DARWIN'S CAMERA

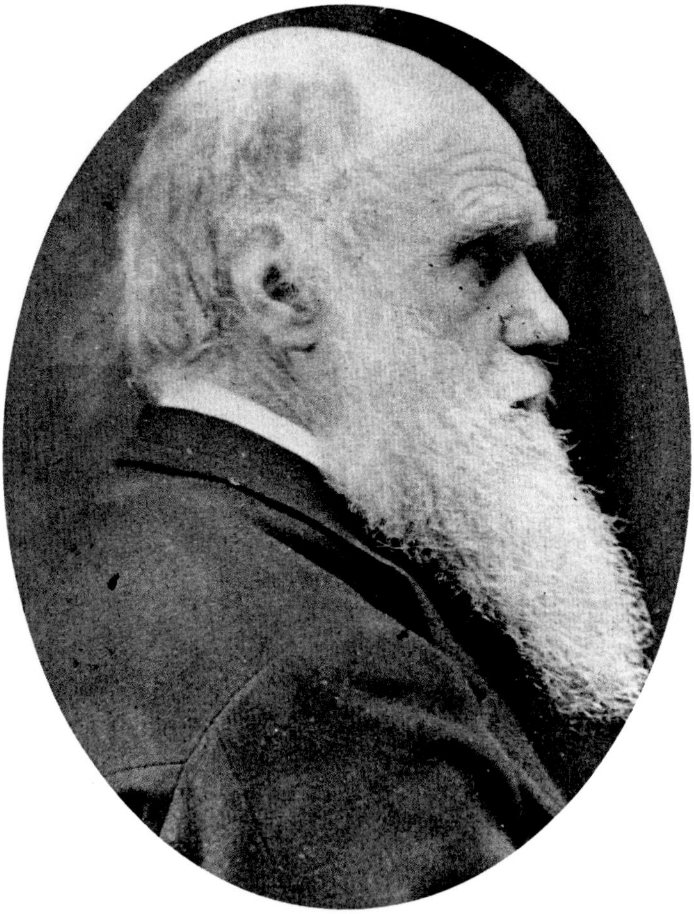

PHILLIP PRODGER

DARWIN'S CAMERA

Art and Photography in the Theory of Evolution

OXFORD
UNIVERSITY PRESS
2009

Frontispiece: Oscar Rejlander, *Profile portrait of Charles Darwin*, c.1871–72

OXFORD
UNIVERSITY PRESS

Oxford University Press, Inc., publishes works that further
Oxford University's objective of excellence
in research, scholarship, and education.

Oxford New York
Auckland Cape Town Dar es Salaam Hong Kong Karachi
Kuala Lumpur Madrid Melbourne Mexico City Nairobi
New Delhi Shanghai Taipei Toronto

With offices in
Argentina Austria Brazil Chile Czech Republic France Greece
Guatemala Hungary Italy Japan Poland Portugal Singapore
South Korea Switzerland Thailand Turkey Ukraine Vietnam

Published by Oxford University Press, Inc.
198 Madison Avenue, New York, New York 10016
www.oup.com

Library of Congress Cataloging-in-Publication Data
Prodger, Phillip.
Darwin's camera : art and photography in the theory of evolution /
Phillip Prodger.
p. cm.
Includes bibliographical references.
ISBN 978-0-19-515031-5
1. Darwin, Charles, 1809–1882.—Illustrations. 2. Darwin, Charles,
1809–1882—Art collections. I. Title.
QH365.Z9P76 2009
704.9'495—dc22 2008033854

LO, here a CAMERA OBSCURA is presented to thy view, in which are lights and shades dancing on a whited canvas, and magnified into apparent life!—if thou art perfectly at leisure for such trivial amusement, walk in and view the wonders of my ENCHANTED GARDEN.

— ERASMUS DARWIN, *The Botanic Garden*

TIMELINE

1806 Charles Bell's *Essays on the Anatomy of Expression in Painting* published

1809 Birth of Charles Darwin

1816 Fitzwilliam Museum, Cambridge, founded

1824 National Gallery of Art, London, founded

1825–27 Darwin studies at University of Edinburgh, attends Charles Bell's lectures on expression

1827–31 Darwin studies at University of Cambridge, visits Fitzwilliam Museum

1831–36 The voyage of the H.M.S. *Beagle*

1839 *Journal of Researches,* also called *The Voyage of the Beagle*, published; Invention of photography announced

1851–52 Wet-plate negatives available

1857 Oscar Rejlander, *The Two Ways of Life*, exhibited

1859 *On the Origin of Species* published

1862 G. B. Duchenne de Boulogne, *Mechanism of Human Facial Expression*, published

1865 Louis Agassiz, photographic survey of Brazil to disprove evolution

1869 Thomas Henry Huxley begins photographic survey of peoples of British Empire

c. 1869 Darwin begins to collect photographs of emotional expressions

1871 *The Descent of Man* published

1872 *Expression of the Emotions* published

1872–73 Muybridge photographs of galloping horses

1874 First gelatin dry-plate negatives available (not mass-produced until 1880)

c. 1878 Francis Galton begins to make composite photographs

1882 Death of Charles Darwin

CONTENTS

PREFACE

THIS BOOK BUILDS on the achievements of the "Darwin industry," that large international community of scholars who have dedicated themselves to studying Charles Darwin's life and work. Darwin is an intriguing subject, and not just because his theories were important and controversial. He wrote letters to nearly everyone he knew, kept copies of virtually all incoming and outgoing correspondence, and made organized notes about his ideas as they developed. So he is not just an extremely interesting scientist, he is also an ideal test case for how scientific theories developed and spread in Victorian England.

In 1948, largely as a result of the efforts of his son Francis, the majority of Charles Darwin's archive was transferred to Cambridge University Library, where it is currently held in the library's department of manuscripts. Although many interesting items are scattered in other locations—there are smaller concentrations, for example, at Darwin's last home, Down House, in Downe, Kent, and at the John Murray Archive in London—the material at Cambridge is authoritative. Cambridge boasts approximately 9,000 of the 14,500 known letters sent to and from Darwin in his lifetime, as well as his notebooks, nearly all of his personal library, and numerous related items such as handwritten manuscripts, edited proofs, newspaper clippings, and even a handful of specimens from the *Beagle* voyage.

The Cambridge archive also includes an amazing collection of original photographs, drawings, watercolors, and prints commissioned, collected, and in some cases even drawn by Darwin, which are the heart

C

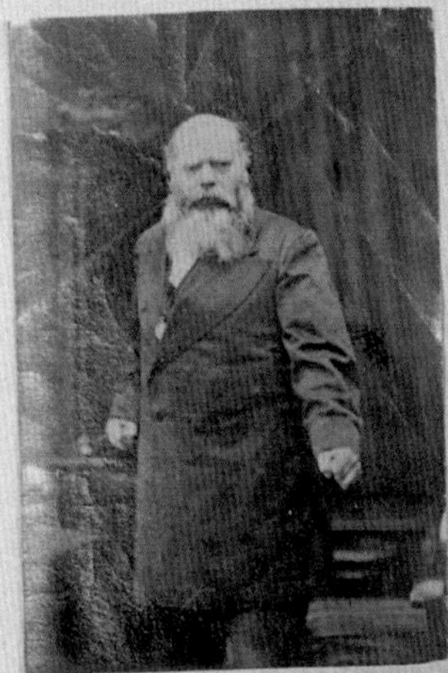

71

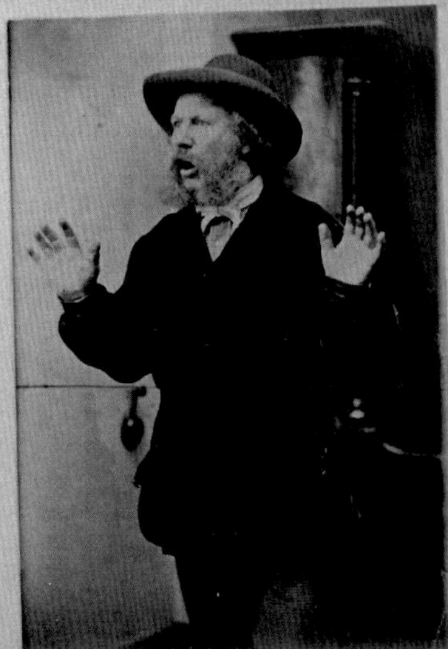

72

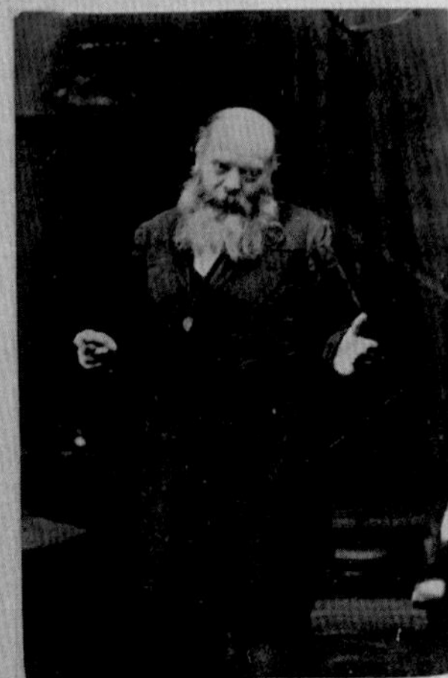

73

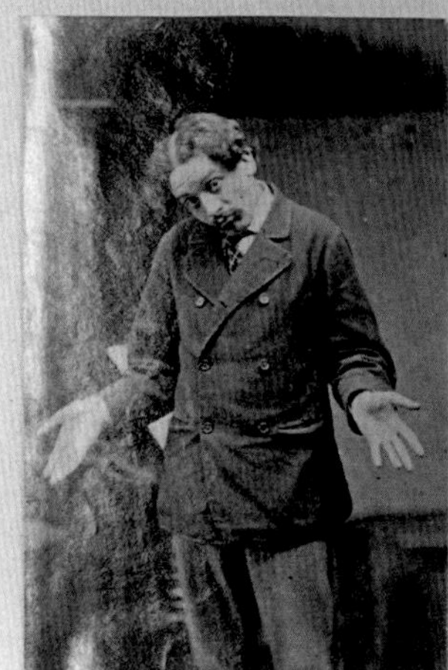

74

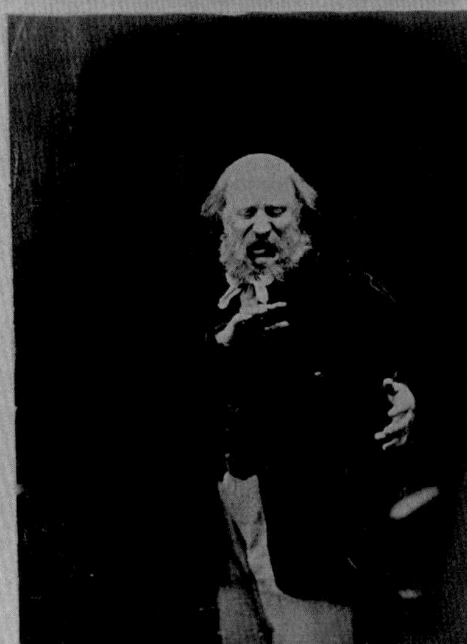

75

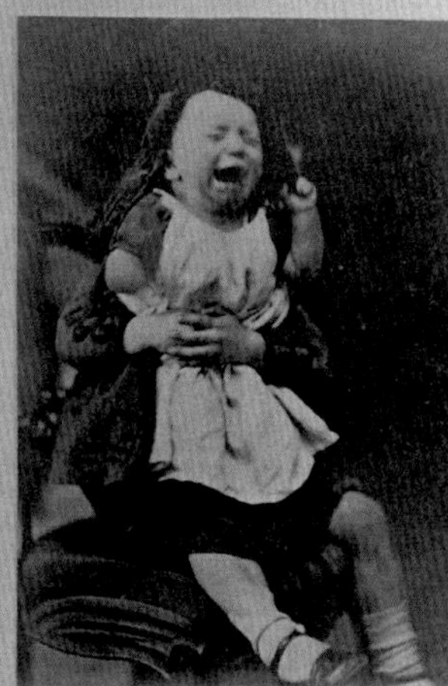

76

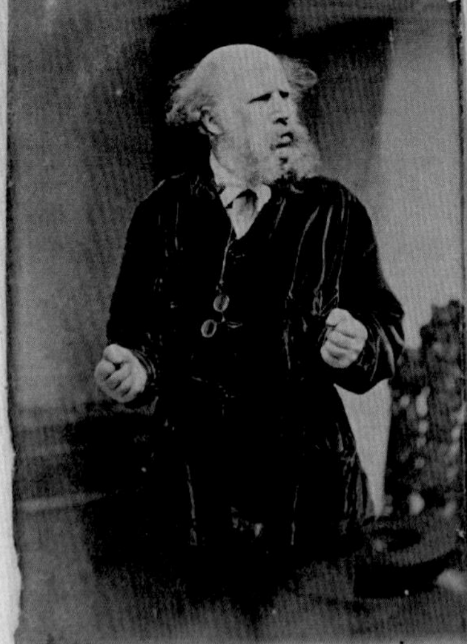

77

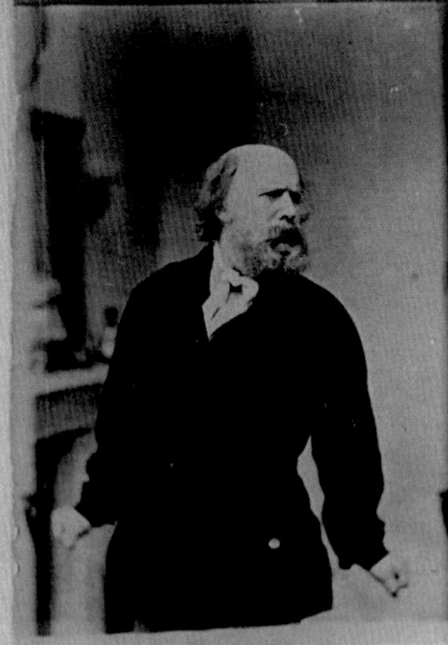

78

of this book. Most are kept in posthumously bound albums, where they roughly reflect the order in which Darwin organized them (figure P-1). A Herculean project to catalogue and transcribe the letters in the archive, called the Darwin Correspondence Project, began in 1974 and is ongoing. There is an enormous amount of work to do—transcribing, checking, and cross-checking. As of this printing, the project has published bound transcriptions of all known letters through 1867 and has made most other letters available online. At the time I performed the research for this book, the letters from the late 1860s and early 1870s, when Darwin began to look seriously at art and photography, had not been published. The Darwin Correspondence Project team was an enormous help in my research, sharing early drafts of key letters and pointing out interesting connections. Nevertheless, because published versions of most relevant letters did not exist, the transcriptions contained herein, unless specifically noted, are my own. I should also note that the Darwin Correspondence Project has for the time being set aside visual material such as prints, drawings, and photographs for later study. As a result, all identifications of artists, titles, dates, and media are also my own.

Darwin identified his notebooks by letter. Notebooks "M" and "N" record Darwin's ideas about emotional expression; I cite these often in this book. He drew heavily on these notes while writing *The Expression of the Emotions in Man and Animals*, published in 1872, which is discussed at length herein. *Expression* was written when Darwin was about sixty-three years old, after several other landmark books, including *The Voyage of the Beagle* (more properly titled *The Journal of Researches*, 1839) and *On the Origin of Species* (1859). The notebooks, however, were begun much earlier. Darwin began notebook "M" in 1838, just before the publication of the *Voyage of the Beagle*. "N," which he started when "M" was full, began in 1856, three years before the publication of *Origin of Species*. In other words, Darwin began thinking about emotional expression early in his career, long before publishing his famous theories of evolution. As the notebooks show, the questions Darwin addressed in *Expression* nagged at him for at least thirty-four years.

Expression contains wood engravings and photographs. The wood engravings, which Darwin referred to as "figures," appear as vignettes embedded in the main body of the text. Photographs were collected on separate pages, which Darwin called "plates." Darwin also referred to his photographs as figures, but referenced them with the plate number to distinguish them from the engraved figures: plate 1, figure 1; plate 2, figure 1, and so forth. I have adopted this convention in this book, using Darwin's figure number alone when a wood engraving is mentioned, and using plate and figure number together to refer to *Expression* photographs. To distinguish Darwin's illustrations from the ones herein, my illustrations are numbered with both chapter and figure numbers.

Opposite:
P-1 A page of photographs by Oscar Rejlander from the Darwin Archive, 1871–72. Albumen prints.

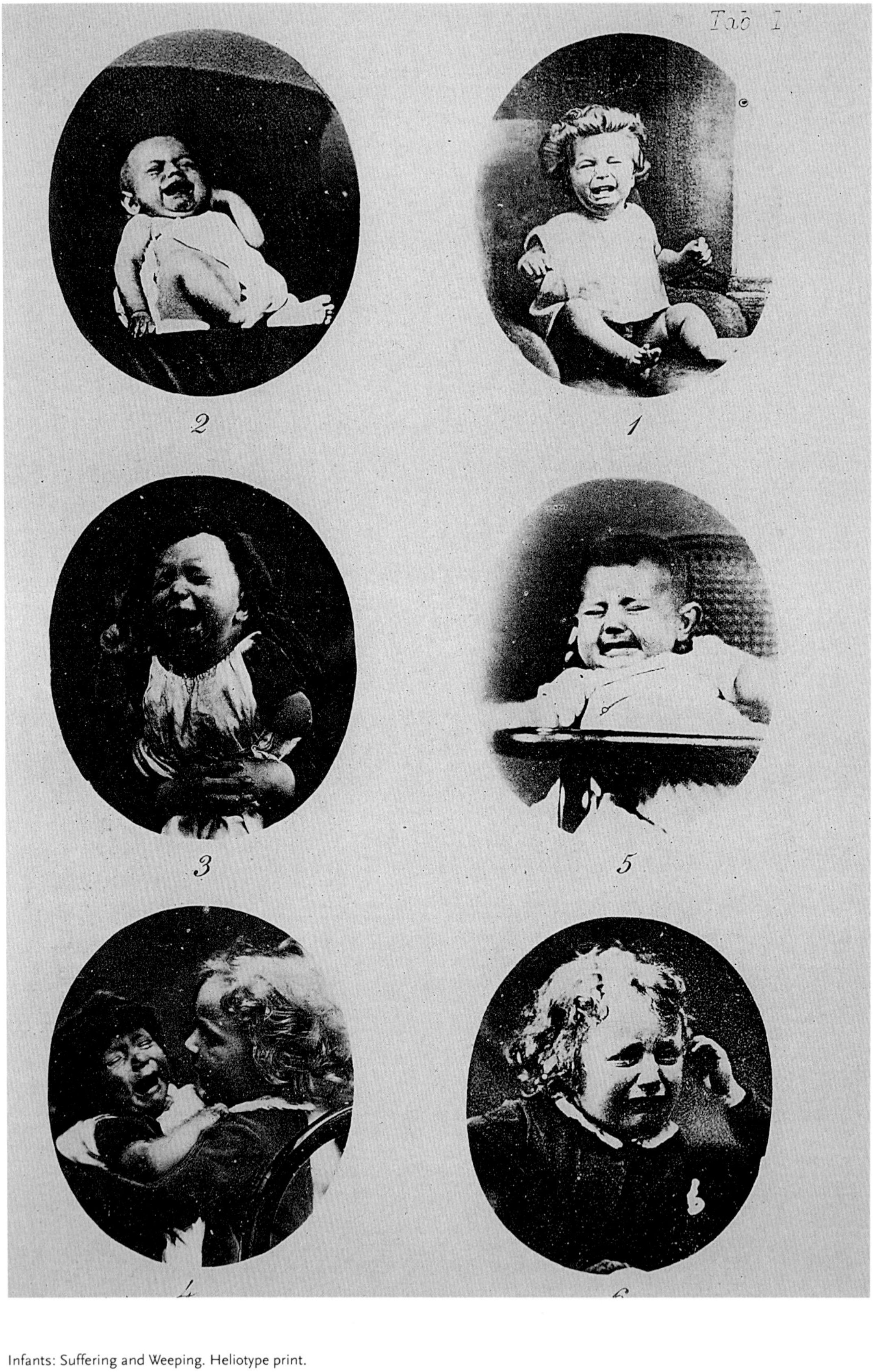

Infants: Suffering and Weeping. Heliotype print.

Plate 1, *The Expression of the Emotions in Man and Animals.*

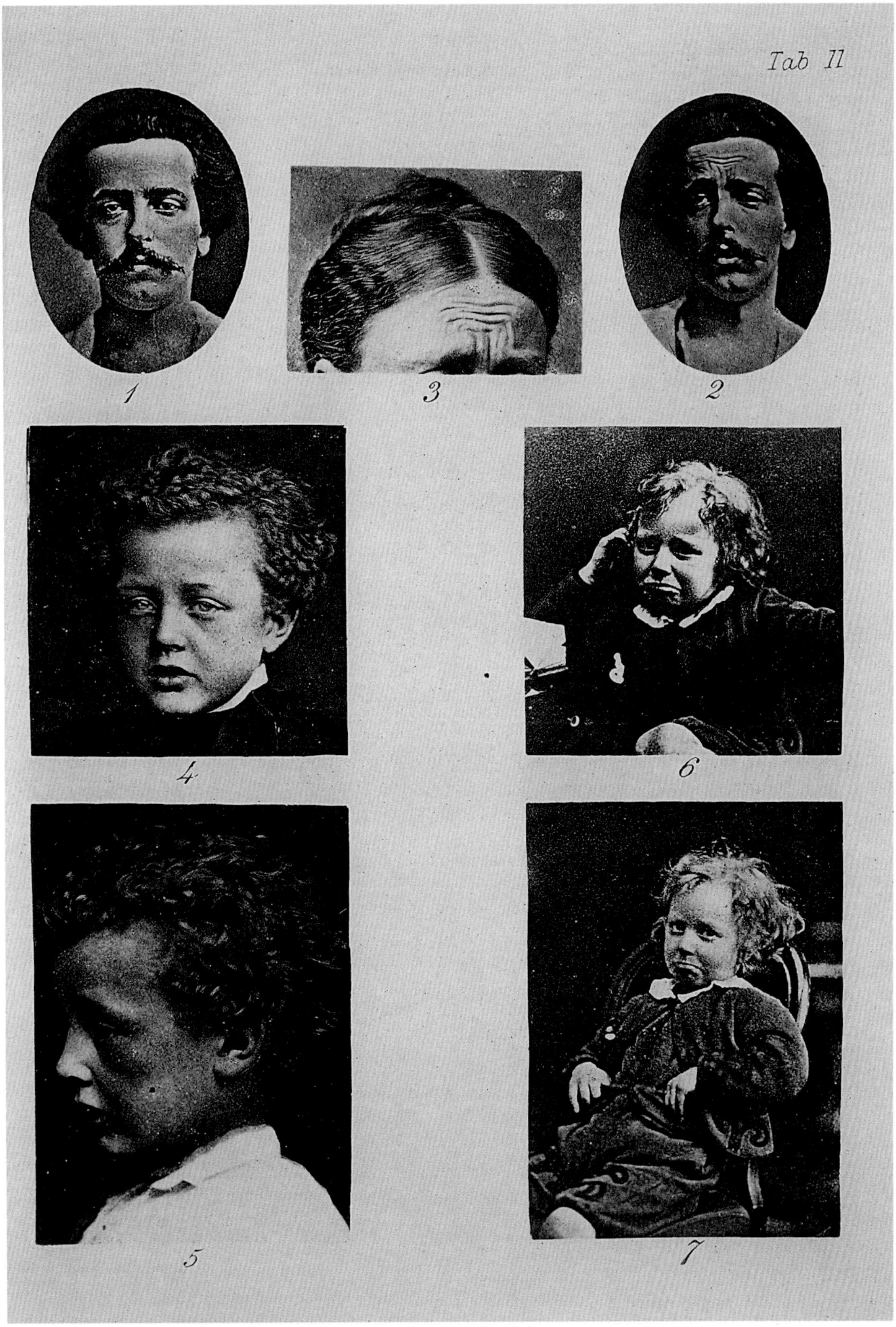

Low Spirits, Anxiety, Grief, Dejection, and Despair. Heliotype print.
Plate 2, *The Expression of the Emotions in Man and Animals.*

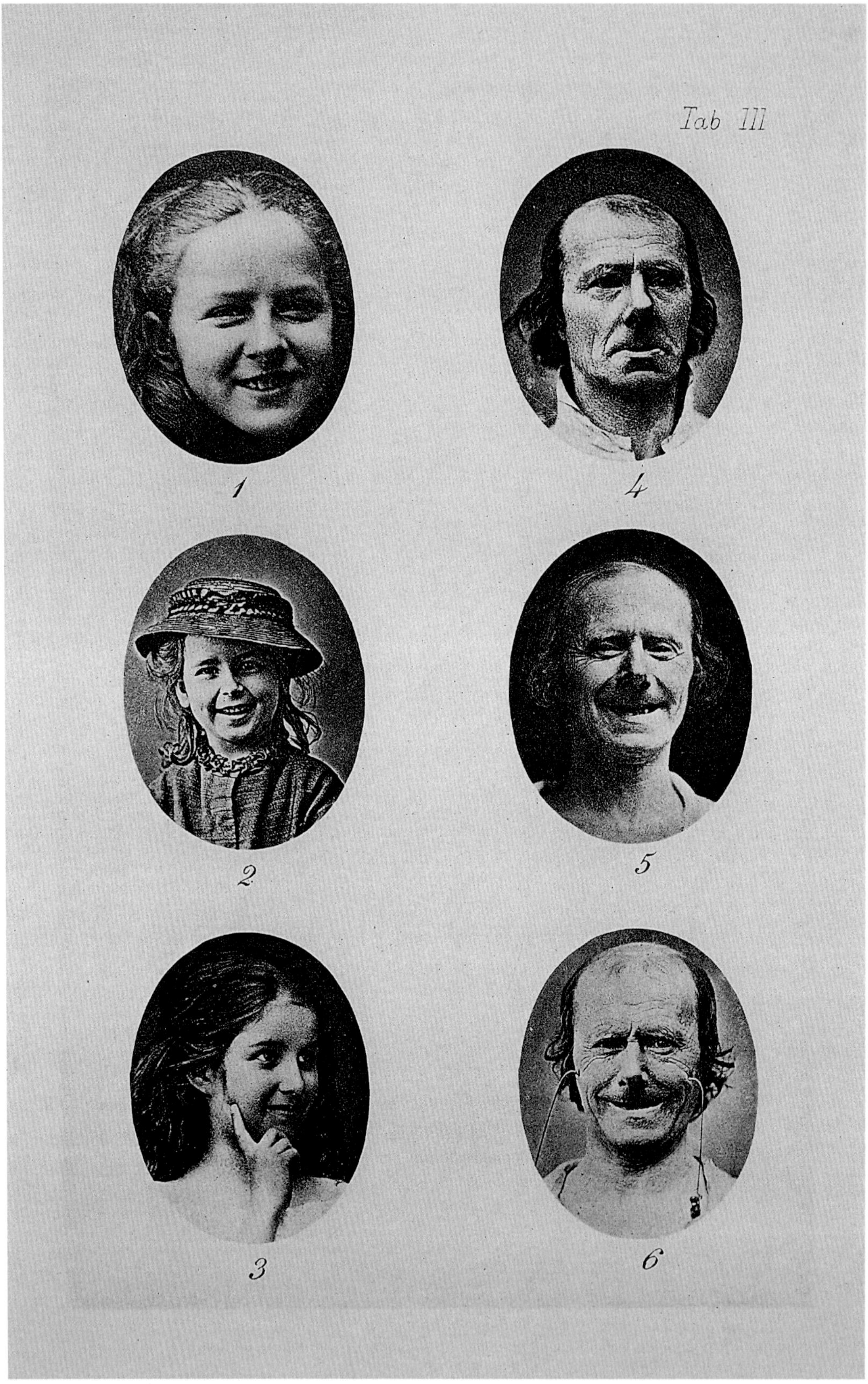

Joy, High Spirits, Love, Tender Feelings, and Devotion. Heliotype print.
Plate 3, *The Expression of the Emotions in Man and Animals.*

Tab. IV

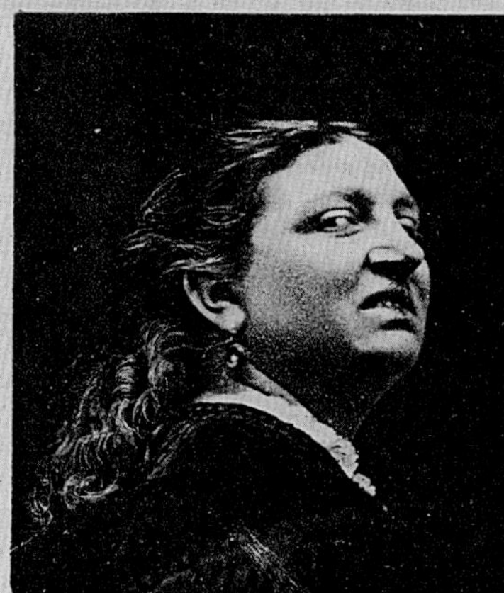

1

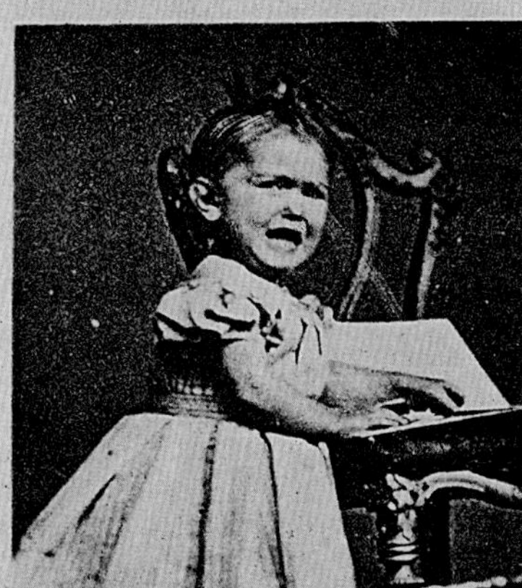

2

Hatred and Anger. Heliotype print.

Plate 4, *The Expression of the Emotions in Man and Animals.*

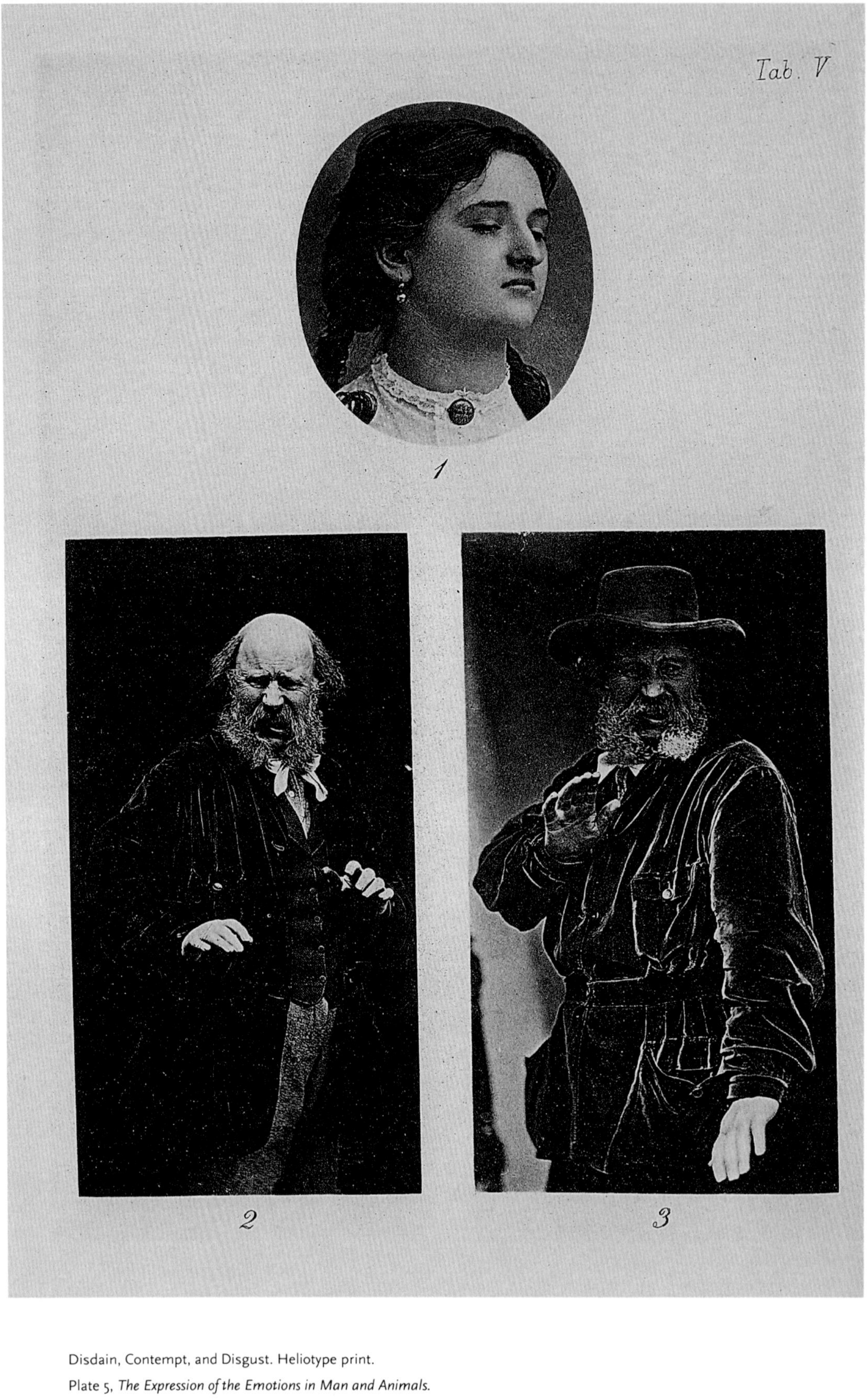

Disdain, Contempt, and Disgust. Heliotype print.
Plate 5, *The Expression of the Emotions in Man and Animals.*

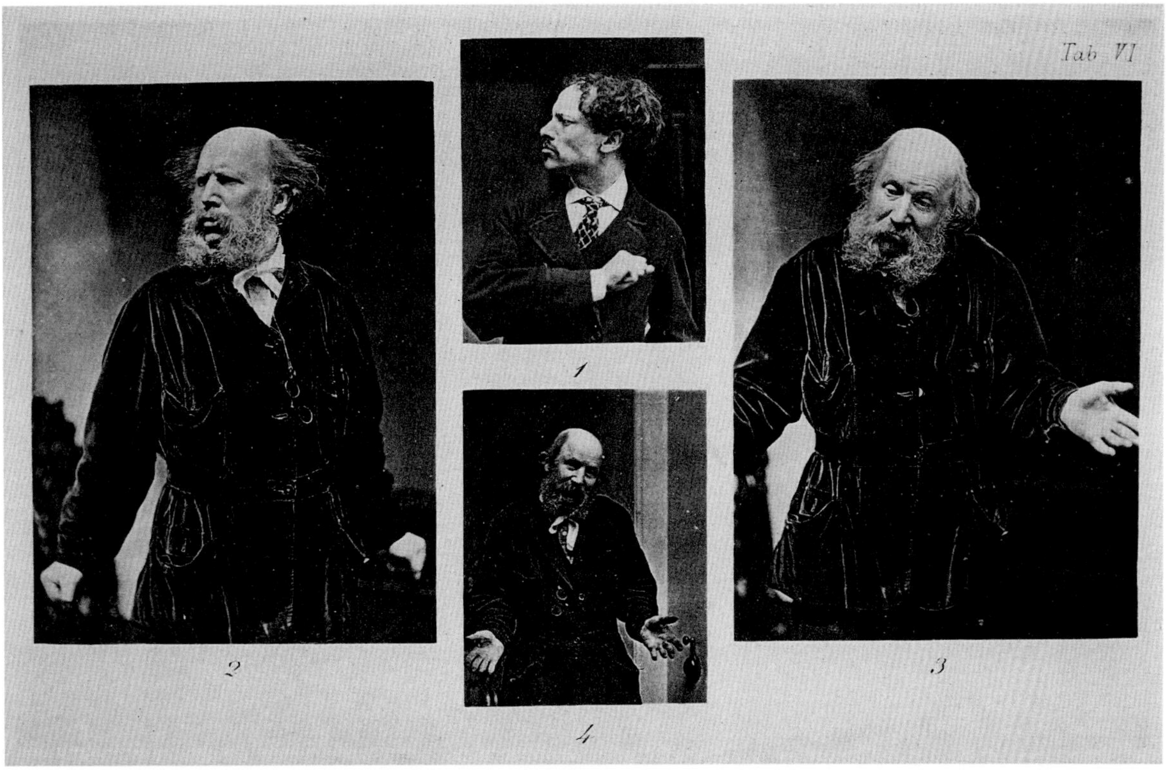

Indignation and Helplessness. Heliotype print.
Plate 6, *The Expression of the Emotions in Man and Animals.*

Tab VII

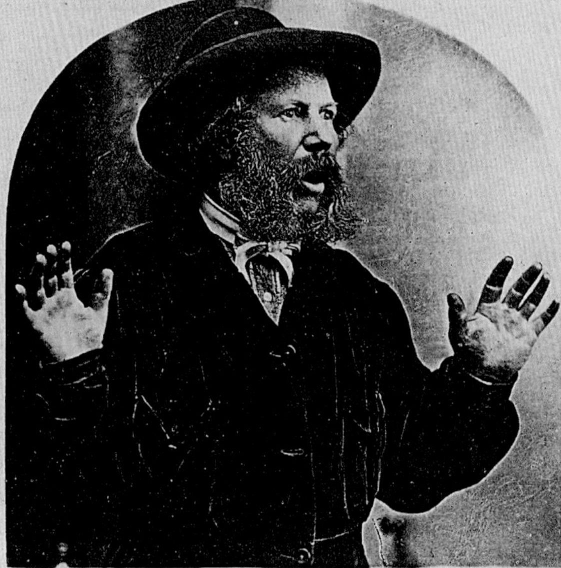

1

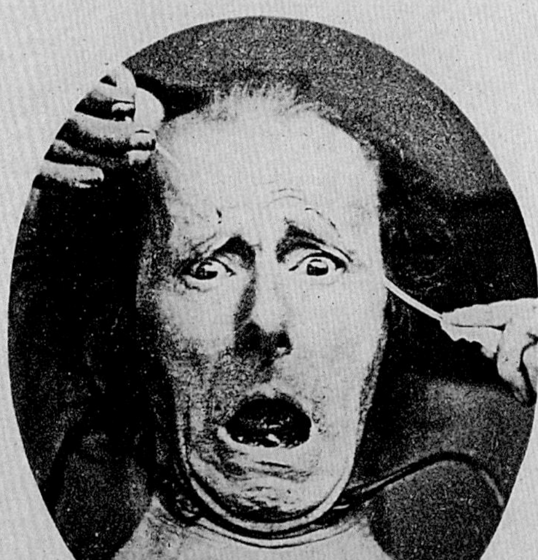

2

Surprise and Astonishment, Fear and Horror. Heliotype print.
Plate 7, *The Expression of the Emotions in Man and Animals.*

INTRODUCTION

We tend to think of photography as having been born whole, with all the ease and authority it currently commands. But the rival processes that the Frenchman Louis Daguerre (1787–1851) and the Briton William Fox Talbot (1800–77) invented around 1839 bore little resemblance to photography as we now know it. Daguerre's photographs were made directly on metal plates, and Talbot's with paper negatives, so there were physical differences between the way early photographs looked. Each process also had its strengths and weaknesses, but they shared similar limitations. Chemistry was slow, requiring long exposure times. Ready-to-use materials were not available, so photographers were required to make their own plates and negatives, sensitize them before use, and develop them shortly thereafter. The whole ordeal was expensive, fraught with error, and required specialist training. Success was hard-earned.

The qualities we now associate with photography—consistency, accuracy, economy, and convenience—were unknown to the first practitioners. This meant that photography itself was viewed in very different terms. The idea that it could be objective—that the camera might be used to record things impartially better than a human can—was several decades away. Some commentators imagined photography would one day achieve objective status, but no one argued that the young medium was inherently privileged. Exciting, yes, and distinctive—photographs looked great, they were reproducible (though not nearly as easy to reproduce as they would later be), and they enabled the more-or-less simultaneous recording of lots of detailed information. But they did not enjoy any sort of special standing. They were seen principally as an alternative to drawing or printmaking, complete with the limitations associated with those media.

In the nineteenth century and before, viewers believed prints and drawings if they met several mutually reinforcing standards. The information they conveyed had to be plausible; in other words, they had to fit with viewer's expectations, evaluating whether a picture resembled a subject or situation that might reasonably exist. The authority of the presenter—in the case of natural history, the scientist who commissioned it—was weighed, and the skill and authority of the artist who rendered it was sometimes considered, too. The best artists were highly coveted, not just because of the quality of their pictures, but also because retaining the right artist added credibility. The participation of a famous illustrator in a natural history work, such as a member of the Sowerby family in England (1700s through 1900s), Georg Ehret (1708–70) in Germany, or Pierre-Joseph Redouté (1759–1840) in France and Belgium, instantly lent authority because those artists were widely regarded as particularly skilled and insightful. And the pictures themselves had to look convincing, regardless of who made them. A crude sketch was viewed with more suspicion than a fine polished rendering—an artist's ability to make realistic-looking pictures had a large role in his or her success.

In all of these criteria, the judgment of the viewer was paramount; essentially, it was up to the reader to determine whether to accept a picture as accurate. This was a negotiated stance—author and artist tried to convince the viewer that what they depicted was real, while the viewer had to decide whether to believe them. This paradoxical position is addressed in a landmark paper published by the historians of science Lorraine Daston and Peter Galison in 1992, "The Image of Objectivity," and elaborated in their splendid book *Objectivity* in 2007. Daston and Galison describe the increasing clamor for objectivity in scientific illustration over time, as well as the strategies authors and artists used to cultivate the perception of authority and, ultimately, of human noninterference in their works. As Daston and Galison have shown, the drive to achieve objectivity in scientific illustration has a long history, independent of the introduction of photography. Thus, when photography arrived on the scene in 1839, it did not enter a vacuum. It became part of this rich history, and the first authors to use photographs as scientific illustrations acted according to conventions that had been hammered out over previous decades and even centuries.

At first, photographs were judged in exactly the same way as prints and drawings. The same standards that applied to them—plausibility, authority, skill, and convincingness—applied equally to photographs. But photographic technology improved rapidly. From the 1830s to the 1880s, exposure times were shortened from minutes to fractions of a second. Commercially prepared dry plates were developed for handy use, handheld cameras were invented, and laboratories were created to process and print pictures. Photographs became affordable to make, and photomechanical printing techniques made them increasingly easy to reproduce in books. It took approximately fifty years, but during the latter half of the 1800s photography moved into territory traditional drawing and printmaking could not. Once it became capable of taking pictures faster than what the naked eye could see, it began to affect measures of scientific integrity.

Fast and efficient photography rendered useless the old system of evaluation based on plausibility and authority. No one could judge whether something was right or wrong anymore. Who could decide the way a lightning bolt should look mid-strike? Or the way horses' hooves look as they gallop? Or what a solar flare looks like, or a speeding bullet? Once the vision threshold was breeched, new thinking was required. Photographs assumed a dual role. They illustrated something, but they were also experiments in their own right. They became more than mere pictures—they became data. At that point, scientists became concerned about exactly how their photographs were made. They developed self-contained protocols to enable like-minded scholars to reproduce their results.

This was a time of profound transition in visual culture. It was also precisely the time during which Charles Darwin lived and worked. Darwin published his famous account of the *Voyage of the Beagle* in 1839, the same year Talbot announced photography in England. The maturation of Darwin's theories paralleled the development of increasingly sophisticated photographic technologies. Year after year, photography got better, and year after year, Darwin extended and refined his theories.

This might have been little more than a historical coincidence, except that Darwin eventually found advancements in photography irresistible and began to use photography in his work. Consequently, he became a prominent voice in scientific photography, and his efforts in the field helped shape photo history. He was not a photographer, nor is there any evidence he ever made photographs of his own. Wide-ranging in his interests and concerned with how his books were illustrated, Darwin took great pains to find just the right pictures to make his books compelling. In his first books he used familiar techniques such as lithography and wood engraving; in the late 1860s and early 1870s he began to use photography. At the time, the medium still suffered from pronounced technical limitations, and practitioners were only just beginning to have success photographing moving subjects. Yet with the publication of his sensational book of 1872, *The Expression of the Emotions in Man and Animals*, Darwin became one of the first scientists ever to publish photographs in a scientific treatise and made significant inroads in action photography. As with his science, Darwin was ahead of his time.

Darwin's Camera tells the extraordinary story of how Darwin not only changed the course of science but also forever changed the way pictures are seen and made. Using photographs to depict fleeting expressions of emotion—laughter, crying, anger, and so on—as they flit across a person's face, he managed to produce dramatic images at a time when photography was famously slow and awkward. The things he wanted to photograph changed too quickly to be photographed easily, and he struggled to get the pictures he needed. So he scoured the galleries, bookshops, and photographic studios of London, looking for pictures to satisfy his demand for expressive imagery.

He finally settled on one of the giants of photographic history, the eccentric art photographer Oscar Rejlander, to make his pictures. It was a peculiar choice. Darwin was known for his meticulous science, while Rejlander was notorious for altering and manipulating photographs. Their remarkable collaboration, and the lengths they went to in order to create the pictures Darwin needed, is one of the astonishing revelations in this book.

Darwin never studied art formally, but he was always interested in art and often drew on his knowledge of the field as his work unfolded. He studied art as a student and befriended the artists on the voyage of H.M.S. *Beagle*; he visited art museums to examine figures and animals in paintings and read art history books and artists' training manuals. He befriended the celebrated animal painters Joseph Wolf and Briton Riviere and accepted the Pre-Raphaelite sculptor Thomas Woolner as a trusted guide. He corresponded with legendary photographers Lewis Carroll, Julia Margaret Cameron, and G.B. Duchenne de Boulogne, as well as many lesser lights. This book provides the first examination ever of these relationships and their effect on Darwin's work, and how he, in turn, contributed to the way scientific illustrations are made and understood.

Photographic illustration was an inexact process. Because there were no preset rules for using photographs in books, Darwin attempted to create them. Working at a time when printmaking still dominated scientific illustration, he internalized prevailing notions about authority and authenticity in picture making. In this regard, he was a transitional figure, with one foot firmly in the past—lessons learned from the books he knew and admired—and one foot in the future, with the enormous potential he recognized in photography. He was somewhat inconsistent in his approach, and the way in which he compromised between these two poles demonstrates the incremental nature of paradigm shifts, such as photography's rise as a dominant form of scientific illustration. Darwin contradicted himself at times, but he never meant to deceive. Even his foes conceded his scrupulous honesty and commitment to seeking truth; in shaping his illustrations, he attempted to convey accurate information pertinent to his theories.

This is not a book that weighs the virtues of Darwin's theories of evolution by natural and sexual selection, which have been treated at length by many capable authors. Rather, it is a journey, funny and amazing at times, into a time, a place, and a culture. Darwin worked at a decisive time in the histories of science and art. The British Empire was at a peak, incredible advancements in technology and medicine were announced almost daily, and the world of ideas was developing at a dizzying pace. Major discoveries occurred almost routinely, and entire disciplines were created, debated, and defined. When Darwin went to school, the idea of becoming a professional scientist was almost inconceivable—the word "science" as we now know it had not even been coined. Yet he became the archetypal scientist of his day. Information flowed fast and furious, and Darwin, arguably one of the most inquisitive men who ever lived, soaked it up like a sponge. His contributions to many fields—evolution, botany, geology, and psychology among them—are widely known. Now we must add his decisive influence on the history of art to his impressive list of accomplishments.

DARWIN'S CAMERA

1

Darwin's Art Collection

The prints, drawings, and photographs Darwin collected in the 1860s and 1870s

Collecting, Charles Darwin wrote in his *Autobiography* in 1876, was his passion. It "was very strong in me, and was clearly innate," he stressed, a side of his personality that fueled his scientific undertakings.[1] Having made a name for himself with the publication of *Voyage of the Beagle* in 1839, he continued to study natural phenomena throughout his career. Over time, his research into the flora, fauna, and geology seen on the voyage of H.M.S. *Beagle* gave way to other interests. With them grew needs for new specimens, from orchid flowers to earthworm castings. Yet his basic method did not change. It was typical of Darwin that when embarking on a new inquiry, he would first acquire, classify, and study as many examples of his subject as he possibly could.

In the late 1860s and 1870s, Darwin turned his attention to emotional expression, a subject different from any he had studied previously.[2] For much of his earlier work, it was enough to look at dead specimens. Expressions, by contrast, are fleeting, complex behaviors, not simple things that can be sampled in a conventional way. Although they typically come together in a recognizable posture—a smile or a frown, for example—most are composed of dozens of associated muscle contractions. One contraction may precede another, causing an expression to unfold in phases over a brief period. The scientist's job is complicated by the need to identify which phases of activity are critical, singling out moments when related motions coalesce. Once that is settled, there are degrees of expression to consider (a chuckle is not the same as a guffaw, for instance), and humans have the uniquely frustrating ability to feign expressions. Darwin, however, wished to study only expressions as they occurred naturally. Acting was a good way to approximate behaviors that could not be seen any other way, but only genuine expressions of happiness, sadness, terror, and ecstasy were suitable for study. What was Darwin to do?

Expressions themselves can be observed, but not collected; the best he could hope for were pictures. One promising source was artists' renderings, so Darwin began by studying expressions in paintings, drawings, sculptures, and prints. Contemporary and ancient art were equally useful. Modern art might better reflect modern sensibilities, but older examples could confirm the existence of behaviors across generations. Yet, manually produced pictures were prone to all sorts of error. The hand of the artist was central to the result; skill in capturing the truth of a situation varied. Besides, works of art are made to communicate ideas, not facts. When artists resort to visual shorthand to represent certain behaviors, as they frequently do, they defeat meaningful analysis. Artists trade in viewers' perceptions, not accuracy for its own sake. It was no good to Darwin if a painter made a subject green with envy, or showed him growling like a lion. Darwin required realism. The comparatively new medium of photography offered a possible answer to these problems, so Darwin began to collect photographs.

In the 1860s and 1870s, with rapid improvements in technology, photography began to be accepted as a practical alternative to manual illustration. Unlike drawings, which are limited by the skill and integrity of the artists who produce them, photographs were thought to afford an objective means of recording empirical events. The question of what constitutes objectivity in the history of science is a subject of lively scholarly interest, discussed by the historians Lorraine Daston and Peter Galison in their book *Objectivity* (2007) and in their landmark 1992 paper "The Image of Objectivity." In retrospect, it is clear that early photographs were at least as contrived as the drawings they were supposed to replace, but scientists increasingly came to equate photography with accuracy in representation. Like many of his colleagues, Darwin accepted the apparent superiority of photography in certain applications. As a result, the photographs he collected assumed a dual role: as specimens for developing his theories and as evidence to demonstrate the validity of his ideas.

Darwin described the ephemeral nature of expression as one of the "main sources of error" inherent to the study of emotional expression:

> From the reasons above assigned, namely, the fleeting nature of some expressions (the changes in the features being often extremely slight);

our sympathy being easily aroused when we behold any strong emotion, and our attention thus distracted; our imagination deceiving us, from knowing in a vague manner what to expect, though certainly few of us know what the exact changes in the countenance are; and lastly, even our long familiarity with the subject—from all these causes combined, the observation of Expression is by no means easy, as many persons, whom I have asked to observe certain points, have soon discovered.[3]

Photography enabled Darwin to capture behaviors for analysis without depending on artists. Artists attempting to render expression are constrained by their ability to accurately absorb and recall the intricacies of the events they witness. Darwin's annotations on the drawings he collected reveal that he frequently disagreed with the details of artists' renderings. While not yet capable of producing truly instantaneous images, photography enabled him to record action without relying on others to witness, visualize, and transmit complex information.

Unlike living subjects, photographs provided a fixed record of activity that could be examined at leisure. They were solid, stationary samples, much like the natural history specimens he so treasured. They were valuable in that they represented discrete moments in complex events and could be used to isolate constituent elements of behavior. This capacity of images of any kind to stand in place of scientific specimens is what the theorist Bruno Latour has described as the "immutable mobile."[4] Darwin collected photographs of expression as substitutes for expressions themselves, since they are difficult to observe effectively with the naked eye. Yet the photographs did not stand on their own. Darwin also used them to supplement his first-hand observations of expression in friends and associates.

Darwin's book *The Expression of the Emotions in Man and Animals*, published in 1872, extended Darwin's theory of evolution by natural and sexual selection to the realm of the mind; it was arguably his boldest extension of evolutionary theory. In *On the Origin of Species* (1859) Darwin described the natural selection of organisms best adapted to compete with others for scarce resources. He explained how physical adaptations determine whether living things will survive and breed in a given environment. This is the book in which Darwin laid out his so-called survival of the fittest doctrine, though he actually never used that phrase in the book.

The Descent of Man, and Selection in Relation to Sex, published in 1871, a year before *Expression*, extended Darwin's evolutionary theories to consider how organisms attract mates. It was also the first book in which the evolution of people was addressed. "I have been led to put together my notes," he wrote, "so as to see how far the general conclusions arrived at in my former works were applicable to man."[5] Darwin examined human beings, their physiology, and behaviors, and discussed in detail the secondary sexual characteristics of animals. *Descent* is best known for its analysis of attractive patterning in animals, for example, in butterfly wings, birds' feathers, and the colors of tropical fish. It also contains Darwin's most direct claims for the animal ancestry of *Homo sapiens*, a subject with which he had been identified ever since the publication of *Origin*, but which he had diplomatically avoided. Conceptually, *Descent* is an ambitious and at times difficult work, designed to address some of the criticisms leveled against *Origin* by considering not just how animals find food and survive in harsh climates, but also how they interact with members of their own species.

Expression emerged from that effort. In it Darwin explored emotional expression as a special type of behavior, describing how expressions affect survival. Originally, Darwin thought it might form a single chapter in *Descent*, but, he explained, "as soon as I put my notes together, I saw that it would require a separate Treatise."[6] Unlike many scholars of his generation who considered emotion to emanate from the soul and expression a gift from God, Darwin viewed expressions as evolutionary phenomena. He considered them as signals: the vestigial result of something useful in earlier times. For example, he argued that the expression of anger, characterized in humans by baring the teeth and snarling, was related to the ancestral instinct to bite in self-defense. This type of convergent behavior seemed to Darwin to prove the common ancestry of humans and other life forms. According to Darwin, people owed the language of emotional expression to the survival needs of their animal progenitors—we can literally see our animal origins in the way we act.

Expression was among the first scientific books ever published with photographic illustrations, and the only photographically illustrated book produced by Darwin. A handful of other books with scientific themes were published with photographic illustrations prior to or at approximately the same time, but none was so widely circulated, and none so immediately influential.[7] Darwin's publisher, John Murray, at first opposed the idea of using photographs to illustrate the book, informing Darwin that it would make *Expression* a money-loser, costing an additional £75 per 1,000 sets.[8] After startup costs and sales of foreign editions were taken into account, Darwin calculated it would ultimately add one shilling and six pence to the cost of each volume,[9] at a time when "the public are accustomed to words at 2 shillings."[10] But Darwin, then one of the most powerful authors in the world, insisted, and the photographs went ahead.

Darwin knew photographs could have special appeal, not just in selling the book, but also in selling his theories. By the time he began assembling notes for *Expression*, photographers had already begun to challenge the accuracy of other media. The French neurologist Guillaume-Benjamin Duchenne de Boulogne (1806 – 75), for example, had written at length about the physiological inaccuracies of sculptures such as the *Laocoön* in the Vatican, revealed by comparing the original object to photographs of models in identical poses.[11] Similarly, Oscar Rejlander (c. 1813 – 75), who eventually became Darwin's main source for photographs of expression, had attempted to judge the accuracy of Italian Renaissance paintings by photographing reenactments of the scenes they represented.[12]

Technical problems slowed the adoption of photography in science. The first photographic materials were complicated and awkward to use—to produce even a simple photograph in the late 1860s, negatives had to be made by hand, usually on wet glass plates.[13] This was tricky, requiring patience and manual dexterity. To make wet-plate collodion negatives, a sticky, noxious solution had to be poured on glass plates before exposure and sensitized immediately, else they would dry and become impervious to the chemicals used to develop them. As a result, preparation and processing had to be performed at the same time that the photographer was taking pictures. Photographers had to be chemists, craftsmen, and time managers as they shuttled from darkroom to studio and back. Evidence of this is visible in *Expression* itself. In one of the pictures Darwin used, the hands of the photographer Oscar Rejlander, who posed for the pictures himself, are stained with the chemicals he used to prepare his negatives (figure 1-1).

To complicate matters, normal exposure times ranged from several seconds to a minute or two, depending on conditions, so that subjects had to remain absolutely still while their photographs were taken. Action, or "instantaneous" photographs, as they were popularly called, were not possible.[14] Many photographers attempted to produce images that *appeared* to have been taken instantaneously, or at least conveyed the illusion of movement frozen in time, but they were limited in what they could do. In the late 1860s and early 1870s, instantaneous photography was an elusive goal. Prizes were awarded in competitions for the most successful instantaneous photographs, and a wide range of photographic paraphernalia, including lenses, light reflectors, and even whole studios were developed with the express purpose of reducing exposure times.[15] Rejlander condemned the practice. "To award a medal for the quickest exposure seems disrespectful. I mistake much if I believe [a photographer would not be] more gratified by being considered the most artistic photographer . . . than being pronounced the quickest at capping his lens."[16]

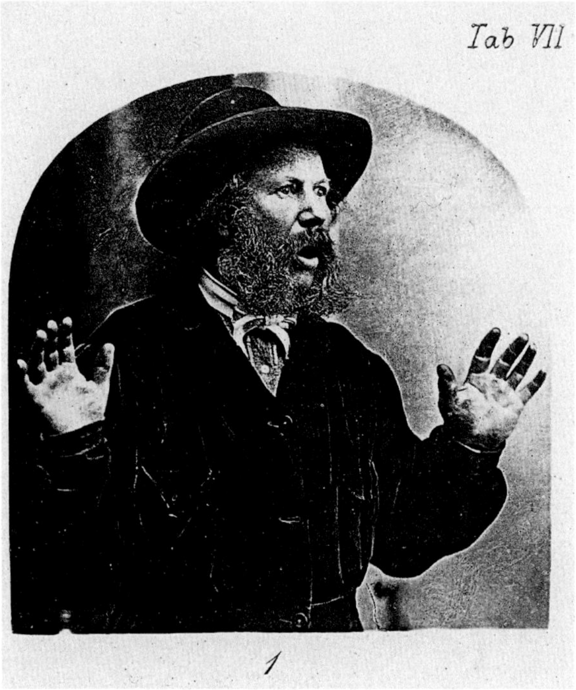

1-1 Oscar Rejlander, Surprised Man (Rejlander), 1871–72. Heliotype. Plate 7, figure 1 of *Expression.* A self-portrait, with fingers stained from silver nitrate.

For Darwin, the problem of producing photographs of rapidly occurring action was paramount. To illustrate *Expression* convincingly, he required images that captured sometimes subtle and ephemeral muscle activity. As he explained in the introduction:

> The study of Expression is difficult, owing to the movements being often extremely slight, and of a fleeting nature. A difference may be clearly perceived, and yet it may be impossible, at least I have found it so, to state in what the difference consists. When we witness any deep emotion, our sympathy is so strongly excited, that close observation is forgotten or rendered almost impossible. . . . Our imagination is another and still more serious source of error; for if from the nature of the circumstances we expect to see any expression, we readily imagine its presence.[17]

Photography offered a vantage from which Darwin might observe discrete moments in the occurrence of expressions with accuracy and detachment. The unassisted eye was incapable of recording the kind of detail he wanted to study.

There was no obvious single source for the material he needed, so when he began looking for photographs, Darwin tried a variety of possible suppliers, including a number of London shops and studios. He bought the pictures he thought best; many of these are preserved in the Darwin Archive at Cambridge. Some forty-one photographs in the archive appear to have been bought specifically for their expressive content, including images by the London Stereoscopic Company (1854–1908), the Leicester firm of John Burton & Sons (fl. 1860s), the Italian photographer Giacomo Brogi (1822–81), the French firm of Fernand Vadon & Cie. (active 1860s) and the American photographer James Landy (1838–97).[18]

Like an entomologist hunting for beetles, Darwin rifled through hundreds of pictures seeking useful examples of expressive imagery. Inscriptions on the pictures he collected reveal that in the two years before *Expression* was published, he traveled throughout London looking for pictures, from Victoria to Baker Street and Strand to South Kensington. This is remarkable in itself, as Darwin suffered from frequently debilitating illness and preferred to stay home whenever possible. At this stage of his life he generally avoided public outings, retreating to the comfort of his own home. Yet the desire for pictures of expression apparently reinvigorated him.

Any photograph that revealed aspects of the physiology of expression seems to have interested Darwin, regardless of how and why it had been created. He continued to collect prints, drawings, and photographs right up to the time *Expression* was published. He prized photographs most. Although truly instantaneous photography was not yet possible, he examined as many as he could for clues. They provided a broad overview of how expressions were popularly portrayed and perceived. Ultimately he considered most of them unsuitable for inclusion in his book, but they helped shape his ideas.

Lessons from Old Masters

Darwin's search for expressive imagery probably began at the British National Gallery around the late 1860s, where he studied depictions of people in paintings. Sebastiano del Piombo's *Raising of Lazarus* (1517–19) was a particular favorite (figure 1-2), "exciting in me a sense of sublimity," he would later recount.[19] Of all the pictures he could have singled out for praise, this highly mannered composition was a peculiar choice. The painting depicts the critical moment in John 11 when Jesus, prefiguring his own resurrection, brings the pious Jew Lazarus back from the dead. The central figure of Jesus raises his right hand skyward, palm open, simultaneously commanding attention and metaphorically acknowledging the divine power that fills him. His left arm is extended, with his hand pointing to Lazarus and bidding him back to life. Behind Jesus stand the twelve apostles: "doubting" Thomas, who was said to resemble Christ, appears behind Jesus cloaked in a robe nearly identical to the one he wears; Judas is framed in the "U" created by Jesus' raised

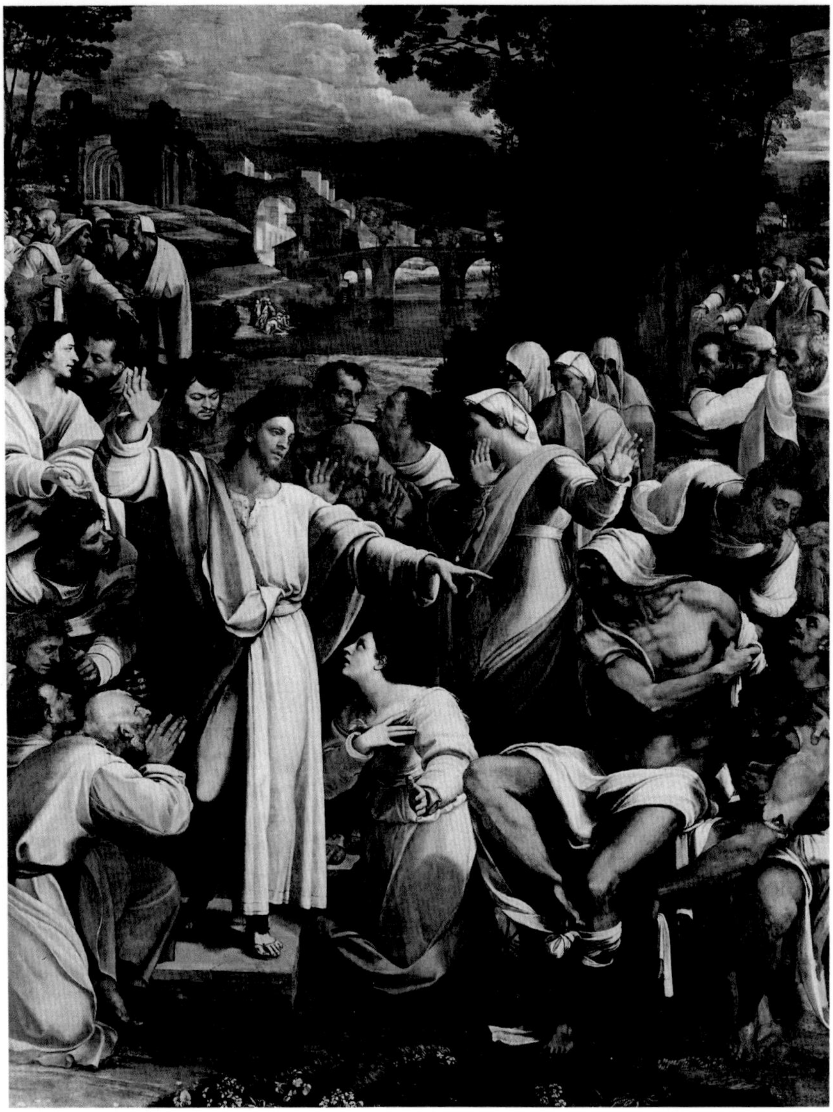

1-2 Sebastiano del Piombo, *The Raising of Lazarus*, 1517–19. Oil on canvas transferred from panel. Sebastiano's painting excited "a sense of sublimity" in Darwin.

arm, isolating him from the other apostles in a clever foreshadowing of his betrayal. At Jesus' feet kneels Lazarus' sister Martha. It is the moment at which Jesus pronounces "I am the resurrection, and the life: he that believeth in me, though he were dead, yet shall he live: And whosoever liveth and believeth in me shall never die" (John 11:25).

Darwin's reference to the painting's "sublimity" may be an allusion to its potent religious theme. However, he also would have been mindful of the scholarly meaning of sublimity associated with the eighteenth-century philosopher Edmund Burke. Burke divided aesthetics into two categories: the sublime and the beautiful. Beauty, according to Burke, is composed of those qualities of things that relax the viewer: picturesque elements pleasing to the eye. Sublimity, by contrast, is made up of fierce and terrible qualities, elements that excite wonder, astonishment, and even horror. Darwin owned a copy of Burke's famous *Philosophical Enquiry into the Origin of Our Ideas of the Sublime and Beautiful* (1757), in which Burke explained, "The passions which belong to self-preservation, turn on pain and danger; they are simply painful when their causes immediately affect us; they are delightful when we have an idea of pain and danger, without actually being in such circumstances. . . . Whatever excites this delight, I have called sublime."[20] He was among the first to analyze aesthetics in terms of physiological effects on the viewer. Sublimity, for Burke, was not an abstract concept, but a fact that could be observed in very real reactions in people. For all its philosophical content, *Enquiry,* with its detailed descriptions of various emotions ranging from curiosity to revulsion, is one of the pivotal books in which the passions were analyzed. It was certainly one of the sources Darwin consulted as he began to write *Expression.*

Darwin's real interest in Sebastiano's painting will never be known. However, the painting does convey Burke's notion of sublimity in the reactions of some of the characters it portrays, a number of which are based on original drawings by Michelangelo. Mary, for example, standing behind Lazarus with her eyes averted from him, raises her hands in astonishment. Saint Peter crouches at Jesus' right foot, his hands clasped in supplication. Perhaps most remarkable, the figure of Saint Andrew, who appears just behind Jesus' left shoulder, raises both hands to the sides of his face in surprise, a behavior Darwin described as characteristic of astonishment in his book (detail, figure 1-3). Indeed, in

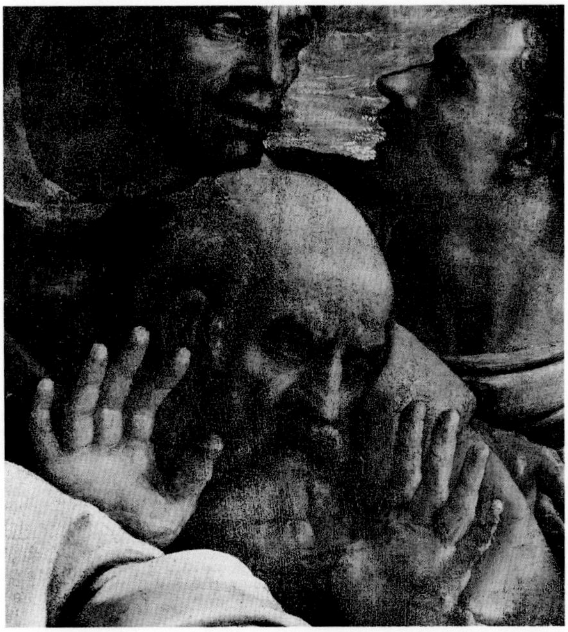

1-3 Sebastiano del Piombo, *The Raising of Lazarus* (detail: Saint Andrew). Saint Andrew's pose resembles Oscar Rejlander's.

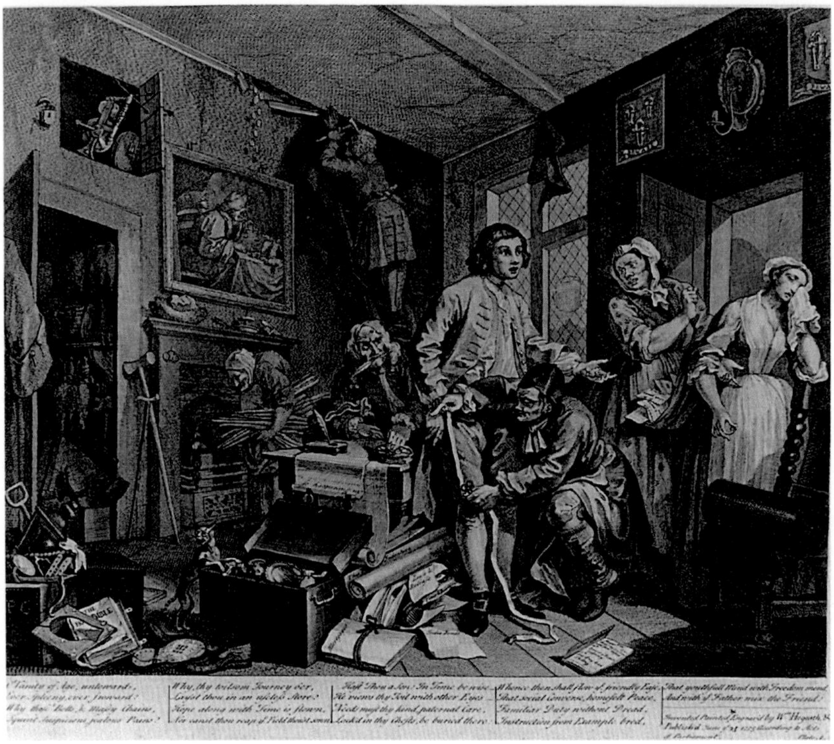

1-4 William Hogarth, *Rake's Progress* (third state), 1735. Engraving. Darwin noted the effective portrayal of the woman baring her teeth.

Sebastiano's painting, Saint Andrew resembles Rejlander's pose in the photograph Darwin published of him illustrating surprise (figure 1-1). Could it be that Darwin directed Rejlander to Sebastiano's painting for inspiration?

Darwin clearly was looking for ideas in Old Master paintings. In *Expression* he also cited a picture by William Hogarth in which "a drunkard raises his eyebrows, and this gives to him a puzzled, foolish look."[21] The picture is not identified, but it may have been *Shortly after the Marriage* (1743), the second of six paintings in the series *Marriage à la Mode* at the National Gallery. Darwin also described a "wonderful" Hogarth in which "passion is represented in the plainest manner by the open glaring eyes, frowning forehead, and exposed grinning teeth."[22] This is an apt description of the charwoman depicted in the doorway on the right side of the first plate of *Rake's Progress* (1735), which Darwin would probably have known from engravings (figure 1-4). Similarly, his only exposure to Leonardo da Vinci's *Last Supper* (1498) would also have been through reproductions. Of this he noted: "[T]wo of the Apostles have their hands uplifted, clearly expressive of their astonishment."[23]

Darwin's family suggested other paintings for him to look at. His cousin Hensleigh Wedgwood, for example, recommended the works of Benozzo Gozzoli (1420 – 97), particularly "a picture in the Campo Santo at Pisa of Noah's daughter looking at her drunken father, called the Vergognosa de Campos Santo," which he reported contained a mock depiction of modesty.[24] Overall, though, Darwin confessed disappointment with such historical depictions in *Expression*. "I had hoped to derive much aid from the great masters in painting and sculpture, who are such close observers. Accordingly, I have looked at photographs and engravings of many well-known works; but, with a few exceptions, have not thus profited."[25]

Darwin also studied natural history illustrations, including those in John James Audubon's *Birds of America* (1827–38).[26] Audubon was celebrated for his contextual realism; his highly detailed birds were frequently shown interacting with other animals, surrounded by appropriate habitat. Darwin, who owned the text to Audubon's *Birds* but did not possess the plates (though he surely had access to them), quoted Audubon in *Expression*, describing the tendency of barn owls to extend their wings and tail when confronted, hissing and "clacking the mandibles with force and rapidity."[27] Darwin may have had Audubon's *Barn Owl* in mind; it is one of the most famous pictures in the book, and Audubon's depiction shows an animal in exactly the pose Darwin described, raising its wing to strike a rival and defend its meal of a freshly killed chipmunk.

Collecting Photographs

As early as 1866, Darwin started to purchase commercial photographs containing details of emotional expression. He may have begun earlier, but it was in that year that he made the first datable reference to his collection, writing to his friend the Yorkshire psychiatrist James Crichton-Browne (1840–1938), "[S]ometime ago I went into several shops in London to try to buy photographs of the insane, but failed."[28]

There is also evidence that Darwin's friend, the scientist Thomas Henry Huxley (1825–95), accompanied him in his first attempts to find photographs. A letter sent to Huxley from an official of the London Stereoscopic Company describes a group of ethnographic photographs jointly intended for Huxley and Darwin: "Enclosed are some extraordinary photographs of which I beg your acceptance. They were (or rather the originals) brought over by Sergeant Catherall from the spot. I think you & Mr. Darwin will feel interested in them. I do not know the latter gentleman's address or I would send him a set."[29] Although there are no further direct mentions of photographic collecting in the correspondence between Huxley and Darwin, other connections may be inferred. For example, one of the first photographers Darwin contacted in a scientific context was medical doctor turned professional photographer George Charles Wallich (1815–99), whose studio was at

2 Warwick Gardens in South Kensington, near the newly created Royal College of Science and School of Mines,[30] which Huxley directed.

Wallich and Huxley almost certainly knew each other. Huxley's appointment books have not been preserved; however, the diaries of Huxley's colleague Henry Cole, first director of the neighboring South Kensington Museum,[31] and himself a pioneering advocate of photographic reproduction, indicate that Wallich was a visitor to the museum, where he would likely have met Huxley socially.[32]

Darwin did not limit his search to expressions in English culture, nor was he interested in subjects of a specific body type. The arguments in *Expression* were universal in scope. He believed most expressions are innate among humans and that shared expressions are evidence of common descent.

Along these lines, Darwin tried to obtain written evidence that different races express themselves in the same way. In 1867 he devised a questionnaire, titled "Queries about Expression," which he sent to correspondents at outposts throughout the British Empire. Originally titled *Expression of Savages* in Darwin's notes (the word "savage" was widely accepted to describe preindustrial societies at the time), the text consisted of sixteen questions later reprinted in *Expression*. It is unclear whether the published questionnaire is the same as the one he sent to his correspondents: similar versions exist with slightly different wording and an additional question concerning the use of a hissing noise as a signal to be silent.[33] The recipients, most of whom were colonists of European descent, were asked to observe whether the native inhabitants of the countries in which they lived expressed emotions the same way Europeans do. "Observations on natives who have had little communication with Europeans would of course be most valuable," he wrote, "though those made on any natives would be of much interest to me."[34] The questions ranged from whether the jaw dropped open and the eyebrows rose when a person was astonished, to whether children pouted and protruded their lips when sulky. The leading nature of the questions and the fact that the respondents varied in training and experience rendered the information he gathered suspect; the exercise lacked objective criteria by which results could be compared. In any case, Darwin used the responses to substantiate his claim that expressions were common to all peoples.

Darwin did not request photographs of the observations he sought. He may have wished to have such photographs, but "Queries" was apparently drafted before he had decided to gather photographic evidence. Moreover, if he had appealed for photographs, it is unlikely he would have received them, given the extreme difficulties involved in producing action shots at the time.

Darwin seems to have been as interested in the perception of expressions as he was in the physiology of the persons displaying them. Although he carefully avoided discussing the communicative function of expressions in *Expression* (voluntary expressions, as opposed to instinctive reactions), he acknowledged their two-part importance as signals first displayed by an originator and second understood by a recipient. Darwin showed photographs to others, asking them to judge the persuasiveness of certain depictions and hoping to gauge their authenticity. Usually the targets of this study were friends and family. In *Expression*, he reported the result of one of his private experiments:

> Dr. Duchenne has given us a photograph of an old man with his eyebrows well elevated and arched by the galvanization of the frontal muscle; and with his mouth voluntarily opened. This figure expresses surprise with much truth. I showed it to almost twenty-four persons without a word of explanation, and one alone did not understand what was intended. A second person answered terror, which is not far wrong; some of the others, however, added to the words surprise or astonishment, the epithets horrified, woeful, painful, or disgusted.[35]

The anthropologist Roslyn Poignant has argued that Darwin had three purposes in using photographs: to freeze transient expressions, to distance himself from the subject, and to provide representations that could be used to prompt comments from others.[36] The psychologist Paul Ekman has noted that the last purpose was particularly innovative.[37]

In a few cases Darwin seems to have sent some of the photographs he collected along with his "Queries" to gather reactions in foreign cultures. In a letter for Darwin's attention relayed to him by the editors of *Nature*, one J.P. Mansell Weale commented on a sample photograph he had received: "I have been unable to get an answer to this question on grief although I have shown the photograph of the girl sent to me by Mr. D. to several intelligent natives & have had no opportunity of personally determining it."[38] There is no way to know with certainty which image Darwin sent; it may have been the photograph of his niece Katherine, which in edited form became the figure 2 of plate 2 in *Expression*.

When Darwin began his search for pictures of expression, the shops and studios of London were brimming with inexpensive photographic portraits and genre scenes. The photographic industry in London had peaked. According to British Post Office records, in 1860 there were some 150 photographic studios in the city. By 1867 the number had nearly doubled to 275. The total changed little until the mid-1870s, when it decreased slightly.[39]

To Darwin, it was helpful to have many photographs to choose from.

Effective photographs capturing expressions were rare; however, it was possible to find useful images by sifting through the hundreds on offer. Though partly suspicious of the circumstances in which they were made, Darwin valued commercially made photographs because they were made without precise ideas about the correct form of expressive postures. They were of questionable scientific pedigree, but they were at least relatively impartial.

Darwin recognized that certain expressions are enacted differently when they occur spontaneously than when they are voluntarily produced. Duchenne, whose work he carefully studied, wrote in 1862 that genuine and false smiles could be differentiated by the contraction of the *orbicularis oculi* muscle. Darwin noted in his copy of Duchenne's *The Mechanism of Human Facial Expression, Or an Electro-physiological Analysis of the Expression of the Passions Applicable to the Practice of the Fine Arts* that one of the illustrations represented "natural" laughter, while another represented "sham laughter on the right side."[40]

Ultimately, most of the pictures Darwin found in London failed to satisfy his needs. There were two problems. First, the information they contained was unreliable: because he had not managed their creation, he was unable to evaluate how accurately they conveyed the circumstances they purported to depict. They might have *seemed* accurate, but having been produced without experimental controls, any conclusions that could be drawn were tenuous at best. Indeed, the images Darwin obtained had been created for dramatically different purposes than those he intended. He collected the American photographer James Landy's *Expressive Pets*,[41] for example, a collage of photographs of dogs mimicking the expressions of crying babies, which was sold as a comic novelty.

Another album of novelty photographs by the German actor Ernst Schulz was made in a similar, light-hearted vein.[42] It contained pictures of Schulz pretending, tongue-in-cheek, to be men and women in various occupations, and with his face painted to resemble various racial types (figure 1-5). Unfazed, Darwin analyzed even these insubstantial pictures for their expressive content. On Darwin's copy, many of Schulz's images show evidence of having been scrutinized; translations appear penciled in English in Darwin's hand, below the original German captions. Among those he singled out were images translated as "hypocrite," "a learned man scholar," "(avaricious) miser," "good nature," "lazy man," "a cocky fellow dandy," "silly (weak head)," "happy simpleton," and "morose grumbler."[43] Consistent with his belief in universal expression, Darwin appears to have been less interested in the racial photographs, as they are not annotated.

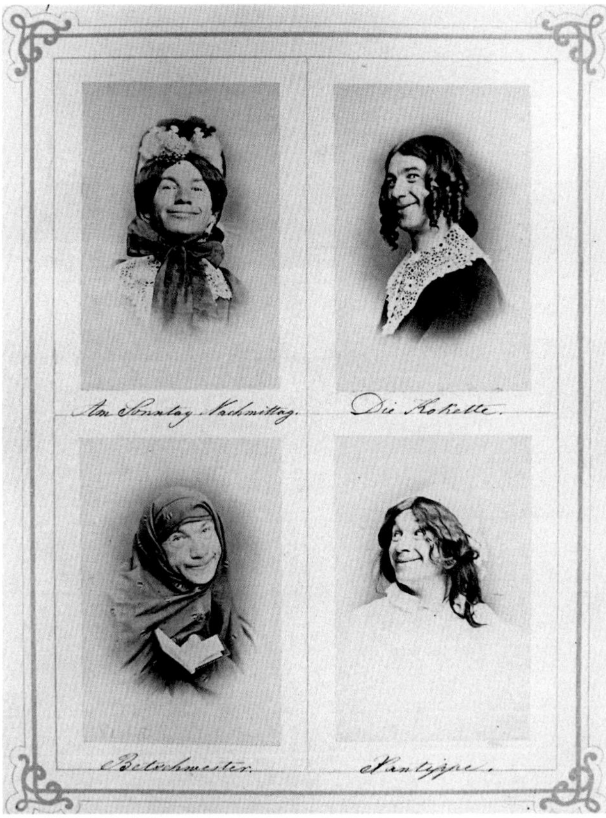

1-5 Studio of Ernst Schulz, *The Actor Portraying Various Character Types*, 1867. Albumen prints.

The second problem with the pictures Darwin found was that, for the most part, they failed to capture expressions as they actually occurred. Much popular Victorian art was mawkishly sentimental. In nineteenth-century Britain, subjects were often portrayed with exaggerated expressions meant to appeal to a popular audience and posed to effect a desired look. Even photographs, which theoretically recorded the actual appearance of things, were often carefully contrived. Because photographic chemistry was extremely slow by modern standards, scenes had to be thoughtfully preconceived. The difficulties Darwin encountered in using commercial imagery is apparent in the many annotations describing the "authenticity" and "genuineness" of photographs in his possession.[44]

In 1870, Darwin adopted a new approach to procuring photographs. While he continued to collect them from commercial sources, in June of that year he wrote to Crichton-Browne that he had begun to *commission* photographs to illustrate *Expression*. Though he knew several photographers, Darwin was unsure whom to approach. His son Leonard, who later became an amateur photographer himself, suggested one possibility:

> There was a man called Conder, an art critic &c. and a friend of the Lushingtons, who has a son here in the Engineers, down here the other day, he said for your expression book Falkner was the best photographer for you as he has done children often, and then you should get them heliotyped [printed using an early photomechanical reproduction process]. Today I got a letter from the heliotype man with some examples of bones and flowers printed this way. I could find out the address of the photographer Falkner. Shall I send you the heliotypes?[45]

There is no record of Darwin having approached Falkner, who was likely too far away for convenient collaboration.

Julia Margaret Cameron (1815–79), the photographer Darwin probably knew best, was at the height of her fame in the late 1860s when Darwin began to collect photographs. However, she generally confined herself to portraits with shallow depth of field and ambiguous expressions. The photographs Darwin sought for *Expression* required as much sharp-focus detail as possible and a spontaneous character at odds with Cameron's aesthetic. Although he was cordial in his dealings with her, there are indications Darwin was uncomfortable with her individualistic approach. When Cameron sent him a portrait she made of him during his stay at her guest cottage in Freshwater, on the Isle of Wight (figure 1-6), Darwin signed it with the inscription, "I like this photograph very much better than any other which has been taken of me."[46] Nevertheless, one year later in a letter to his colleague Alfred Russel Wallace, Darwin described the same portrait as "heavy & unclear."[47] This criticism, which was frequently leveled at Cameron's work at the time, may help to explain her exclusion from the *Expression* project.

Darwin's correspondence with Cameron was remarkably colorless. Two letters exchanged between them are preserved in the Darwin Archive. The first, dated 10 July 1868, urges Darwin to confirm the length of his stay at her cottage. She reveals herself to be preoccupied with her responsibilities as landlord: "[W]e must only ask you as soon as you can decide this point to let us know because in the season months there is such a demand at Freshwater for the few houses that are at all comfortable & good that I am sure you would not like us for the sake of one uncertain week lose a tenant by one or two months if offered."[48] The letter concluded with a warning about the unusually hot dry summer. A second, undated letter from that year is more personable, venturing, "My Bromley people rejoice for us in your coming to be our neighbours & friends as we hope."[49] The rest of the letter, however, explains options for garden and maid service and concludes by advising Darwin to bring his own plates and linen. Such letters demonstrate that at the beginning of their association, at least, Darwin and Cameron enjoyed little more than a landlord–tenant relationship.

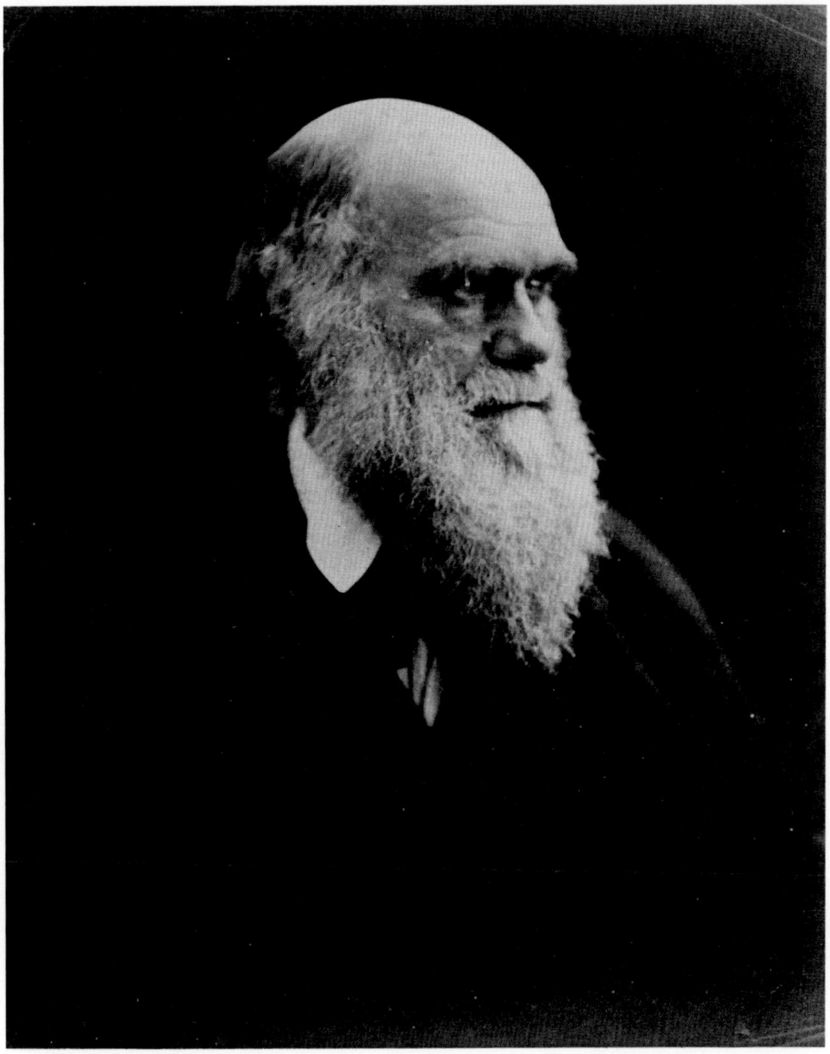

1-6 Julia Margaret Cameron, *Charles Darwin*, 1868. Albumen print. Darwin was of two minds about Cameron's portrait of him.

From 1868 to 1870, Darwin's picture collection grew as an eclectic blend of bought and donated works, obtained ad hoc. The images he acquired during this period were useful in formulating his ideas, and some even served as illustrations, enabling him to test his theories of emotional expression on others. Still, there were substantial gaps in the range of expressions he had managed to obtain as photographs, and none managed to convey the fleeting moments of expression that truly excited him. The completion of *Expression* would require photographs made expressly to Darwin's specifications.

2

Illustration and Illusion

Strategies Darwin used in illustrating his books

At the top of a plate of seven photographs, a fragment of a woman's head forms plate 2, figure 3, of *The Expression of the Emotions in Man and Animals.* Only the area from the base of her eyebrows to the top of her skull is visible, her hair neatly parted and gathered in a tightly curled bun (figure 2-1). Her identity is obscured—the shape of her face, her demeanor, manner of dress, and location; no reliable judgments can be made about her class, age, or occupation. Darwin described the picture as "the forehead of a young lady who has the power in an unusual degree of voluntarily acting on the requisite muscles." It was intended to illustrate the horseshoe-shaped furrows visible on the forehead in acute states of misery, the result of certain muscle groups contracting in opposition to one another. He withheld the rest of the picture, he explained, because the unaltered image was not convincing. The sitter had been so absorbed in enacting the expression for the camera, that "her expression was not at all one of grief." To correct this, he presented only those elements of the picture he considered true, omitting those he thought unsuitable.

This photograph provides a window, though a narrow one, to the way Darwin approached illustration. Of the thirty photographs in *Expression*, only the disembodied forehead shows such a severely edited view of the subject. The other images were not so radically cropped, but many were pared down more subtly before publication. The Darwin Archive helps us to understand just how much Darwin left behind

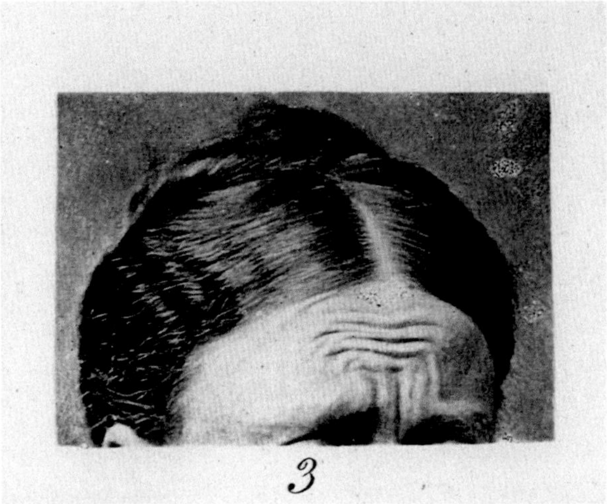

2-1 Attributed to H.T. Melville, The Grief Muscle, 1871. Heliotype. Plate 2, figure 3 of *Expression.*

in preparing his pictures for publication. In some instances he cropped away legs and torsos; in others he instructed artists, photographers, and designers to remove information he considered distracting.

Darwin made these alterations to help strengthen his arguments.[1] He was extraordinarily successful in winning over readers; though his views were at times controversial, they were embraced quickly by many in the scientific community. The effectiveness of his writing is one reason for this. He wrote in clear, lucid language flavored with pleasing anecdotes and clear examples. He used creative metaphors to explain his theories, and he avoided complicated jargon. The strategies he used to win over readers verbally have long intrigued historians of science, who recognize that carefully crafted prose accelerated the acceptance of his ideas.

Much has been written about Darwin's use of language, but what of his illustrations? If words helped popularize Darwin's books, then it stands to reason that pictures may have behaved similarly. This book is one attempt to examine the remarkable control Darwin had over the illustrations he used, and the surprising lengths he went to in order to find pictures to advance his theories. Darwin knew far more about art than we give him credit for, and he used that knowledge to create persuasive and visually compelling books.

In the 1970s, the historian of science Edward Manier was one of the first to analyze Darwin's written innovations.[2] Darwin chose words thoughtfully, using new terms for innovative concepts. Not only were his theories innovative, but crucial terms used to explain them such as "natural selection," "struggle," and "chance" were unprecedented in evolutionary science and may account for the relative ease with which his ideas gained acceptance. By developing a vocabulary specific to his theories and his alone, Darwin avoided challenging established scholars on their terms; by devising new ways of talking about nature, he facilitated new ways of thinking about it. As Manier saw it, Darwin's inventive prose became a neutral framework within which his observations could be considered without prejudice.

Extending Manier's position, literature professor Dame Gillian Beer has argued that the use of literary devices such as analogy and metaphor were also critical to his success: "It is essential for Darwin's theory that the multitudinousness and variety of the natural world should flood through his language."[3] As a result, Beer argues, readers engaging a Darwin book were required to assume a rhetorical omniscience comparable to that found in written fiction. Put another way, they were invited to think about evolution abstractly.

The historians Adrian Desmond and James Moore provide another piece to the puzzle of Darwin's rapid acceptance, commenting on the remarkable success of his books as best-sellers, eagerly consumed by an admiring public. Imaginative and accessible, they sometimes read more like first-rate novels than scientific prose, leading readers through the twists and turns of theoretical arguments like so many developments in the plot of an amazing tale. Not only was persuasive writing a sound tactic for swaying doubters, Desmond and Moore contend, but it also helped boost sales. Writing about *Descent of Man*, they explain:

> It told an arm-chair adventure of the English evolving, clambering up from the apes, struggling to conquer savagery, multiplying and dispersing around the globe. In Darwin's early anxious jottings such a story seemed dangerously implausible; his secret assault on man's ancestry had been a brazen act of faith, fit only for radicals and their ilk. But now, habituated to material progress, social mobility, and imperial adventure, the *arriviste* reading classes lapped it up.[4]

The same logic that scholars such as Manier, Beer, and Desmond and Moore have applied to Darwin's writing can also be applied to Darwin's illustrations. Darwin wanted his pictures to be content rich, without being too specific; to be metaphorical and abstract, and for their unique visual appeal to help his books sell well. The same strategic logic he used to construct his writing he also used to select and edit his pictures. This meant choosing pictures to help him make his case, but it also meant leaving out pictures, or elements of pictures, he viewed as counterproductive.

Layers of Meaning

Darwin wanted his pictures to have semantic force: to embody a general truth about human or animal physiology. The disembodied forehead in *Expression* takes on the meaning of *all* foreheads, showing everything

Darwin wanted us to learn from a single example. The furrows above the brow are lines he claimed one can see on any person in the right circumstances. So the illustration gives readers information they can extend and apply to other, similar examples. By editing away unwanted and, he would have argued, distracting information, he created a visual abstraction of the original, a directed and carefully condensed version. This process of abstraction is a tactic he used repeatedly in *Expression*.

Yet even if a picture reveals a greater truth, it still has a unique life of its own. Each image has size, texture, clarity, and tone. The distinctive appearance of an illustration forms syntactical meaning: the stylistic qualities that make a Darwin illustration distinct from others. Syntactical qualities are ones that tether them to a certain time and place. Today, many of Darwin's pictures seem old-fashioned. At the time, however, they were an exciting change from the engravings and lithographs to which readers were accustomed. Now, the process itself can be distracting. Modern viewers may not notice the contractions of facial muscles in his pictures immediately, focusing instead on their sepia tint, small size, and grainy texture, or the manner of dress of the sitter. With the passing of time, syntactical information becomes more pronounced, as dated materials and techniques stand out more.

Pictures may also have emotional associations. This is one reason Darwin cropped the forehead picture as severely as he did. He was concerned that if readers saw the rest of it, they might feel pity or humor, which were unwanted associations that detracted from his message. He feared the creep of these elements into his pictures, so often he edited out elements one could react to emotionally. At the same time, he occasionally used personal associations to help the reader relate to a picture, as a means of establishing a picture's authenticity.For example, he sometimes allowed children to be shown in recognizable domestic settings, and in one instance he changed a picture of a cat rubbing against a chair leg to the same cat rubbing against a person's leg. Tension between scientific detachment—what might cautiously be called objectivity—and this kind of calculated emotional weight can be seen in a number of Darwin's illustrations.

Fortunately, we can surmise some of Darwin's thinking about his illustrations from clues left in the Darwin Archive. In the case of the forehead photograph, the discarded portion of the published figure is preserved (figure 2-2). On the back, in colored pencil, is inscribed: "Plate II, Figure 3, the forehead & eyebrows alone / K.E.W. 1871—Grief Muscle."[5] The archive also contains two variant photographs that were never used (one is shown in figure 2-3). These smaller prints are glued on studio mounts with the name of the photographer, H.T. Melville of Bromley, printed in letterpress. They may have been taken at a different

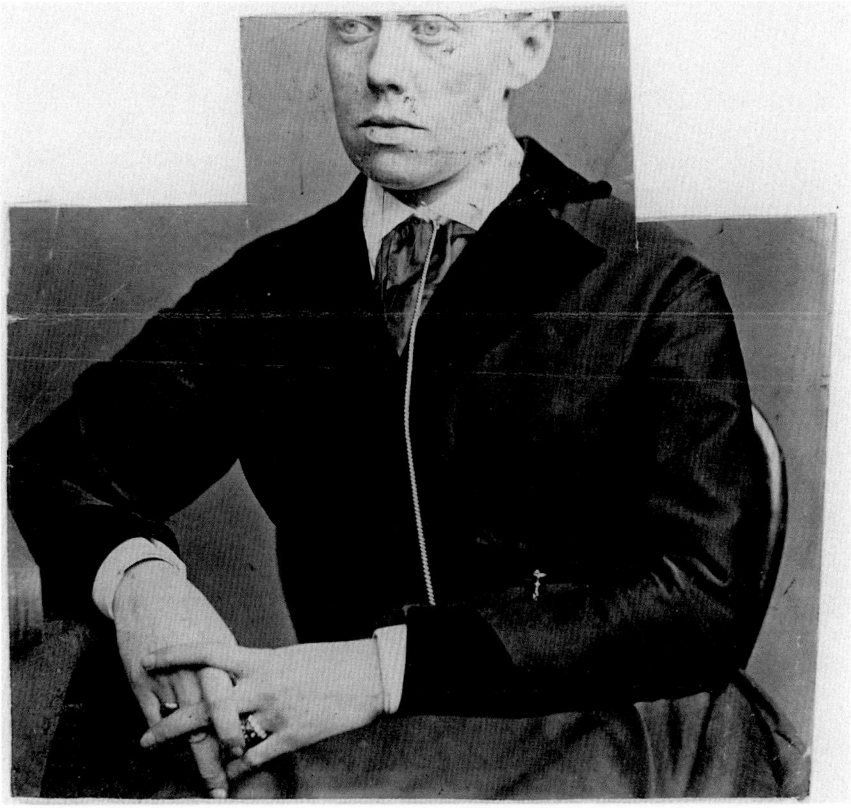

2-2 Attributed to H.T. Melville, The Grief Muscle (K.E.W.), 1871. Incised albumen print. What remained after the forehead had been cut away.

time than the published version, since they depict the subject against a different backdrop.

Darwin did not identify the sitter, but considering the annotation "K.E.W." on the verso of one, and given how close Melville's studio was to his home at Down House, it is likely she was Darwin's niece, Katherine Euphemia Wedgwood. Initials alone would have been enough to identify a family member, and since K.E.W. seems to have sat for her portrait on more than one occasion, it is reasonable to imagine she was someone Darwin knew well—the picture may even have been made at Down House. Concerns for the woman's privacy would have prevented him from labeling the photograph with her full name. Omitting the rest of her face in the published illustration obscured her identity further.

Fragmentation is a powerful theme in contemporary art theory. It can make self-contained images enigmatic; removed from context, mere slivers of pictures can assume iconic status. A sculpted foot, when connected to a statue, is a means to an end—it serves only to support the figure above it. But a foot alone standing on a pedestal is an object for contemplation. Many scholars have explored this fertile ground in art. In the writings of theorists Walter Benjamin, Roland Barthes, and their

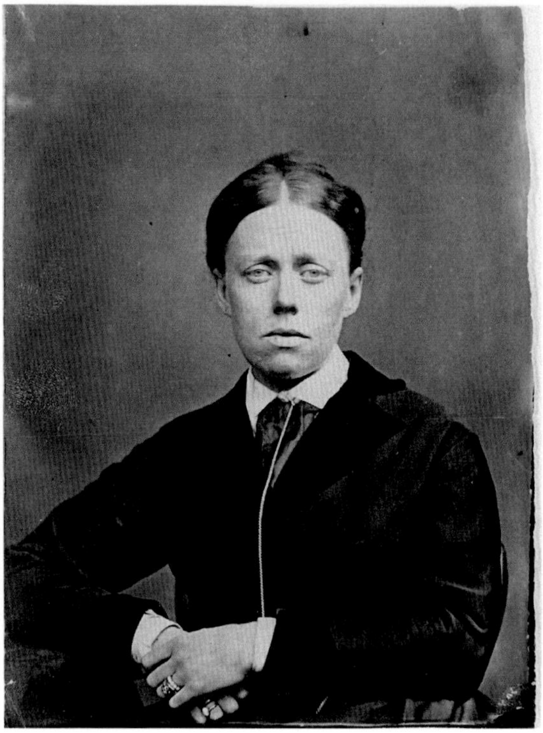

2-3 H.T. Melville, The Grief Muscle (K.E.W.), 1871. Albumen.

successors, what constitutes completeness in an image has long been a matter of animated discussion. Without revisiting these positions, it may be sufficient to acknowledge the theoretical potency of Darwin's scissors. He could not have realized it, but by cutting away the nose, mouth, cheeks, and ears of the subject in this modest figure, he tapped into one of the enduring preoccupations of postmodern thought.

Darwin's primary concern was with archetypes. He wanted his illustrations to prove universal truths, and he did not want readers to become preoccupied with the stylistic decisions he made in editing his pictures. In this sense, he was suspicious of syntactical meaning, those elements that draw attention to the way a picture is made. And yet his readers could not help but notice his pictures were *photographs*, a new technology increasingly (but not yet fully) associated with objectivity and authenticity. When *Expression* was published in 1872, photographs in a book were fresh and enticing: they ensured that it would receive attention as a novelty, if nothing else.

Darwin was also sensitive to the question of emotional meaning. His illustrations were designed to stand alone, without explanations of context. But if viewers saw something they recognized in them, this might make them seem more real to readers unaccustomed to photographic illustrations in science books. Convincing the viewer was important; after all, one had to believe that what one was looking at actually happened.

These are the tightropes Darwin walked in illustrating his texts. He wanted them to serve as factual examples to make generalizations about natural phenomena. At the same time, he wanted his pictures to win over readers and to sell more books. A picture revealing too much information could tip the balance toward the syntactical and expressive, and his science might not be taken seriously. Too little, and the pictures would be dry and uninviting. As a result, the process of editing pictures became extremely important. He had to decide what to keep and what to discard. This process of refinement, of whittling away at meaning until the right balance was struck, is seen repeatedly in the steps he took to find and prepare pictures for *Expression*. Consciously and unconsciously, he developed ways of integrating pictures to support his arguments.

In designing his books, Darwin had to choose the number of illustrations to use and the technique with which to print them. For some books, such as *Expression*, *Power of Movement in Plants* (1880), and *Contrivances by which British and Foreign Orchids are Fertilised by Insects* (1862), plentiful illustrations were appropriate; in others, it was more effective to leave illustrations out completely. Leaving the text to speak for itself was appropriate when the subject was more abstract, as in *Origin of Species*. To illustrate the various adaptations described in *Origin* would have been counterproductive. Those who believed in divine creation needed abundant illustrations to support their cause: the more seemingly inexplicable complexity readers encountered, the less likely they were to think that evolution could have been achieved through random selection alone. By withholding illustrations from *Origin*, Darwin kept the discussion on a theoretical level.

Decisions also had to be made about the style of the pictures. Not a single color illustration appears in all of Darwin's published work, even in *Descent of Man*, in which the colorful sexual displays of animals is a major theme. Color illustrations were available; though expensive, they might have enhanced the aesthetic appeal of his books. But it was not worth the risk, because they would contribute to the perception of complexity in the organisms depicted. Darwin would then have been forced to defend them, explaining how pleasing colors can result from random mutation and variation.

Darwin had considerable control over the appearance of his books. From individual pictures and plates to the design of the covers and spine, Darwin decided many design elements personally. For example, the Darwin Archive contains tracing-paper sketches for four of the seven photographic plates in *Expression* (e.g., figure 2-4). The sketches show evidence of having been reworked as Darwin attempted to find the right order to convey his ideas most effectively, working out how images might play off one another, and which one would introduce the next. In one notable case (plate 6), he wove the images into a narrative: one portrait seemingly reacts to the indignant expression of another. This is purely a result of how the images are arranged on the page; the circumstances leading up to this mock confrontation are invisible in the images themselves. Interestingly, despite such careful contrivances, Darwin was often casual about referring to pictures in his texts. This is surprising, because without being clearly referenced in the text, the purpose of an image is inherently unclear. Images without explanation are cast adrift, free to assume a role assigned by the reader. This does not necessarily mean they are useless, however; Darwin may have hoped they would contribute an air of authenticity without being directly connected to the points he hoped to make.

Naturally, as in the case of the detached forehead, Darwin also had to vet each picture for internal content. For example, a photograph of a crying child (plate 4, figure 2) was useful to reproduce, but not in its raw state (figure 2-5, bottom). The chair the child leant against in the original version was extraneous, while a poodle cowering beneath the chair was distracting. To solve these problems, Darwin drew a line across the image above the chair and dog, indicating to the publisher that the image should be cut down (figure 2-5, top). Variations of this

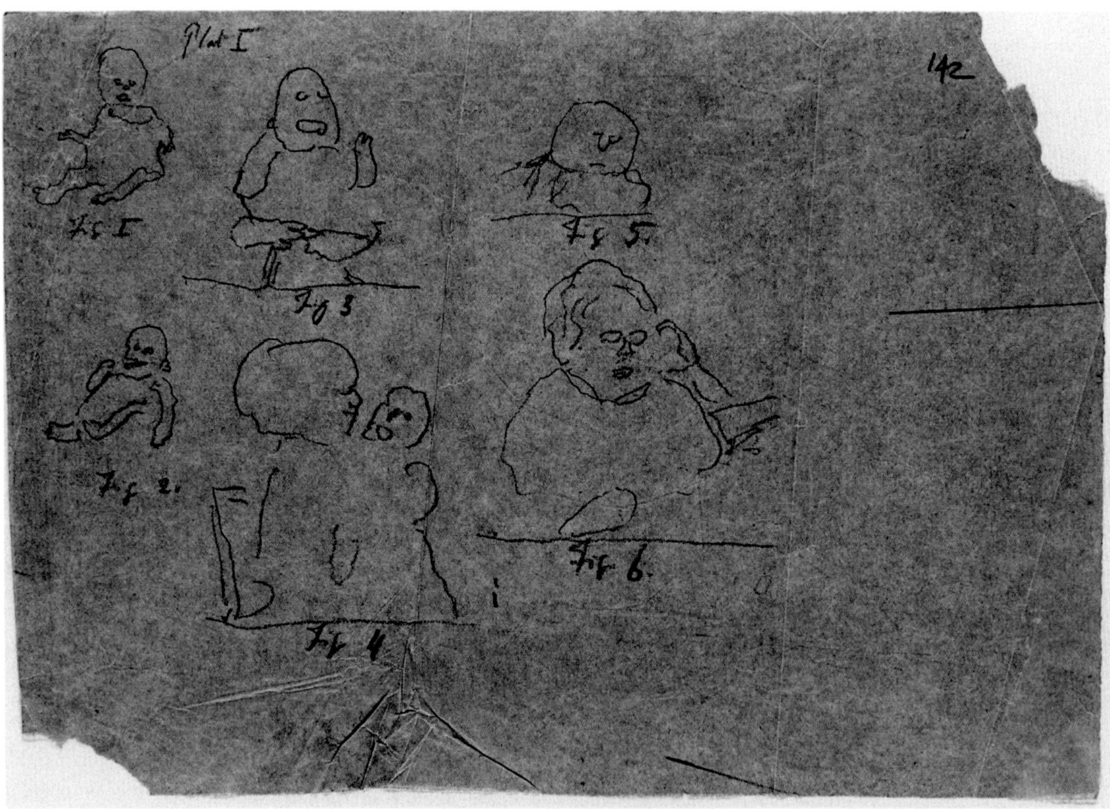

2-4 Charles Darwin, Layout for plate 1 of *Expression*, 1872. Pencil and colored pencil on tracing paper.

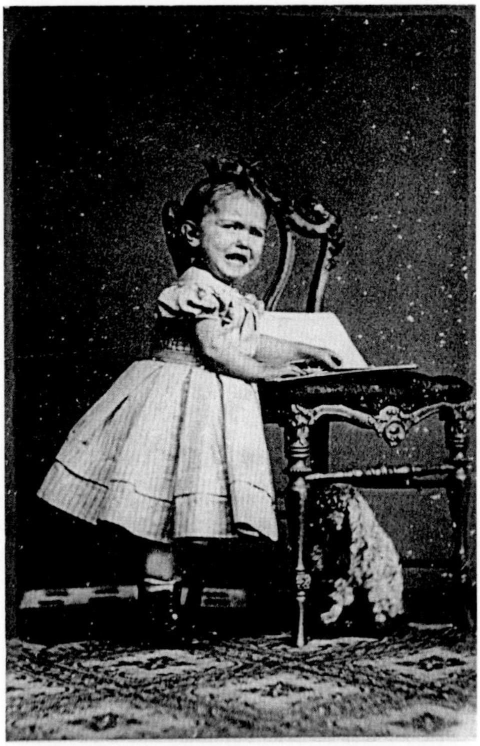

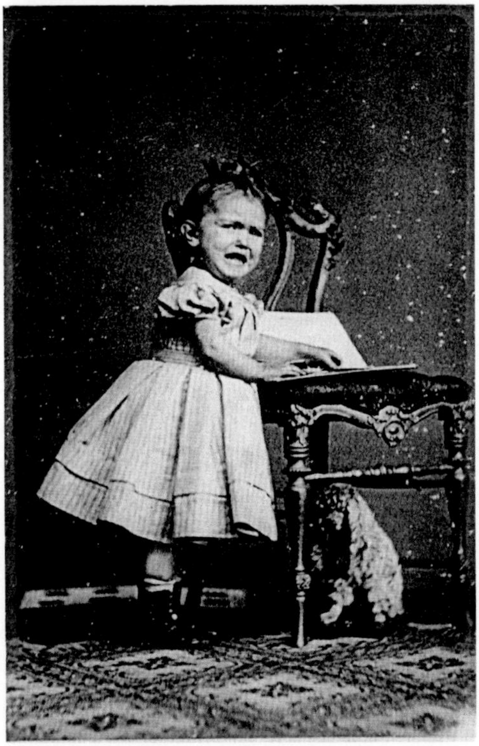

2-5 Adolph Diedrich Kindermann, *Defiant Girl*, c. 1865–70. Albumen prints.

process played out continually as Darwin assembled pictures for *Expression*. The goal was to find a balance between the compelling and the marginal, to choose a level of information most effective with readers.

Illustration as Strategy

For authors to succeed, they must attract and persuade readers. These are substantial obstacles to the acceptance of any new scientific theory, but they were especially challenging for Darwin. The broad shift in belief necessary to the acceptance of Darwinian theory was radical. It was particularly troubling to those who sought to explain biological phenomena in religious terms. It was also a substantial departure from much of the scientific work that preceded it. Darwin's ability to gain acceptance for his theories is remarkable given the explosive nature of his research and its cultural implications.

Numerous scholars, including Manier, Beer, Desmond and Moore, and others have suggested that the assimilation of Darwinian theory into Victorian culture was due in part to Darwin's rhetorical abilities. Building on their accomplishments, it is now possible to address how Darwin's visual acumen also contributed to his success. Like the text they partner with, illustrations can help convince readers of an author's points. Or, as Lewis Carroll's *Alice* put it: "[W]hat is the use of a book . . . without pictures or conversation?"[6]

Yet illustrations are also potentially hazardous. The sociologist of science Bruno Latour puts this point well: "Bringing pictures, figures, numbers and names into the text and then folding them in is a source of strength, but it may also turn out to be a major weakness. Like references, they show the reader what a statement is tied to, which also means the reader knows where to pull if he or she wishes to unravel the statement."[7]

Most of the figures in *Expression* are abstracted versions of larger pictures. As metaphor is to text, abstract images are an indirect means of communication; they force the reader to fill in gaps using personal knowledge. The art historian Ernst Gombrich referred to this phenomenon as "projection," the way in which an incomplete painting or sculpture "can force a beholder's imagination to project what is not there."[8] Darwin recognized the phenomenon of projection, although he never labeled it as such. In his notes he wrote:

> This unwillingness to consider creator as governing by laws is probably that as long as we consider each object an act of separate creation [*sic*]. We admire it more. Because we can compare it to the standard of our own minds. Which ceases to be the case when we consider the forma-

tion of laws invoking laws. & giving rise at last even to the perception of a final cause.[9]

At the same time, Darwin was concerned that his photographs could be abstracted too much and, as a result, be vulnerable to competing interpretations. The Darwin Archive reveals various instances in which he showed photographs to friends and colleagues to gauge their reactions. By presenting pictures to his colleagues and asking "What does this mean to you?" he was checking to be sure that they conveyed enough information to be useful. Their effectiveness was not a foregone conclusion. Darwin went to considerable effort to obtain clear photographs, and he was keen to ascertain whether his friends understood them to convey what he believed them to.

This was both a test of his pictures and an experiment in its own right. Darwin wished to know whether particular expressive features, such as the arrangement of muscular contractions of the face associated with smiling or crying, could be recognized without the visual and aural cues experienced in daily life. He sought to isolate particular elements of expression for comparison without the interference of corroborative information. He was genuinely interested in knowing the minimum amount of information required for an individual to recognize expressions. The understanding of expression was another one of the subjects Darwin was attempting to investigate.

Expressions themselves are communicative and visual. Consequently, his depictions of expression had a double role Darwin could not escape. For each photograph, there were at least two elements critical to their function: the meaning of the expression and the way it was received by those who saw it. In giving his photographs to others for evaluation, Darwin was learning, in a loosely deductive way, both what constitutes an expression and how such an expression is understood. In other words, he attempted to use his pictures not just as illustrations, but also as data. As such, the photographs in *Expression* are incomplete, however, because they were not made according to strict experimental controls. Instead, they invite the reader to evaluate them as true or false based on personal experience. This was a crucial part of Darwin's method—to invite readers into the anecdote, bringing them along as observers.

In her marvelous book *Nature Exposed: Photography as Eyewitness in Victorian Science* (2006), the historian Jennifer Tucker chronicles photography's emerging acceptance as an authoritative medium in the middle to late nineteenth century. Photography was not always considered superior to drawing and printmaking. Faster, yes, and potentially more convenient, but given the long exposure times early photography required, there was nothing that could be depicted photographically that could not be shown equally well with paper and pen. It was only when photography began to push into spheres beyond the limits of human vision—for example, rapid actions that happen too quickly for the naked eye to track—that photography came to assume a privileged place in the visual realm.

At the time *Expression* was published, photographs were judged on how real they looked, not on how scrupulously they had been produced. Later, as scientists began to use photography as evidence of events invisible to the unassisted eye, viewers began to demand proof that the photographs they witnessed were accurate. Scientists began to look more carefully at the circumstances in which photographs had been produced and devised protocols for collecting reliable photographic data. *Expression* was produced at the cusp of this change in attitude to photographic information and is an important part of the story of how photography came to be accepted as "objective." As the following chapters show, Darwin took liberties with some of his photographs that might be considered unacceptable in scientific work today. He did not do this to be deceptive or tricky—he simply tried to make the best of an imperfect medium at a time when there were no rules about what could or should be done photographically. He was, in a very real sense, making up rules as he went along. His work, and the way in which it was received, became part of the history of how photography achieved its now widely accepted status as authoritative.

Interestingly, Darwin was looking for pictures right at the threshold between what could be seen with the unassisted eye and what could only be seen photographically. What he really wanted became routine about a decade later, with the invention of speedy gelatin dry-plate chemistry of the kind Eadweard Muybridge (1830–1904) in California and Etienne-Jules Marey (1830–1904) in France used to analyze the gaits of galloping horses. Darwin hoped to find instantaneous photographs to enable him to differentiate between discrete phases of emotion, breaking behaviors lasting several seconds into intervals of 1/10th of a second, for example. Although one of his photographers, Oscar Rejlander, experimented with sequential imagery, he was unable to produce sequential pictures suitable to Darwin's purposes.

Ultimately, photography became a dominant means of representation in nineteenth-century Britain. Even before its status as an objective medium was established, the relative economy of the medium proved irresistible to Victorian consumers. By the 1860s, commercial mass production of photographic materials ensured that even individuals of modest income could afford to buy photographic prints. Even then, the quantity and quality of imagery that could be produced using photography was unprecedented.

Indeed, it is through photographs, as well as his writings, that Darwin is himself remembered. To those who knew him well, Charles Darwin was an expressive man, prone to dramatic gesticulations and facial movements. "When he was excited with pleasant talk his whole manner was wonderfully bright and animated," wrote his son Francis, "and his face shared to the full in the general animation." This is a little-known side of Darwin, whose character the public have deduced largely through austere portrait engravings, and photographs imbued with the air of intellectual gravitas that his peers thought befitting a scientific celebrity. In life, however, his laugh was a "free and sounding peal, like that of a man who gives himself sympathetically and with enjoyment to the person and the thing which have amused him." Francis also remembered his lively use of his hands, which he would wave to indicate some nebulous concept, or bring down on his leg with a slap. "He was given to gesture and often used his hands in explaining anything," he remarked, "in a way that seemed rather an aid to himself rather than the listener. He did this on occasions when most people would illustrate their explanations by means of a rough pencil sketch."[10]

Portraits made of Darwin in his lifetime show the appearance of the man, but like the disembodied forehead in *Expression*, they show only a narrow slice of him, omitting elements their makers could not or did not wish to convey. For a scientist who routinely used his own expressions to illustrate ideas in conversation, this leaves them substantially incomplete. Darwin was a man whose revolutionary efforts to illustrate his books prompted new ways of thinking about pictures. What we can glean of his working methods can only be reconstructed from his magnificent legacy of published and unpublished works and the amazing pictures he left on the cutting-room floor.

3

Art, Experience, and Observation

Darwin's knowledge of art history and use of illustration in his books

When Charles Darwin arrived as an undergraduate at the University of Cambridge in 1828, he must have felt relief. He had escaped the sickening lectures he endured as a medical student at the University of Edinburgh, with their grisly investigations into the anatomy of cadavers. For a young man with a weak stomach, nineteenth-century clinical procedures were difficult to endure. Dissections took place without refrigeration, and operations were performed without anesthesia. Three years earlier, in 1825, he had enrolled at Edinburgh at the insistence of his father. He stayed for two years, studying anatomy, chemistry, and other medical matters. Some of the lectures Darwin considered boring, but others were simply horrifying. His experience at Cambridge was much more pleasant. Now a cleric in training, he may still have been subjected to dull lectures, but they were at least inoffensive, and his new course of studies provided opportunity to pursue leisure interests. Afternoons were spent hunting beetles in the neighboring town of Cherry Hinton or strolling the grounds of King's College Chapel listening to the sounds of the daily choral anthem. It was also during these years at Cambridge, from 1828 to 1831, that Darwin began to develop an interest in the visual arts.

Unlike many of his colleagues, Darwin had little talent as a draftsman. He knew almost nothing about the proper handling of line or the effective manipulation of tone, texture, form, perspective, and color. His research notes include some pen-and-pencil sketches, but for the most part they are crude and tentative. He had no formal training in visual art; nearly all his drawings were traced from books or copied under the microscope. They are mainly thumbnail sketches of ideas or observations more easily recorded in pictures than in words.

Darwin—the dedicated scientist, obsessed with beetles, shells, and finches—is well known from standard biographies. But there is another side of the man. Quietly, he could be found studying paintings in the National Gallery, collecting prints for the walls of his dormitory, debating evolution with John Ruskin (1819-1900) and the Pre-Raphaelite sculptor Thomas Woolner (1825–92) and building a library that included some of the seminal art books of his time. There are many more surprising and consequential connections between Darwin and the arts. He was a man who became fascinated by and deeply immersed in the relatively new realm of photography; who commissioned the most controversial photographic artist of his time, Oscar Rejlander, to create pictures to illustrate his book on facial expressions; who hunted for interesting photographs in shops and studios across London; and who published one of the first science books illustrated with photographs. Visual art was a small part of Darwin's life, but it was nevertheless important. Darwin had omnivorous tastes, and art was just one of the things he consumed. There was little in intellectual life that did not intrigue him, and he never ceased to crave new information. Yet few realize how profoundly Darwin shaped the visual culture of his time, and how it, in turn, shaped him.

There were at least two factors at play in Darwin's encyclopedic interests. First, and most obvious, he had an extremely lively mind. When he was not out gathering information personally, as he famously did on the voyage of H.M.S. *Beagle*, he sat in his study like a spider at the center of an immense web, gathering morsels of information as a spider does flies. He gathered as much information as he could through correspondence, and he pored over what he found, discarding the irrelevant and following up promising leads. His genius lay not just in the quantity of information he obtained, which would have overwhelmed a lesser mind, but in the way he processed it. Large amounts of observations were digested, and every piece of evidence was assigned its proper place. Darwin had the rare talent of making order out of the obscure: distilling thousands of facts and assertions to their vital essence and seeing the links between them.

The second reason Darwin was so wide-ranging in his interests is the age in which he worked. The twentieth and twenty-first centuries are characterized by increased specialization of research, by the funneling of ideas into narrowly defined fields of study. It is now commonplace to draw distinctions between closely related fields, as one does, for example, between biochemistry, molecular biology, and genetics. Darwin bridged these fields and many more. His work drew on the accumulated knowledge of disciplines now labeled as geology, geography, population biology, psychology, systematics, anthropology, animal husbandry, and anatomy, among others. However, he worked at a time when the very idea of "science" as an enterprise was new. There were few preconceived ideas of what a scientist should do, because science was itself a fundamentally nebulous undertaking. No rules existed, written or otherwise, to discourage Darwin from exploring diverse fields of study. Why not turn to Plato and Aristotle, or Shakespeare and Milton for inspiration? Why, indeed, not examine paintings?

Art and Cambridge

"Oh Lord what a jolly place Cambridge is," Darwin wrote to his friend Charles Whitley shortly before his departure on the *Beagle* in 1831. "But it is all over," he sighed, "there is no use thinking about it."[1] Their university days behind them, Darwin embarked on the surveying mission that would help define his career, while Whitley began a quiet life as a vicar. A classmate at the Shrewsbury School and a friend of Darwin's brother Erasmus, Whitley had become one of his best friends at the university. He was a founding member of a group its members lovingly described as the Glutton Club, and Darwin's principal guide to the mildly hedonistic delights of the city, including art, music, and the sensation "which some call sublime, but the wise know to be the full round feeling from a contented dinner."[2] Whitley also cultivated Darwin's taste in prints, escorting him to view works from the Fitzwilliam Bequest at Free School Lane, precursor of the Fitzwilliam Museum, and it was under his influence that Darwin purchased several "first class line-engravings," as the historian Janet Browne has noted. In December 1828, Darwin wrote to Erasmus, telling him of his latest hobby: "I imbibed your tastes about prints, and put it into practice, and have bought some very good prints, which I long for you to see."[3] According to Browne, these included a 1793 study after Raphael by Florentine engraver Rafaello Morghen, a portrait of Leonardo, and an unidentified print by Johann Mueller, whose chief engraved work was the *Sistine Madonna*.[4]

The Fitzwilliam Museum did not move into its current premises on Trumpington Street in Cambridge until 1848, when George Basevi's imposing neoclassical structure was completed. Housed in the earlier space on Free School Lane was an impressive group of 144 paintings of the Italian, Dutch, and Flemish schools, willed to the university by the 7th Viscount Fitzwilliam of Merrion (1745–1816).[5] Among the finest of these were such paintings as Titian's *Venus and a Lute Player*, Rem-

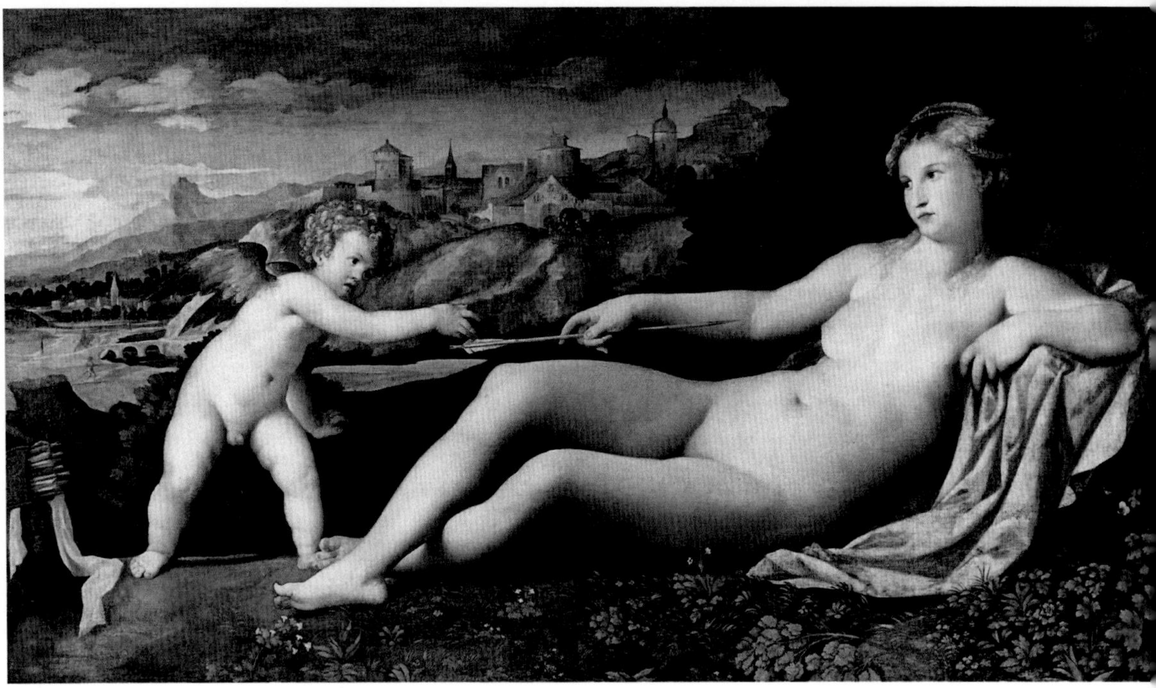

3-1 Palma Vecchio, *Venus and Cupid in a Landscape*, 1522–24. Oil on canvas.

brandt's *Portrait of a Man*, and Lodovico Carracci's *Christ Appearing to the Virgin*. Palma Vecchio's High Renaissance *Venus and Cupid* (figure 3-1) is typical of the style of pictures Darwin would have seen: naturalistic in its portrayal and imbued with narrative content, but not obviously scientific. The print collection, which Darwin and Whitley especially enjoyed, consisted of bound albums that could be viewed by appointment. Some seven years after he left Cambridge, Darwin noted in his Notebook "M," in which he recorded observations on expression, that the Fitzwilliam bequest was still having an effect on him. "When in the National Institution & not feeling much enthusiasm, happened to go close to one & smelt the particular smell of Picture. Association with much pleasure immediately thrilled across me, bringing up old indistinct ideas of Fitzwilliam Museum."[6]

After Darwin left Cambridge for his five-year voyage on the *Beagle*, his interest in natural history intensified and his friendship with Whitley weakened. In 1835, as the *Beagle* explored the coast off Valparaiso, Chile, Whitley wrote to Darwin, anticipating his return and reminiscing about their mutual interests. He invited Darwin to visit him in London: "There are plenty of new things for you to see and do here by way of music & painting. I have made some superb additions to my family of prints, and there are some glorious treasures added to the National Gallery. So let us look forward to intellectual pleasures yet to come."[7] Darwin, however, was thousands of miles away. Despite Whitley's hopes, he had become preoccupied with natural history and had set aside his interest in fine art.

Darwin's exposure to art at Cambridge was a significant part of his education. The interest he had displayed in it as a student reflected his strong collector's mentality, a trait that distinguished him throughout his life. Yet the effects may have been more profound. "Ultimately more significant," Browne has argued,

> Whitley seems to have taught Darwin how to look—to search out the meaning behind the artist's composition and patiently to follow through the layers of technique and allusion. Though he learned a good deal about how to look at nature from Robert Grant [Darwin's associate at the University of Edinburgh], this cultural experience of "looking" gained through aesthetics was equally important in the development of his sensibilities.[8]

Darwin himself provides one of the strongest endorsements of Browne's view. Writing about William Whewell's *History of the Inductive Sciences* (1837), he noted, "All Science is reason acting (systematizing) on principles, which even animals practically know (art precedes science—art is experience & observation)."[9]

Some early documents in the Darwin Archive hint at an aesthetic interest predating his matriculation at Cambridge. With his rich imagination, restlessness, and eye for visual meaning, the engraving printed prominently on the front of his membership card for Edinburgh's Plinian Society, for example, would not have escaped his notice. Depicting a ship voyaging at sea, it could as easily have served as an invitation to the voyage of the *Beagle*. The society's slogan, emblazoned below the engraving, was taken from Virgil's *Fourth Eclogue*: "*Terrasque traetusque maris coelumque profundum*," it teased; or, "Earth and wide ocean, and the heavens profound."[10] The archive also contains notes for a talk Darwin gave at Edinburgh University in 1826, titled "Zoological Walk to Portobello" (a beach resort east of the city), in which he critiqued the city's architecture. Young and sentimental, he romanticized the royal palace of Holyroodhouse, challenging: "Who can see this venerable pile without at the same time connecting in his mind the various scenes and changes it has witnessed?"[11] Not all his comments were so sanguine. Later in the same lecture he expressed growing disaffection with Edinburgh, at the same time revealing his Whig leanings. He railed at violations of the city's natural landscape. "The far-famed Salisbury Crag," he scoffed, "another striking specimen of Scottish taste—not of picturesque beauty, but of money. At one time this vaunted hill was perchance an ornament to Edinburgh. Now it merely stands a monument."[12]

Illustration in Darwin's Work

Although as a student Darwin admired, collected, and even criticized art and architecture, the importance of imagery of all types in his work is seldom acknowledged. In his landmark essay "The Emergence of a Visual Language for Geological Science," for example, the historian Martin Rudwick goes so far as to describe Darwin as a nonvisual scientist, citing the lack of illustrations in his notebooks and published works.[13] Rudwick differentiates between communicative and conceptual visualization, stressing the relatively low number of printed pictures in Darwin's works. He defines visual communication as writing with pictures: complementing texts with "maps, sections, landscapes and diagrams of other kinds,"[14] distinguishing this from the less tangible notion of conceptual visualization, such as that involved in the formulation of an evolutionary theory. According to Rudwick, Darwin was a nonvisual author, but not necessarily a nonvisual thinker. His works simply contain fewer illustrations than those of his contemporaries.

The idea that Darwin was not particularly interested in illustration may seem beyond doubt. But this view is based on three widespread misconceptions about his work. First is the belief that he used few illustrations in his books, as expressed by Rudwick. Darwin himself perpetuated an additional misconception, by publicly decrying his inability to draw well. And finally, despite his early association with Whitley, he is thought to have had little personal and professional interest in fine art. Collectively, these ideas disguise the true character of Darwin's use of illustrations. Throughout his career, Darwin recognized the importance of supporting his arguments clearly.

The most conspicuous of the contentions about Darwin and art, that he used comparatively few pictures to illustrate his texts, is based on a narrow selection of his works. His two most famous books, *Journal of Researches into the Geology and Natural History of the Countries Visited by H.M.S. Beagle etc.* (1839), and *On the Origin of Species* (1859), are indeed sparsely illustrated. But a different picture emerges when his entire output of seventeen books is considered. Table 1 summarizes these illustrations, as they appeared in the first editions of Darwin's books.

The numbers listed in table 1 are approximate, because many of the listed plates consist of two or more separate illustrations. The seven plates in *The Expression of the Emotions* (1872), for example, actually include thirty individual photographs. Similarly, the individual wood engravings tallied in his other books were in many cases assembled from a number of different drawings. For example, figure 1 of *Fertilisation of Orchids*, listed as one of thirty-three wood engravings, contains six related renderings of the orchid *mascula*, and figure 2 was made from two draw-

Table 1 Illustrations in Darwin's Books (First Editions)	
Book	**Illustrations**
Journal of Researches, or *The Voyage of the Beagle* (1839)	2 charts, 4 wood engravings
The Structure and Distribution of Coral Reefs (1842)	3 maps, 6 wood engravings
Geological observations:	
Volcanic Islands (1844)	1 map, 14 wood engravings
South America (1846)	1 map, 5 plates (1 colored)
A Monograph of the Sub-class Cirripedia (1851)	40 plates (3 colored), 24 wood engravings
A Monograph on the Fossil Lepadidae (1851)	7 plates, 12 wood engravings
On the Origin of Species (1859)	1 diagram
Contrivances by which British and Foreign Orchids are Fertilised by Insects (1862)	1 plate, 33 wood engravings
Variation under Domestication (1868)	43 wood engravings
Descent of Man (1870)	76 wood engravings
Expression of the Emotions in Man and Animals (1872)	7 photographic plates, 21 wood engravings[a]
Climbing Plants (1875)	3 wood engravings
Insectivorous Plants (1875)	30 wood engravings
Cross and Self Fertilisation (1876)	1 diagram
Different Forms of Flowers (1877)	15 wood engravings
Movement in Plants (1880)	196 wood engravings
The Formation of Vegetable Mould (1881)	15 wood engravings

Compiled from Richard Broke Freeman, *The Works of Charles Darwin: An Annotated Bibliographic Handlist*, 2nd edition [Folkestone, England: Wm. Dawson & Sons, 1977].

[a]Freeman neglected to include these 21 wood engravings in his count.

ings of orchid pollen masses. The quantity of illustrations is amplified if the increased numbers of pictures appearing in later, revised editions is considered. The first printing of *Coral Reefs*, for example, was published with three maps and six wood engravings. The second edition contained the same number of maps but fourteen wood engravings. The number of illustrations in other books also increased. As a result, the numbers in table 1 show, at most, only the quantity of plates rather than the much larger number of illustrations. Together, the seventeen volumes listed account for at least 561 maps, plates, and wood engravings.

Table 1 also excludes the sumptuously illustrated, nineteen-part *Zoology of the Voyage of the Beagle* (1838–43), which Darwin edited. Although each volume in the series had its own author, Darwin oversaw this extraordinary publication and had a hand in selecting and commissioning its illustrations. Arguably one of the masterworks of natural history illustration, *Zoology* is composed primarily of richly detailed, full-plate lithographs and engravings, including many spectacular images, such as the British artist Sir George Scharf's (1788–1860) massive slip-cased lithograph *Base of a Skull of Toxodon platensis*, and some fifty full-page, hand-colored lithographs illustrating part XI, legendary British ornithologist and illustrator John Gould's (1804–81) celebrated volume on birds.

Darwin's first published book, *Journal of Researches*, or *Voyage of the Beagle*, was originally the last part of the three-volume *Narrative of the Voyages of H.M. Ships Adventure and Beagle* (1839), edited by Captain Robert FitzRoy. Later that year *Voyage of the Beagle* was rereleased as a separate volume. The close relationship between the nineteen-volume *Zoology*, the three-volume *Narrative*, and the singular *Voyage* complicates analysis of the illustrations in each. Although *Voyage* has enjoyed a substantial life of its own, the number of pictures it contains was determined by the other two sets. As a part of FitzRoy's *Narrative*, the illustrations in Darwin's *Voyage* were chosen to avoid duplication. FitzRoy's diary of the *Beagle* voyage, which constitutes the second volume of the *Narrative*, includes 25 plates and charts. This reduced the need for illustrations in Darwin's account. The *Narrative*, in turn, was intended to complement *Zoology*, and this also affected the number of illustrations it contains.

Further complicating matters, various editions of Darwin's *Voyage of the Beagle* contain different numbers of illustrations. The first edition contained just four wood engravings; the second, fourteen. And in 1890, John Murray released a posthumous edition with two maps and ninety-three wood engravings. However, in modern editions there has been a tendency to reduce the number of illustrations.[15] The popular Penguin edition, for example, an abbreviated version of the original, includes updated renditions of Darwin's maps but omits the wood engravings. This belies the considerable number of images historically associated with the book and fuels the perception that Darwin was not concerned with illustration.

On the Origin of Species includes just one drawing, Darwin's schematic diagram of evolution, usually referred to as the "Tree of Life," but referred to in his notes alternately as "branching seaweed" and "the coral of life." The Tree of Life is arguably the least appropriate of these names; the metaphor implies that each leaf and branch remains vital and is connected to a single root. Darwin's diagram, by contrast, represents both extinct and surviving lineages and is designed to show discontinuous phases of evolution. Darwin argued that living creatures may not have contemporary ancestors because the species from which they evolved may have become extinct. The diagram, rendered with broken lines and discontinuous branches, is intended to convey the erratic course of evolution. Thus, he wrote in Notebook "M": "The tree of life should perhaps be called the coral of life, base of branches dead; so the passages cannot be seen."[16]

It would be wrong to conclude that Darwin was not interested in illustration on the evidence of *Origin* alone. The book is unique among his works, and somewhat unusual for scientific texts generally for its sweeping theoretical scope. Its central message is difficult to illustrate. As the scholar Howard Gruber has noted, the Tree of Life serves to convey the major elements of his evolutionary theory. It illustrates the fortuitousness of nature, the explosiveness of growth and its limitations, and "most important, the fundamental duality that at any time some must live and others die."[17] Although it is only one illustration, the Tree of Life thus serves as an index to the entire theory of evolution by natural selection and was arguably the most effective way to illustrate the essence of Darwin's ideas.

Scientist and Draftsman

A second misconception is that Darwin seldom drew pictures himself. He perpetuated this belief in his *Autobiography*, identifying his two greatest failings as the "irremediable evil" of his inability to dissect (which he blamed on the poor quality of the lectures at Edinburgh) and his "incapacity to draw."[18] Later in the book he expressed regret that his research on the *Beagle* voyage was not as effective as it would have been if he had been better able to record what he had seen. "From not being able to draw and from not having sufficient anatomical knowledge a great pile of MS. which I made during the voyage has proved almost useless," he wrote.[19] Privately, he repeated the regret. Writing to his close friend the scientist Thomas Henry Huxley, he complimented him on two drawings of barnacles he had sent: "I think your diagrams will do very well—What an advantage to be able to sketch easily! No one has a right to attempt to be a naturalist who cannot."[20] At the end of his career his self-deprecation had not wavered. Writing to John Collier (1850–1934), the artist who painted his portrait for the Linnean Society in 1881, he wrote: "I must thank you for the gift of your art primer, which I have read with much pleasure. Parts were too technical for me who could never draw a line, but I was greatly interested."[21]

Darwin suffered in comparison to Huxley, who was a talented draftsman and who frequently illustrated his own works. Huxley also supplied drawings for others, including numerous depictions of barnacles that Darwin included in his two volumes on barnacles (or *cirrepedia*) (1851-1854).[22] As Huxley's grandson Julian wrote, "Huxley was a great scientist and a great man of letters; if he had had time and opportunity he would have been a great artist as well."[23]

Although Darwin downplayed his drawing skills, he was able to make basic pictures and occasionally used them to illustrate his books. For example, many of the engravings in *Insectivorous Plants* (1875), attributed to his sons George and Francis, are based on preliminary sketches he made himself. This was not unusual; Darwin frequently roughed out his conception of a drawing for a more talented hand to clean up later. His sketch under the microscope of the vascular tissue in a small leaf of *Drosera rotundifolia* (a kind of sundew; figure 3-2) is typical of the process.[24] Once complete, his son George rendered the drawing more clearly, as the note "see George's drawing" on the lower margin of the sketch indicates. The new drawing was then passed to the engraver, James Davis Cooper (1823–1904).[25] Cooper's engraving is figure 11 of the book (figure 3-3).

Darwin usually confined himself to drawings like this section of *Drosera*—simple images that could be drawn in two dimensions, such as microscopic sections or geologic strata. Unsure and slightly imprecise, they betray Darwin's struggles as he tried to transmit what he saw to paper. A critic would agree with Darwin's professed lack of talent: the *Drosera* sketch is shaky and uncertain, and some of the original pencil lines are drawn over in ink to clarify details unclear in the underlying strokes.

Yet Darwin was exacting in the demands he made of the various artists he employed. Like the figures in *Insectivorous Plants*, several of the illustrations in the barnacle volumes were adapted from sketches he supplied personally. One of Darwin's collaborators was the distinguished artist and naturalist James de Carle Sowerby (1787–1871). Sending him corrections to some of the plates in 1850, Darwin was firm in his request for clarity:

> I yesterday went carefully over some of your drawings. The outlines appear to me very accurate; but yet a few alterations are wanted in most of them; but these are trifling & refer chiefly to lines of growth.—The scale is not at all too large. What I now write for, is to beg you to do them a little harder & with the lines of growth more distinct. Some

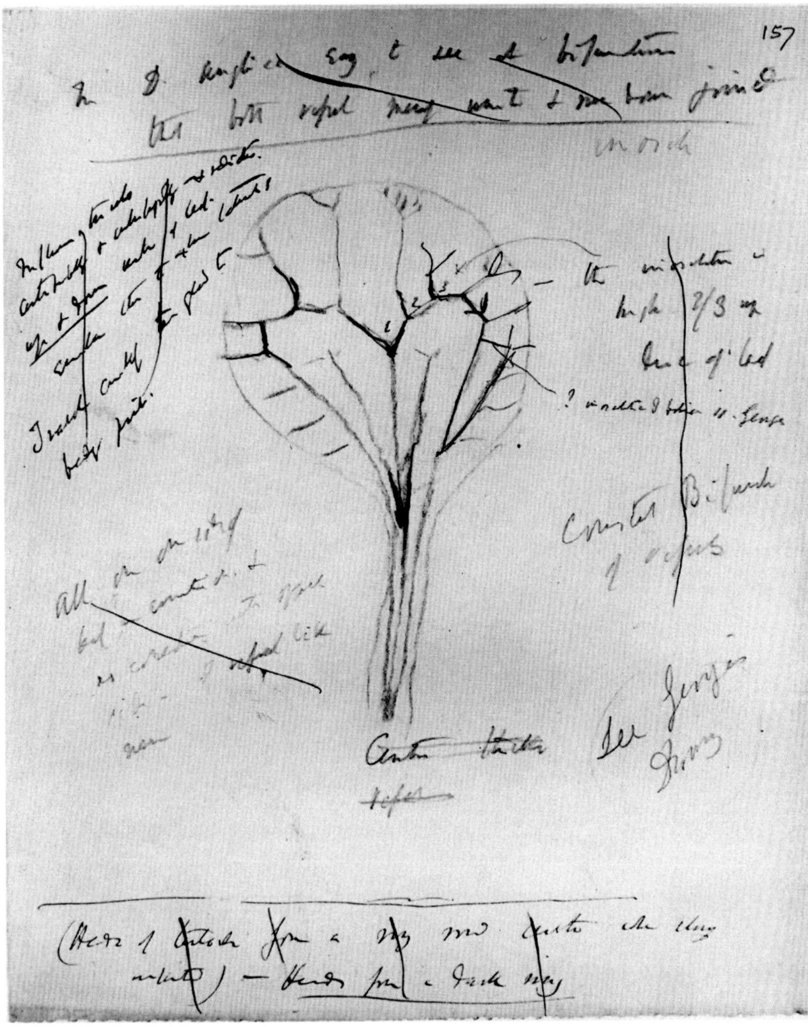

3-2 Charles Darwin, Vascular Tissue in a Small Leaf of *Drosera rotundifolia*. Ink and pencil. Darwin sometimes drew his own illustrations.

> of the drawings have the muzziness of Lithography,—a style of art (viz. Lithog.) which in my opinion has been highly injurious to Nat. History—I do not care for artistic effect, but only for hard rigid accuracy.[26]

Notwithstanding the success of the lithographic plates in *The Zoology of the Beagle*, Darwin's aversion to lithographic printing persisted throughout his career. Of his seventeen books and numerous articles, none was illustrated with lithographs. Moreover, he persisted in his wish for "hard rigid accuracy." It is one of the main factors that drove his use of photography in later works, both in direct form in *Expression* (1872) and copied as engravings in *The Variation of Animals and Plants under Domestication* (1868), *The Descent of Man, and Selection in Relation to Sex* (1870), and *The Formation of Vegetable Mould through the Action of*

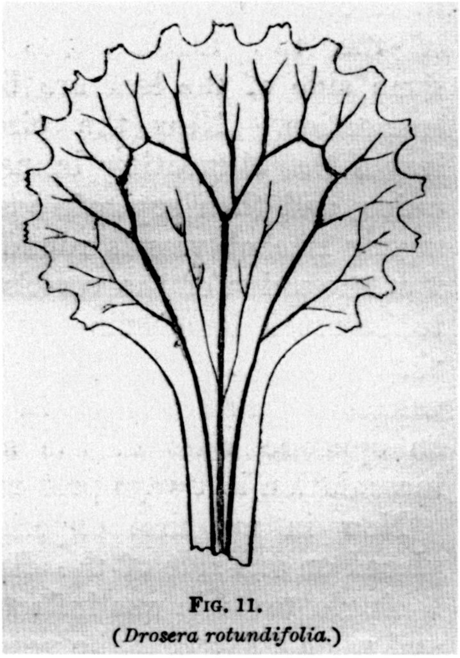

3-3 George Darwin after Charles Darwin, Vascular Tissue in a Small Leaf of *Drosera rotundifolia*. Wood engraving. Figure 11 of *Insectivorous Plants* (1875). Darwin's drawing, drawn more clearly for publication.

Worms (1881).[27] The earliest of these appeared in *Domestication* in 1868. The subject was a large Yorkshire pig named Golden Days, which appeared as figure 3 in the book; the photograph on which it is based is now lost (figure 3-4).

Art World Contacts

Darwin's bookshelf contained a surprising number of books both directly and indirectly related to theories of art, indicating comfort with developments in the field. In addition to Edmund Burke's *Philosophical Enquiry into the Origin of Our Ideas of the Sublime and the Beautiful* (1757), there were George Harris's *The Theory of the Arts* (1869), William Hogarth's *Analysis of Beauty* (1753), Augustus Pugin's *Contrasts or Parallels between the Noble Edifices of the Middle Ages and Corresponding Buildings of the Present Day* (1841), and Gustav Zerffi's *Manual of the Historical Development of Art* (1876).[28] Such books constituted a small portion of the books Darwin owned, but they represented an active dimension of his interests.

Darwin was also personally connected to a significant number of influential figures in the art world. From his grandfather Josiah Wedgwood, founder of the Wedgwood pottery in Etruria (Stoke-on-Trent), he inherited a sizable collection of ceramics and maquettes, including a group of thirty-three early wax reliefs by Giuseppe Angelini

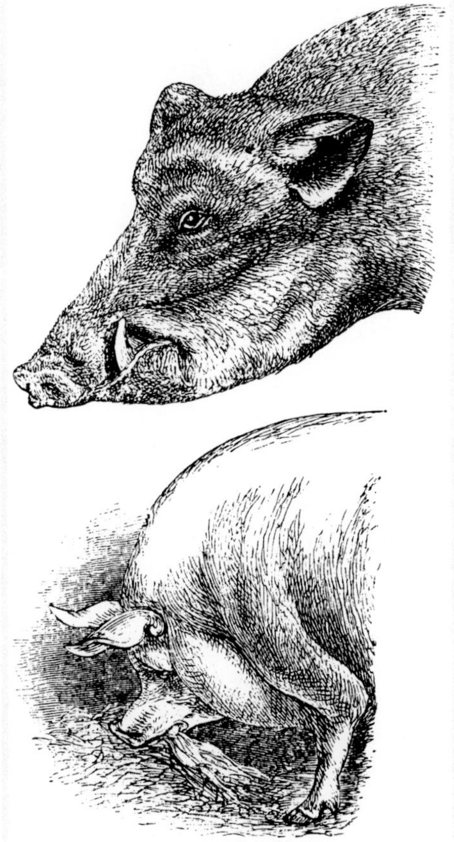

3-4 Maker unknown. Head of a Wild Boar and of "Golden Days," a Pig of the Yorkshire Large Breed, the Latter from a Photograph, c. 1860. Wood engraving. Figure 3, *The Variation of Animals and Plants under Domestication* (1868). A well-fed pig was the subject of Darwin's first photographic illustration.

(1735–1811), John Flaxman (1755–1826), Camillo Pacetti (1758–1826), and others (figure 3-5), which he eventually sold.[29] He also owned an early Wedgwood copy of Sir William Hamilton's renowned Portland Vase, the exquisite Roman first-century B.C. cameo glass vessel that inspired countless artists after its arrival in Britain in 1778. His other grandfather, Erasmus Darwin, was famously acquainted with the artists Joseph Wright of Derby (1734–97) and George Stubbs (1724–1806). Erasmus's collection of works by these artists was spread throughout the family. John James Audubon (1785–1851) was also a family friend. Audubon's wife, née Lucy Green Bakewell, was a daughter of William Bakewell, another of Erasmus Darwin's associates.[30] In 1826, Audubon lectured in Edinburgh, and as a student Darwin saw him speak. Shortly thereafter, Audubon visited the Derbyshire town of Matlock to pay homage to Erasmus Darwin's tomb, signifying his admiration for the Darwin family.[31]

Like his grandparents, Darwin was acquainted with a number of prominent artists. Among them was John Ruskin, whom Darwin first met at the home of Oxford natural history professor William Buckland

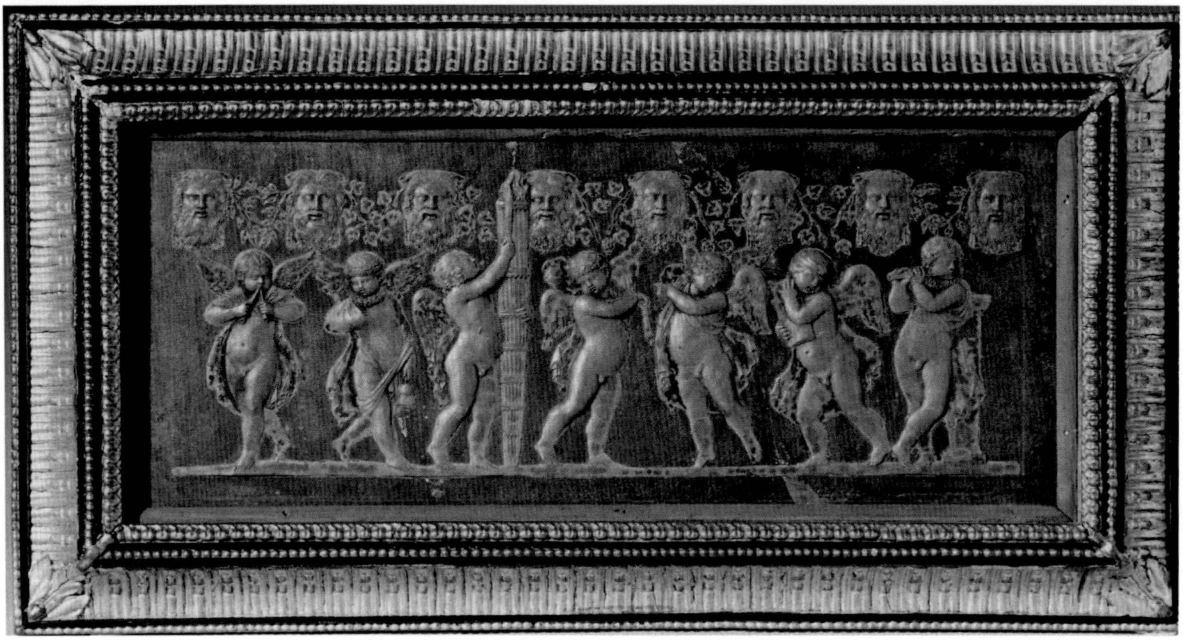

(1784–1856) in 1837.[32] After university, the two lost touch until reunited by the American art historian Charles Eliot Norton (1827–1908), who took up residence at Keston Rectory, near Down House, in 1868.[33] As Darwin became one of England's preeminent celebrities, he was introduced to numerous other influential artists, critics, and collectors. One week, for example, he and his wife Emma might dine with Philip Stanhope (1805–75), director of the National Gallery; another they might stay with the photographer Julia Margaret Cameron in one of her cottages on the Isle of Wight. Though the influence these figures had on Darwin's thinking is difficult to gauge, it is nevertheless apparent Darwin had access to current thinking about art and illustration.

3-5 Giuseppe Angelini, *Geniuses of the Elysian Fields*, 1789. Wax on slate. One of thirty-three wax reliefs Darwin inherited from Josiah Wedgwood and later sold.

These were the resources Darwin had at his disposal as he gathered illustrations for his most richly illustrated work, *The Expression of the Emotions in Man and Animals*, in the late 1860s. Though a mature man of sixty-three, the author of *Expression* was a man Charles Whitley would have easily recognized during their days at Cambridge. In his memoirs, Darwin recalled his frequent visits to the Fitzwilliam with his college companion. "My taste must have been pretty good," he wrote, "for I certainly admired the best pictures."[34]

Expression is the book around which Darwin's ideas about art coalesced. It contained the most ambitious program of illustrations and is also the only one of his books to contain original photographs. The text is relatively short, but the photographs it contains are unlike anything that came before. It also contains a noteworthy selection of wood engravings of animals in expressive postures. Written at a time when Darwin was one of the world's most famous scientists, the book and its illustrations commanded immediate attention upon their publication. Though produced by a man possessing little ability to create illustrations himself, they helped to redefine what was possible, and consequently what was desirable, in scientific illustration.

4

Darwin and the Passions

How passion manuals informed Darwin's research

Expression built on three separate but related traditions of illustrated books: physiognomy treatises, passion manuals, and anatomical studies. If Darwin was occasionally guilty of blurring the distinction between art and science in his book, it was understandable given the diversity of publications about facial structure and expression. He was familiar with books illustrating the "passions," as emotional expressions were habitually called, and owned a number of such volumes himself. The questions of how emotions are expressed and what they represent recur frequently in the history of ideas. One of the earliest known pronouncements on the subject is an Aristotelian text that declares: "[T]he soul and body react to each other; when the character of the soul changes, it changes also the form of the body, and conversely, when the form of the body changes, it changes the character of the soul."[1] The passions were seen as the visible incarnation of the soul manifesting in the body. In the second century, the Greek physician Galen proposed that the study of expression should be considered a medical subject. Arguing that expressions manifest in human bodies, he proposed that to understand them fully one must undertake physiological research.

By the sixteenth century, artists were devoting considerable effort to the study of both physiognomy (the external appearance of inner character) and pathognomy (the external appearance of the passions). As early as 1501, the Italian painter Jacopo de Barbari (c. 1440–1516) complained that contemporary paintings were "improbable and false" because painters lacked the knowledge of astronomy necessary to understand facial expression:

> Physiognomy and chiromancy [palm reading] have their being by the force of the stars, which exert their greatest influence over man's outward appearance in the face and hands. And it is necessary that painters should understand this, in order to record the celestial influences appropriately in their painted histories.[2]

From the Renaissance forward, a variety of books were produced depicting idealized pathognomic and physiognomic types. These were intended to help students learn how to communicate properly the precise character of people in their compositions. Artist's manuals were seen as dependent on close observation, for it was only through accurate understanding of the mechanisms of human expression that an artist could hope to paint a true likeness. Albrecht Dürer, Leonardo da Vinci, and Charles le Brun are just a few of the artists who devoted themselves to documenting ideal expressive types.[3]

Giambattista della Porta (c. 1535–1615) helped to establish the pseudoscience of physiognomy with his *On Human Physiognomy* [*De Humana Physiognomia*, or *Della fisonomia dell'huomo*] (1586), categorizing individuals according to their superficial resemblance to various animals. The character of that animal dictated the personality of its human counterpart. Thus, the man who looks like an ox (figure 4-1) would be brash, stubborn, and dim-witted. According to this theory, one's character is innate and predetermined. A virtual zoo of personalities was described in della Porta's treatise, from owllike and hawklike people to those resembling leopards and dogs. One illustration even likened a

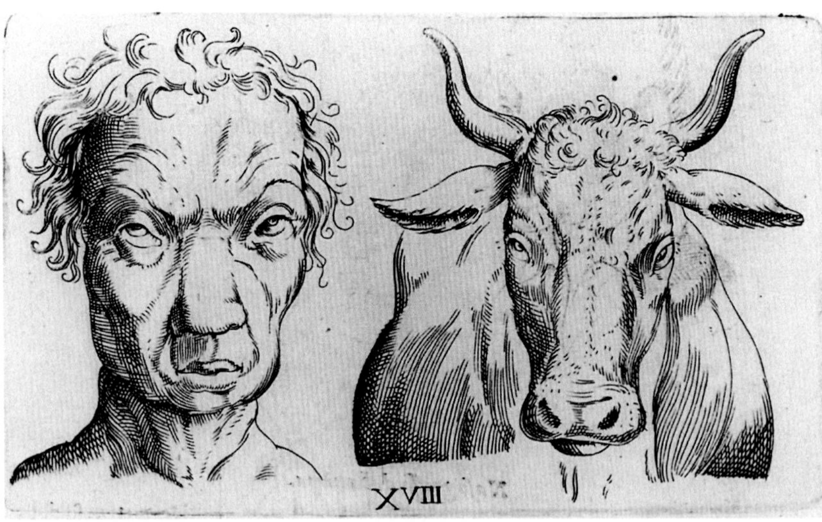

4-1 Giambattista della Porta, Comparing the Broad Nose of an Ox with That of a Similar One on a Man. Plate 18, *Della fisonomia dell'huomo*, 1610 edition. Oxlike features were said to indicate a boorish personality.

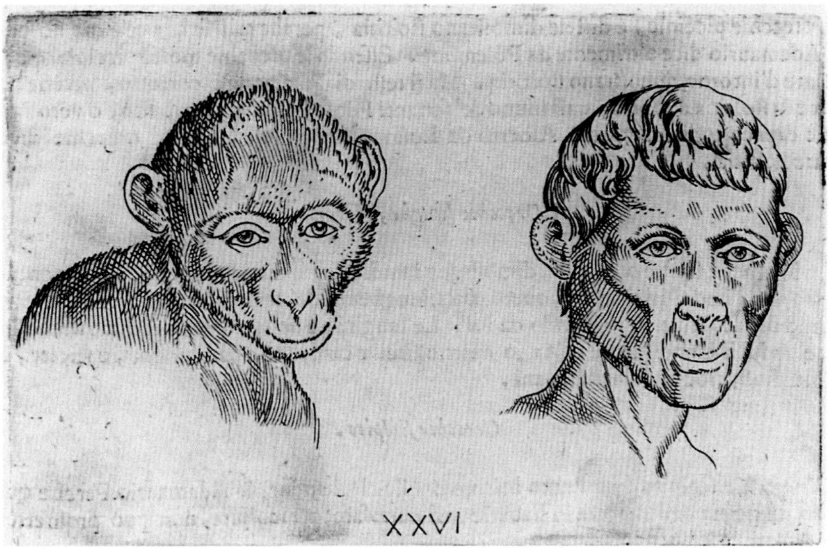

4-2 Giambattista della Porta, Comparing the Extremely Small Ears of a Monkey with Those of a Man. Plate 16, *Della fisonomia dell'huomo*, 1610 edition.

square-jawed man to a monkey, a proto-evolutionary vision that Darwin would surely have appreciated (figure 4-2).

The most popular treatise on expression was published by the Swiss theologian Johann Caspar Lavater (1741-1801) in 1772. His *Fragments of Physiognomy* [*Physiognomische Fragmente*] was elaborately adorned with illustrations by numerous influential artists, including the American Benjamin West (1738–1820) and the Britons Henry Fuseli (1741–1825) and William Blake (1757–1827). The book captured the popular imagination. Followers of Lavater maintained that one could deduce an individual's personality from appearance alone. During the peak of its fashion, educated citizens across Europe paused to assess the physiognomical character of the people they met. *Fragments* went through so many different editions and printings that it is difficult to generalize about its illustrations. Some reappeared with every issue; others came and went. In addition, there were both authorized and unauthorized translations of the book, together with pirate versions in the original German. Seldom has a title so aptly described a book as *Fragments of Physiognomy*: the book existed in numerous discrete sections that occasionally coalesced between the covers of a leather binding. As a result, there are often substantial variations between even closely related printings of the book.

The range of illustrations in *Fragments* was impressive. Included were pictures comparing humans and animals (some clearly indebted to della Porta), silhouettes, catalogues of lips, noses, and other features, portraits of eminent personages, analytical diagrams, and even genre scenes. Most took the form of people dressed in contemporary costume. Ethnicity was strongly emphasized in the work, and manner of dress,

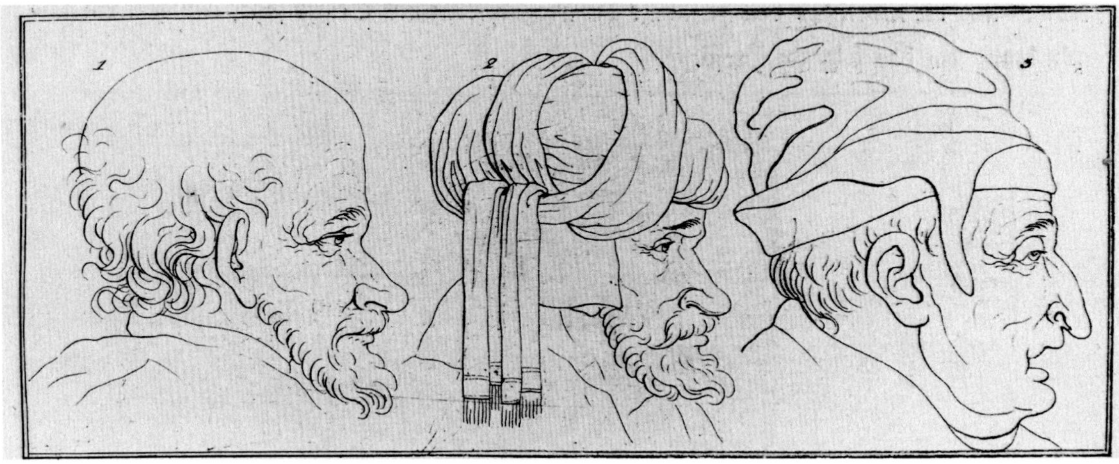

such as headwear (figure 4-3), was used as shorthand for certain national types. At the end of the book, Lavater demonstrated that through the judicious addition of lines, an artist could transform a simple drawing of an animal into a human portrait. He did not intend to suggest that a frog could transform itself into a man exactly, but as with della Porta, his illustrations evoked proto-evolutionary themes (figure 4-4).

4-3 Circle of Johann Caspar Lavater, Three Heads. Etching. From *Physiognomische Fragmente*, 1775.

The philosopher and scientist René Descartes (1596–1650) wrote *The Passions of the Soul* [*Les Passions de l'âme*], published only a few weeks before his death, one of the first philosophical attempts to understand the relation of the soul to the body. Descartes considered the arts subordinate to reason and was one of the first to propose a psychological theory to explain expression.[4] He speculated that the soul, united with the whole body, exercises its functions by the mediation of a "little gland" in the middle of the brain.[5] This gland, he maintained, orients the animal spirits, which in turn affect the nerves and muscles. Although it contained mystical elements, Descartes's conception of the passions was essentially physiological: he argued that emotional expressions are the result of chemical and physical interactions.

Passions of the Soul was illustrated with schematic drawings, probably by the author himself. A more elaborate presentation was devised by the artist Charles Le Brun (1619–90) to illustrate his artist's manual, *Method to Learn to Design the Passions* (1667). Le Brun accepted the Cartesian view that expression should be understood in terms of the relationship between mind and body, but ranked artists as foremost in the physiognomic debate, noting that despite the efforts of philosophers and physicians, "no one has ever thought of making it his particular study with an eye to painting."[6] The book was extensively illustrated with engravings from his own hand.

Opposite:
4-4 Circle of Johann Caspar Lavater, Transformation of a Frog into a Man. Etching. From *Physiognomische Fragmente*, 1775. Meant to exercise drawing skills, Lavater's transformations had evolutionary implications.

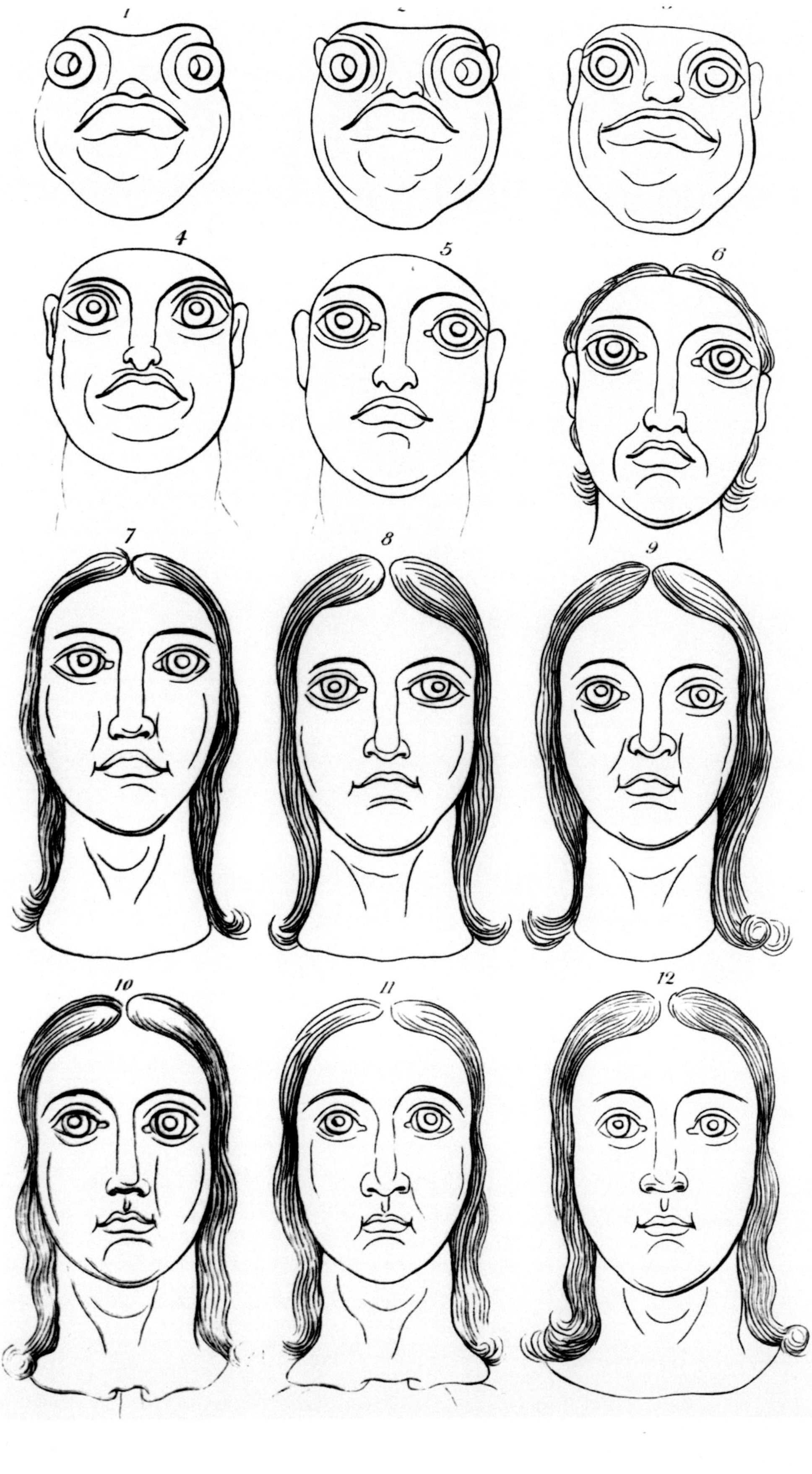
1
4
5
6
7
8
9
10
11
12

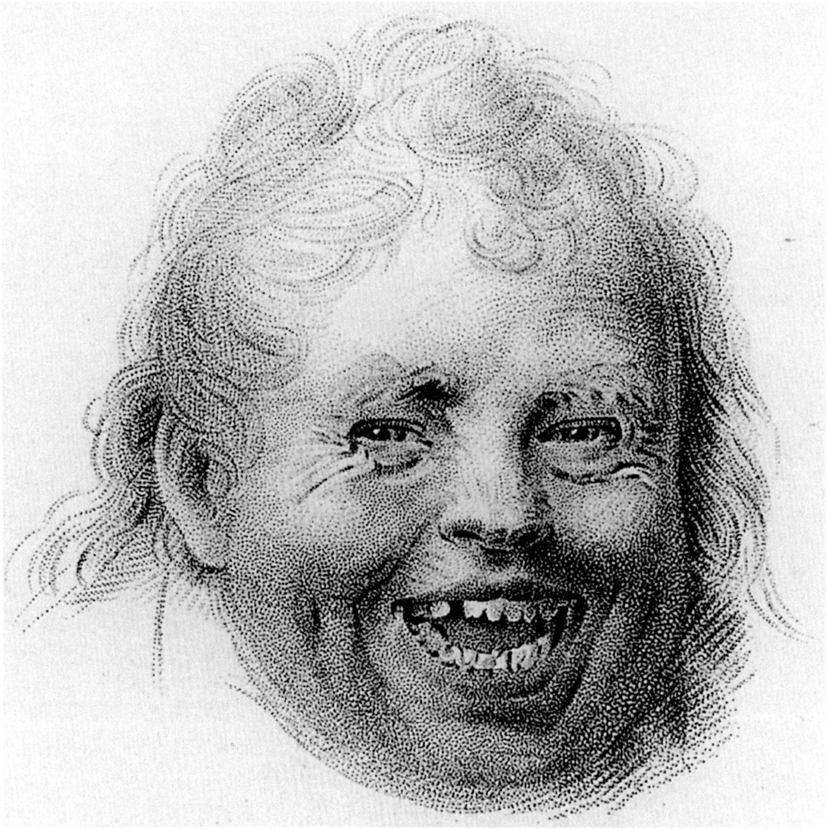

4-5 Charles Bell, Laughter, c. 1806. Stipple engraving. From *The Anatomy and Philosophy of Expression as Connected with the Fine Arts*. Darwin used Bell's illustrations as a model.

The immediate antecedent to Darwin's *Expression* was Scottish anatomist Charles Bell's (1774–1842) *The Anatomy and Philosophy of Expression as Connected with the Fine Arts* (1806).[7] Copiously illustrated with engraved plates depicting Bell's interpretations of human emotion (figure 4-5), the volume focused on the application of human expression to drawing and painting. In the introduction to the third edition (1824), Bell explained:

> I am not without hope that a new impulse may be given to the cultivation of the fine arts, by explaining their relation to the natural history of man and animals. . . . Till he has acquired a poet's eye for nature, and can seize with intuitive quickness the appearances of passion, and all the effects produced upon the body by the operations of the mind, he has not raised himself above the mechanism of his art, nor does he rank with the poet or historian.[8]

This sentiment echoed the aspirations of the artists of the romantic period. In the first half of the nineteenth century, unconscious facility in making pictures became highly prized. Artists tried to become so adept at their craft that they could spontaneously register their feelings in pictorial form.

Darwin had long been acquainted with Bell's work. On 21 November 1826, the night he petitioned to join the Plinian Society at the University of Edinburgh, he attended a lecture debating the book. Darwin's friend William Browne, whose student research focused on the diagnosis of insanity (see chapter 6), led the discussion. Bell claimed that

> anatomy [is] the examination of the structure by which the mind expresses emotion, and through which the emotions are controlled and modified; it introduces us to the knowledge of the relations and mutual influences which exist between the mind and the body. To the painter, therefore, the study is necessarily one of great importance; it does not teach him to use his pencil, but rather it teaches him to observe nature, to see forms in their minute varieties [and] to catch expressions so evanescent that they must escape him, did he not know their sources.[9]

Darwin recorded his reactions to Bell's work in his Notebook "C" (1838). Commenting on Bell's idea that the practice of grinning was designed to expose the canine teeth, Darwin wrote,

> [N]o doubt a habit gained by formerly being a baboon with great canine teeth. . . . Laughing modified barking, smiling modified laughing. Barking to tell other animals in associated kinds of good news [*sic*]. discovery of prey.—arising no doubt from want of assistance.—crying is a puzzler. Under this point of view, expression *of all animals* becomes very curious.[10]

This brief entry foreshadows the book he would write some thirty-five years later.

Bell believed that facial expressions are endowed by God. He explained:

> As the Creator has established this connexion between the mind and external nature, so has He implanted, or caused to be generated, in us, various higher intellectual faculties. In every intelligent being He has laid the foundations of emotions that point to Him, affections by which we are drawn to Him, and which rest in Him as their object. In the mind of the rudest slave, left to the education of the mere elements around him, sentiments arise which lead him to a Parent and a Creator.[11]

This argument paralleled English theologian William Paley's theory of divine design, based on the contention that the rationality of God is evident in the structure of organisms.[12] Bell in fact collaborated with Paley on Paley's seminal book *Natural Theology; or Evidences of the Existence and Attributes of the Deity* (1802).

Darwin learned much from his reading of Bell's *Anatomy and Philosophy of Expression*. Beyond the evolutionary implications of its arguments, the book was a primer on depictions of expression in art. His private copy of the book, now in the Darwin Archive, contains annotations highlighting passages in which he was particularly interested. In the chapter on beauty, Darwin highlighted a passage on classical art:

> With them (the Greeks) the highest effort of art was to represent man deified; as it were, purified from the grosser characters of nature. This they did, as we have already seen, by exaggerating whatever is proper to the human form, by increasing what gives dignity, and bestowing features capable and prone to the finer emotions; representing them, either as still and unperturbed, or as indicating a superiority to the things of this lower world.[13]

Elsewhere Darwin marked passages that described (in Bell's words) the "intimacy and tenderness" of Madonnas in Italian churches, remarked on the success of William Hogarth's renderings of eyes, praised Michelangelo's command of expression, and recommended Francisco de Zurbarán's (1598–1664) painting of Saint Francis in the Louvre.[14] "The finest possible example of [the] condition of suffering is in the picture of Guercino in the gallery of Milan, *The Departure of Hagar and Ishmael*," Bell wrote, and Darwin duly noted. Bell went on: "Those who have only seen the engraving can have little conception of the beauty of the picture, for the perfection is in the colouring." Later in the text Bell explained the importance of eyebrows: "[W]e have only to repeat the experiment of Peter of Cortona (1596–1669); to sketch a placid countenance, and touch lightly with a pencil the angle of the lips and the inner extremity of the eyebrows. By elevating or depressing these, we shall quickly convey the expression of grief or of laughter."[15]

Darwin also underscored a section describing a tour of the Academy of Fine Arts in Bologna. According to Bell, standouts there included Raphael's St. Cecilia in ecstasy, in which "there is not only great beauty, but very fine expression . . . in the picture of the Murder of the Innocents, by Guido Reni, there is an admirable figure of a woman." Raphael and Domenichino, Darwin underlined, "have painted demoniacal boys." He also noted Bell's observation that in Raphael's "great picture of the transfiguration in the Vatican, there is a lad possessed, and in convulsions." Bell's artistic grand tour continues, and Darwin underscored additional passages describing paintings by William Hogarth and Leonardo da Vinci, a sculpture by Gian Lorenzo Bernini (1598–1680), and the Laocoön.[16] Judging from the number of annotations in the book, Darwin scrutinized *Anatomy and Philosophy of Expression*. His interest piqued by the pictures Bell described, Darwin would soon begin to explore London museums, looking for alternative examples of expressive imagery.

Bell's book and others formed part of a continuing historical endeavor to analyze the nature of human expression. It is apparent even from this small sampling of books known to Darwin that studies of human expression involved a mixture of artistic and scientific positions. Some, like Le Brun, argued that the artist's eye should be the guide for the proper expression of the passions. Others, like Descartes, argued that expression was the product of physiological interactions and exists separately from aesthetic ideals. Darwin cites much of this material in *Expression*. His understanding of this work informed his effort to illustrate the book.

Illustration and Narrative Strategies

From the Renaissance forward, books on expression were illustrated with pictures of progressive complexity. The increasing quantity and sophistication of these illustrations may be understood in part as a result of technological advancements in printing methods and in that sense reflect the increasing complexity of books on other subjects. As printed books became more common, illustrating them became more economical. These illustrations also proved popular with the public. Viewing the illustrations to Lavater's *Fragments*, for example, was considered not just informative, but also an entertaining parlor distraction in eighteenth-century Britain. By 1810, twenty different English-language translations of *Fragments* had been prepared, many of which were widely available through newly formed book clubs.[17] To compete in such a marketplace, Darwin had to produce a book that was more appealing to his potential audience. His challenge was to produce illustrations that advanced his theories on expression, but that also persuaded and engaged his readers. The use of photography was one means to accomplish this.

Darwin profited from earlier books on expression and incorporated their best ideas in his work. At the same time, he was also wanted people to buy his book; he competed with other titles for consumers' purchases. It would have been clear to Darwin that existing physiognomical, artistic, and scientific literature was well illustrated and had been for 300 years or more. Illustration was expected in books of this type, so to succeed, of course he would have to use pictures. If his book was to be seen as more appealing to buyers than others in the field, it would be helpful to offer more spectacular images. His decision to choose photography was entirely consistent with the progression in illustration quality evident in previous volumes.

5

Photography and Evolution Meet

Connections between photography and biology in the 1860s

Throughout the 1860s, Harvard University zoologist Louis Agassiz (1807–1873) was one of the main opponents to Darwinian theory in the United States. Always looking for evidence to challenge the concept of evolution by natural selection, in 1865 he arranged an expedition to Brazil to study the country's geology and gather biological specimens. Agassiz believed that the country's spectacular ecology might help him to unravel the mysteries of species development. Organisms as impossibly beautiful and as delicately adapted to their surroundings as Amazonian animals and plants could not have been formed by random mutation and selection, he believed; only divine will could explain their presence. His research on that journey became known as the Thayer Expedition, after the wealthy Bostonian businessman Nathaniel Thayer Jr. (1808–1883) who sponsored it. It was to become a centerpiece of his assault on Darwinian evolution. "The conviction which draws me irresistibly," Agassiz wrote, "is that the combination of animals on this continent, where the fauna are so characteristic and so distinct from all others, will give me the means of showing the transmutation theory is wholly without foundation of fact."[1] Agassiz's research culminated in the book he co-authored with his wife Elizabeth, *A Journey in Brazil* (1867), in which he argued, somewhat predictably, that Darwinian theory was fatally flawed. Tropical plants and animals were not the only evidence of God's divine creation in Brazil; the racial makeup and behaviors of the *people* who lived in Brazil were just as convincing.

Agassiz's analysis focused on speciation. It had long been his contention that crossbreeding between different subspecies of animals could not give rise to new ones naturally, because hybrid offspring are not biologically viable. This inviability can take two forms. In some cases, for example, when a horse and a donkey mate to create a mule, the offspring are simply sterile. In other cases, Agassiz claimed, the offspring may not be sterile but fail to mate because they do not find suitable partners. For example, if an Eastern Swallowtail Butterfly were bred with a Western Swallowtail, what butterfly would want to mate with such strange-looking progeny? And even if one did, he reasoned, at what point would the offspring become so dissimilar from their parents that they could no longer breed with the founding population, for example, as a Swallowtail is dissimilar from a Monarch? According to Agassiz, hybrid animals are not sexually attractive to their relatives because the mating preferences of the standard population favor animals that look like the norm. And even if they did breed, there is no way such "selected" animals could differentiate themselves so much that they are ultimately no longer able to breed with the founding population. Agassiz considered this "evidence" to be Darwin's Achilles heel; if animals cannot mutate, hybridize, and breed in the wild, then the entire theory of evolution by natural selection unravels.

Brazil, Agassiz wrote, provided the perfect laboratory for studying this sort of "speciation" in humans. He believed his theories would be corroborated in the mixed-race children of the Native American, African, and European peoples who lived in Brazil. He considered such offspring sterile crosses, claiming that when parents from different human races have children, their bloodlines are destined to die out because interracial offspring are not attractive sexual partners.

On his arrival in the river port of Manãos in May 1865, Agassiz began to collect the evidence that he believed would prove his argument. Photography, he decided, was the perfect medium to substantiate his theory. Armed with photographs of mixed-race natives, he would illustrate unpleasant features of their anatomy, confirming the unattractive appearance of mixed-race peoples among the general population. With the assistance of Walter Hunnewell, a hastily trained student, he built a photographic studio and, choosing subjects from among Brazilian racial types that he believed would show once and for all that Darwin was wrong, began taking their pictures.

The Thayer Expedition

Although Agassiz's Brazilian project seems misguided now, it was important since it marked the first time photography was used directly in the evolutionary debate.[2] The images he and Hunnewell produced on the Thayer Expedition were never published; the results were inconclusive, and photographs were generally difficult and expensive to reproduce. But three remarkable albums of their photographs were preserved at the Peabody Museum of Archeology and Ethnology at Harvard University. Agassiz maintained that they demonstrated the immutability of species. Most were taken in Agassiz's makeshift studio, which his wife Elizabeth described as a "picturesque barrack of a room, which [. . .] serves as a photographic salon."[3] Another of Agassiz's students, William James,[4] provided a more colorful description:

> I went to the photographic establishment and was cautiously admitted by Hunnewell with his black hands. On entering the room I found Prof. engaged in cajoling three mocas whom he called pure Indians, but who I thought as afterwards appeared, had white blood. They were very nicely dressed in white muslin and jewellery with flowers in their hair and an excellent smell of pripioca. Apparently refined, not at all sluttish, they consented to the utmost liberties being taken with them and without much trouble were induced to strip and pose naked.[5]

The Harvard albums contain more than 100 six- by four-inch studies of people, nude and partially clad, who sat for Agassiz's camera. Nearly all were young adults whom Agassiz considered of prime reproductive age. The majority of them were women, who arrived in the studio wearing fine dresses and showy jewelry, as if ready to pose for a formal portrait.[6] Agassiz requested that the women lower their gowns to expose their uncovered bodies for the camera (figure 5-1). The results are surreal: elegantly clad over her torso, the sitter's clothes are parted or lowered to expose her breasts. To those not familiar with the project, the pictures look more like voyeuristic fantasies than scientific studies. As the historian Gwyneira Isaac has noted, there is a disquieting element to the photographs, as the women seem to have been "dressed up" only to be "dressed down" for the camera.[7] Agassiz seems to have been particularly interested in the forms of women's breasts; the *Journey in Brazil* describes the various manifestations of the feature in extensive detail, in a section titled "Comparative Analysis of Women's Breasts and Inguinal Region." In the albums, Agassiz twinned his photographs of native Brazilian breasts with photographs of classical statuary, including Canova's *Three Graces* and the *Venus de Milo*[8] (figures 5-2, 5-3). He applied a similar technique to his male subjects, contrasting a native male nude with the *Apollo Belvedere*.[9] The vulgarity of mixed-race sexual characteristics compared with classical European ideals was supposed to be self-evident.[10]

Classical statuary was often used as a reference point in nineteenth-

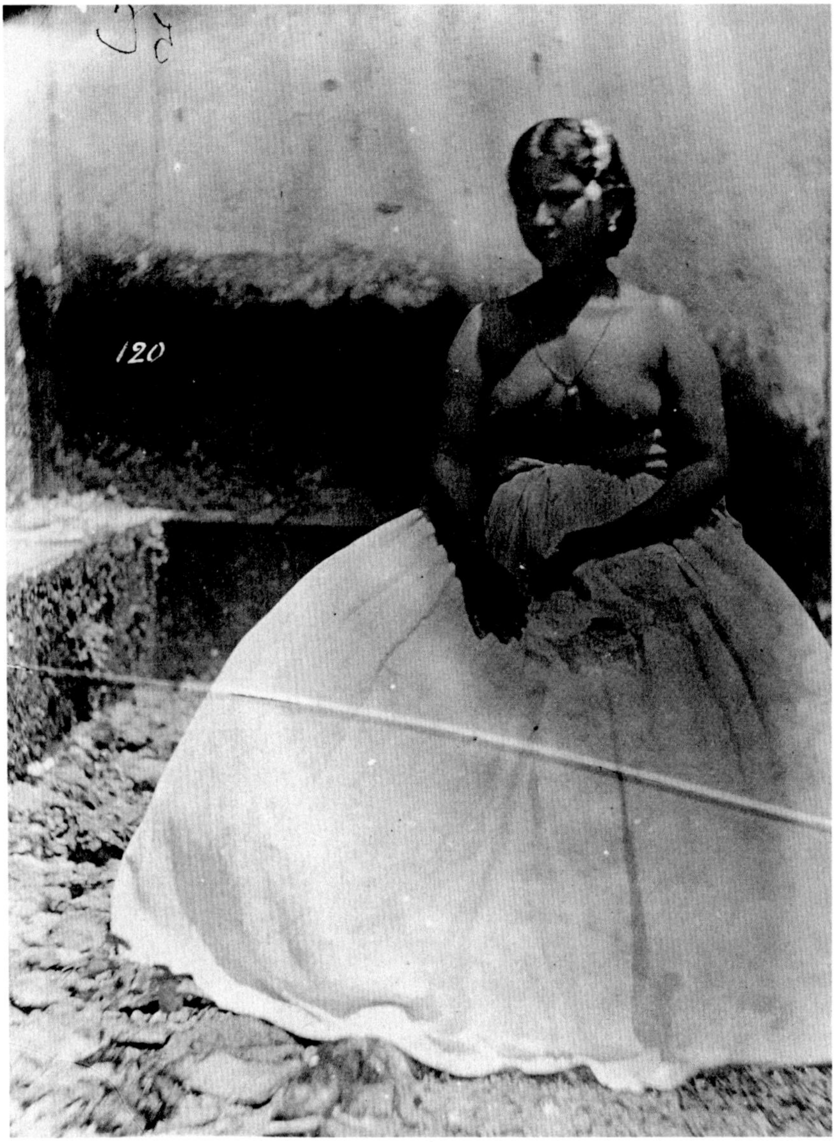

5-1 Louis Agassiz and Walter Hunnewell, *Brazilian Woman with Dress Lowered*, 1865. Albumen print.

century discussions of human anatomy. This is a conceit with which Darwin would certainly have been familiar. Many early writings on physiognomy espoused the perfection of ancient Greek models; these were taken not as idealizations but as actual representations with which living specimens could be compared. The uncritical absorption of this standard owes much to the view of early art historians, such as the Prussian Johann Winckelmann (1717–1768), who praised the "noble simplicity and quiet grandeur" [*Edle Einfalt und stille Grösse*] of Greek art in his *Thoughts on the Imitation of Greek Works in Painting and Sculpture* (1755).

There may not be a single reliable archetype of attractiveness, as

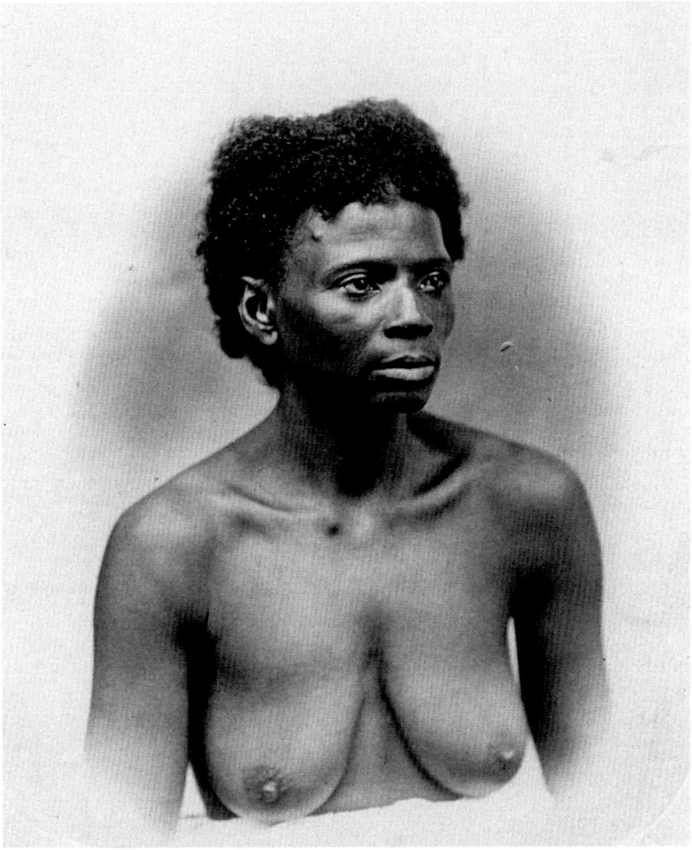

5-2 Louis Agassiz and Walter Hunnewell, *Bust of Brazilian Woman*, 1865. Albumen print. According to Agassiz, much was to be learned from the female breast.

Agassiz suggested in his comparisons of Brazilian women with ossified European figures. And, although one may speculate that the ancient statues faithfully depicted people, it is a stretch to think such works are records of real individuals. A standard that incorporates an artist's conception of the ideal, in which muscles are exaggerated, forms streamlined, and blemishes removed to achieve a desired effect, is theoretical rather than actual. These are issues with which Darwin wrestled while preparing *Expression*. Can one generalize a smile, a laugh, or a frown, from a distinct expression on a particular individual? If so, how should that individual be selected? Would a random person suffice, or one chosen from a cross section of people of different ages and cultures? Darwin chose not to illustrate *Expression* with classical examples.

Darwin may never have seen Agassiz's Brazilian photographs, but he was aware of his work in Brazil and corresponded with him about his journey. Despite the antipathy each had for the other's research, the two remained congenial adversaries. Agassiz wrote to Darwin shortly after his return from Manãos that while he was opposed to Darwin's views, this did not blind him to "the great value of his original researches."[11]

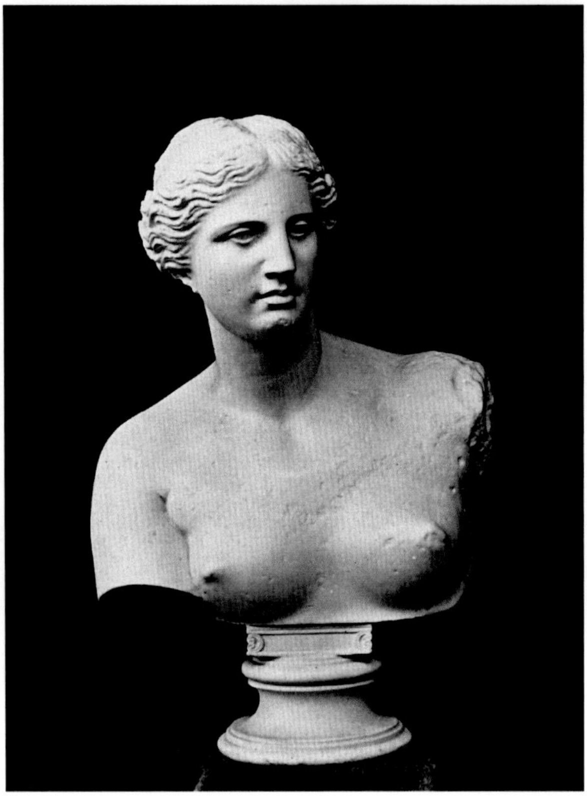

5-3 Maker Unknown, *Cast of the Venus de Milo*, c. 1865. Albumen print.

For his part, Darwin, too, professed respect for the Harvard scientist. Unfailingly diplomatic, he sidestepped Agassiz's more extreme theories, including his stance on miscegenation. Instead, he inquired about less controversial matters such as his thoughts on the sexual coloration of Amazonian tropical fish.[12] Ironically, Darwin included this information in *Descent of Man*, which contains his broadest claims for the shared ancestry of humans of any of his works. Darwin's biographers Adrian Desmond and James Moore have described his exchange with Agassiz during this period as "turning the poor man's anti-Darwinian gains into serviceable spoil, pickpocketing him so gently that he was none the wiser."[13]

Agassiz's use of photography would not have escaped Darwin. From 1869 to 1870, Louis's son, the natural historian Alexander Agassiz (1835–1910), took up residence in England. During this time he made several visits to Darwin at Down House, serving as a sort of unofficial emissary from his father. A specialist in sea urchins and starfish, the younger Agassiz was also employed at Harvard, where after his father's death in 1873 he succeeded him as director of the University Museum of Comparative Zoology. He was also involved in U.S. coastal surveys and shared with Darwin an interest in the formation of coral reefs.

Perhaps for this reason he was less skeptical of Darwin's work than was his father, but he was just as interested in harnessing photography to the cause of science.

Upon his return to Boston in 1871, and shortly after his last visit with Darwin, Alexander published an appeal to scientists in the *Bulletin of the Museum of Comparative Zoology,* urging the use of photography in their research. Perhaps the first article to advocate the use of photography in a scientific journal, it was accompanied by two sample photographs of sea urchins, each printed using a different photomechanical printing technique (figure 5-4 demonstrates photomechanical Albertype printing). He encouraged readers:

> The rapid progress made in carbon printing by the Woodbury and Albert processes[14] promises to furnish us, within a short time, the means for direct application of photography to illustrations of natural history, and these new methods of printing are likely to replace to a great extent the ordinary lithographic plates. The accuracy of photographic printing is of course far beyond that of an engraving or lithograph,

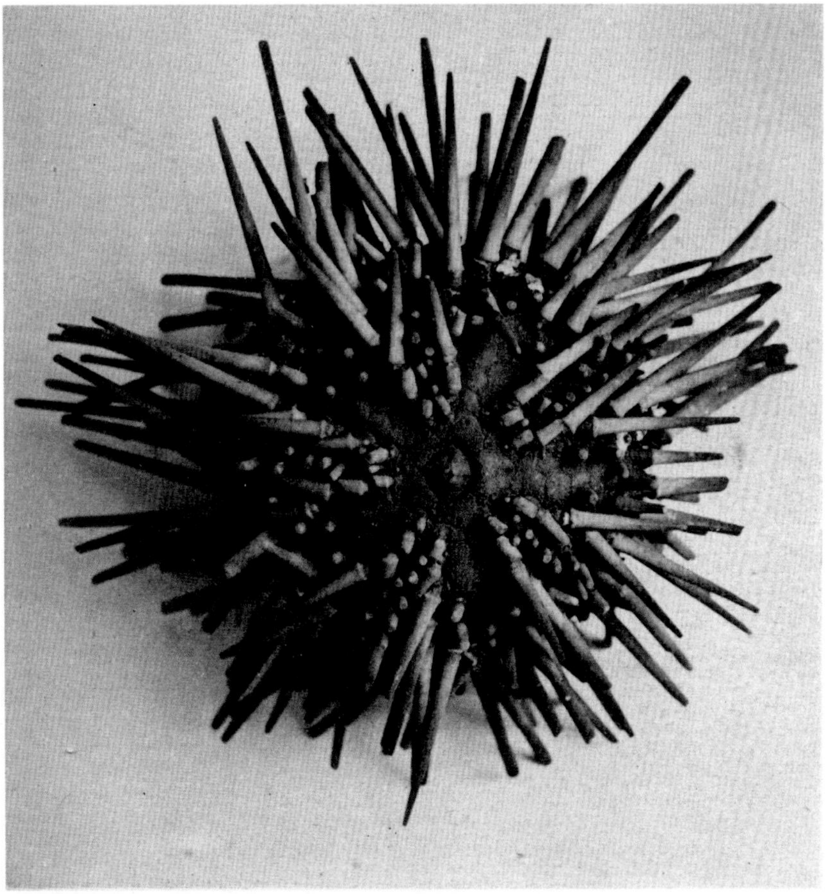

5-4 Circle of Alexander Agassiz, *Echinocidaris punctulata*, c. 1871. Albertype. Alexander Agassiz believed that photographs like his would make lithography obsolete.

and as soon as a few difficulties of printing the separate figures of a plate with one impression are overcome, we shall be able to illustrate our memoirs accurately and economically, and give figures with an amount of detail which the great expense of engraving or lithographing would normally make impossible, even if it were mechanically practicable.[15]

In the winter of the following year, Alexander Agassiz repeated this advice in a letter to Darwin: "I have made pretty extensive use of photographic printing in my book [*Revision of The Echini* (1872)], and from what I have succeeded in obtaining trust it will hereafter be possible to supersede the old lithographic processes, which are wasteful in time and money, and not half as accurate."[16]

Huxley and the Anthropological Debate

Just as Darwin's rivals were seeking to enlist photography in their campaign against the theory of evolution by natural selection, his allies were preparing their own pro-Darwinian photographic campaign. In England, Thomas Henry Huxley, Darwin's staunch ally, had become interested in photographic anthropometry, that is, the application of photography to the mathematical analysis of anthropological specimens. Huxley was a leading member of the London Ethnological Society, a splinter group of the Anthropological Society, which formed in 1863 to fashion a Darwinian approach to anthropological research. The Anthropological Society favored the traditional, static classification of peoples (the approach with which Louis Agassiz was most comfortable), in which researchers aimed to develop a simple topology of various human races. They viewed humans as unchanging, created by God to be separate and distinct. The ethnologists, by contrast, attempted to trace the social evolution of peoples from a common universal ancestor.[17]

In the late 1860s, photography became a centerpiece of the new society's investigations. The anthropologist Roslyn Poignant notes that the *Ethnological Journal* is full of accounts of photographs being exhibited as evidence at meetings of the society, ranging from reproductions of Peruvian antiquities to a picture of a boy in Saigon who grew a tail.[18] Huxley himself began working with photographic evidence at least as early as 1866, when he received a packet of photographs of the legendary Neanderthal cranium from Karl Fuhlrott, the German school teacher who had announced the specimen some nine years earlier.[19]

In 1868, Huxley probably first saw photographs being presented by his ethnologist colleagues, a packet of photographs by the Ellis Studio of a Burmese family with apelike hair all over their bodies. William

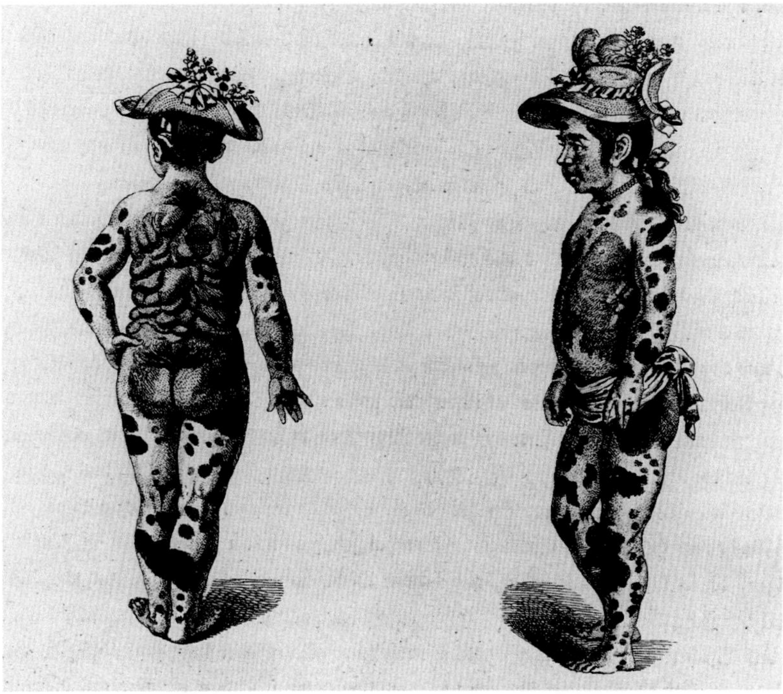

5-5 Circle of Johann Caspar Lavater, Girl With Deer-like Hair All over Her Body. Etching. From *Physiognomische Fragmente*, 1775. This condition had been known since the Middle Ages.

Houghton (1821–95), who sent him the images, explained that they were "photos recently brought home by Captain Houghton—a brother of mine. I read a paper on the subject lately before the Ethnographic Society."[20] These photographs may have been a hoax, as the hair covering the subjects has the look of a costume. Nevertheless Huxley and Darwin were probably familiar with the condition they supposedly depict, which is also illustrated in Johann Caspar Lavater's *Fragments of Physiognomy* [*Physiognomische Fragmente*] (1772). Lavater described the case of a girl who had grown hair "like a deer" all over her body and included two illustrations of the phenomenon in his book[21] (figure 5-5).

It is now known that these accounts are consistent with the symptoms of a rare genetic abnormality known as congenital generalized hypertrichosis.[22] However, there is little evidence that Huxley took such accounts seriously. Darwin, too, seems to have been uncertain what to make of them. In *Descent of Man* he claimed that the thin layer of short hairs that cover the bodies of humans to varying degrees is undoubtedly "the rudiments of the uniform hairy coat of the lower animals."[23] But he did not mention the condition described by Houghton and Lavater. He did, however, possess a set of four photographs similar to those Houghton sent, reproductions of oil paintings photographed by the Austrian studio F.R. Bopp. Portraying a family of four who were said to have grown thick fur, they were described on the mounts as the

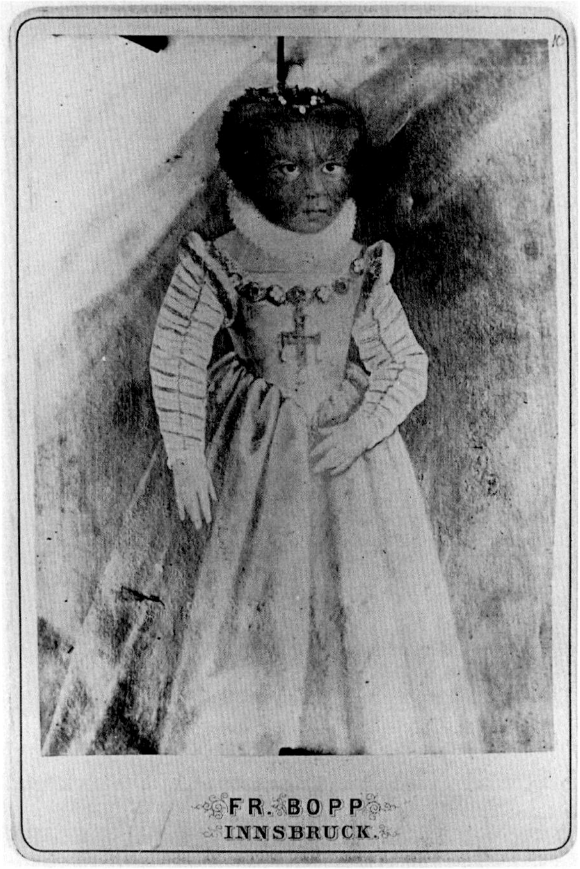

5-6 F.R. Bopp, *Daughter of the "Hairy family of Ambras,"* n.d. Albumen print. Darwin received four pictures of this remarkable family from an unknown source.

"Hairy family of Ambras."[24] The daughter of the family is shown in figure 5-6.

In March 1869, Huxley delivered the keynote address for a series of six lectures before the Ethnological Society on the peoples of India, North America, and Polynesia. The first of the subsequent lectures, delivered by one G. Campbell, was titled "The Peoples of India, as Traced in Existing Tribes and Castes." To illustrate his talk, Campbell used albumen prints from *The People of India*,[25] an ambitious eight-volume collection of photographs of indigenous residents of the subcontinent. The photographs were produced principally by the Indian photographic firm of Bourne and Shepherd.[26] Campbell, speaking less than a year after the first volume was published, would have had access only to the earliest sections of the work. Typically the images feature sitters adorned with selected indications of social status: jewelry, costume, occupational tools, and weapons (figure 5-7). Poignant argues that while the photographs were considered successful, the sitters' accoutrements constituted visual "noise" that worked against anthropological discourse at the lec-

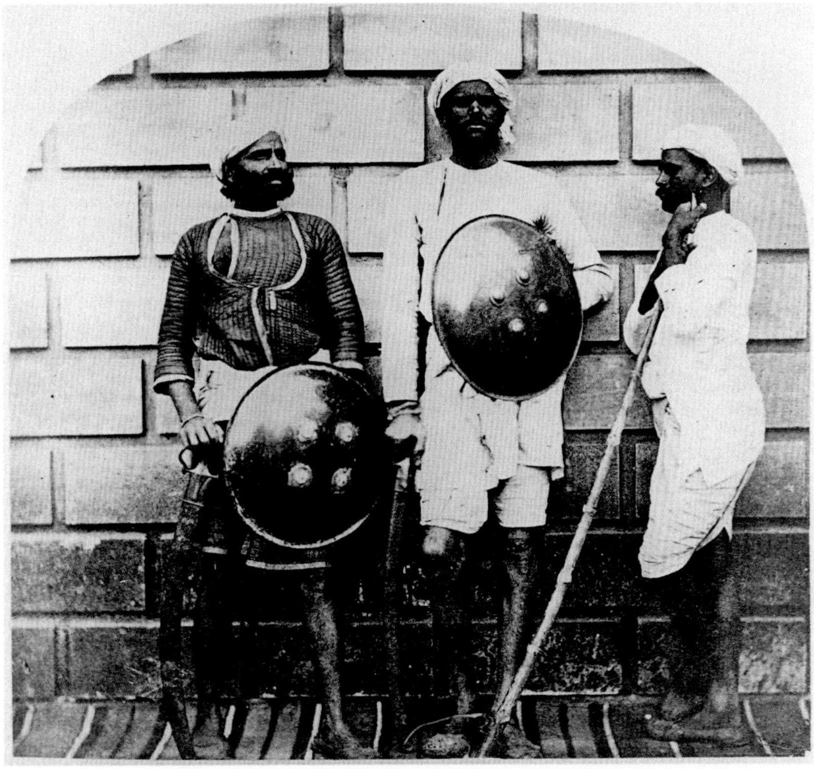

5-7 Attributed to Samuel Bourne, Hindustani Men, c. 1865. Albumen print. From *The People of India.* London anthropologists appreciated Bourne's photographs, but found them unscientific.

tures.[27] After Campbell's lecture, the members agreed that it would be preferable to give a "more scientific character" to meetings in future.

Perhaps with this resolution in mind, some six months later Huxley embarked on a more structured photographic survey of the peoples of the British Empire. Working with the Colonial Secretary George Leveson-Gower, the second Earl Granville (1815–91), Huxley devised a plan for an encyclopedia of "measurable" and "comparable" photographs "of the various races of men comprehended within the British Empire. . . . Great numbers of ethnological photographs already exist," he wrote, "and they lose much of their value from not being taken from a uniform and well considered plan."[28] Under Lord Granville's guidance, the project obtained the official support of the Colonial Office, which conveyed formal requests for the photographs Huxley sought to each of the colonial governors general.

The idea was to create an improved *People of India* on a grander scale, one that encompassed all the tribes and races in the British Empire, to be used for scientific study. In August 1869 Huxley sent Granville the protocol for making such photographs (figure 5-8). Each record was to comprise four parts: two full-figure nudes and two close-ups of the head alone. The photographs were to be no less than three inches long,

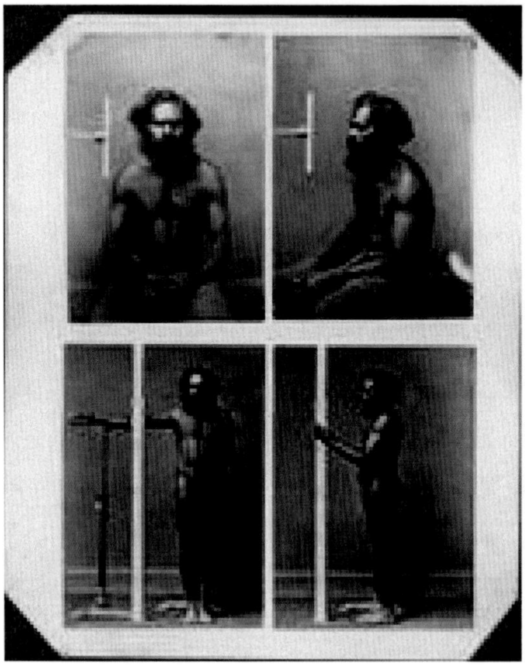

5-8 Maker Unknown, Photographs made according to Huxley's anthropometric instructions, c. 1870. Albumen prints. Huxley envisioned an encyclopedic resource of ethnographic information.

preferably all to the same scale so that the "relative proportions of the different figures could thus be apprehended almost at a glance." The subjects were to be "in a condition of absolute nudity or as near thereto as may be practicable."[29] Both full-frontal and profile views were desired, with the sitter in an attitude of "attention":

> [T]he right arm should be stretched out horizontally—the hand being fully open with the fingers & thumb extended, and the palm turned forwards. The feet should be parallel and with the ankles fairly touching one another. The arm will need a rod to prevent it from trembling and a measuring rod divided in feet and inches may either be fixed to this rod or otherwise included in the plane of the body so as to present a scale. In the profile view, the left side should be turned to the eye of the photographer; and the left arm bent at the elbow and so disposed as not to interfere with the outline of the pectoral region. The back of the hand should be turned toward the photographer and the fingers & thumb extended.[30]

Perhaps mindful of Louis Agassiz's dip into pseudoscience, Huxley eliminated a passage alluding to the female breast. Crossed out, but still visible on his copy of the letter, was the further prescription, "In the female it is desirable that the arm should not be so disposed as to interfere with the contour of the breast, which is very characteristic in some races." Huxley's specifications for photographs of the head were similar

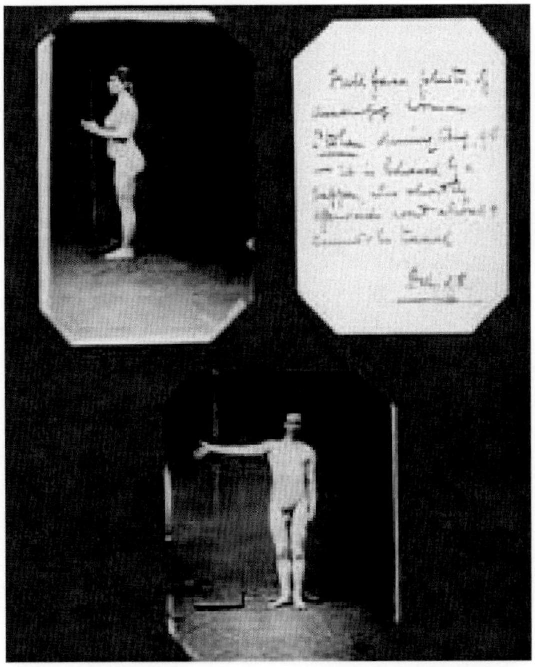

5-9 Circle of Thomas Henry Huxley, Sample photographs of Britons photographed anthropometrically, 1869. Albumen prints.

to those for the full figure. They were to be taken head-on, and again in centered profile, the larger the better, with a scale divided into width and length photographed together with each head.

Along with these instructions, Huxley enclosed sample photographs illustrating his method[31] (figure 5-9). In contrast to Agassiz, who used Grecian statues as models, two Caucasian sitters were used, one male and one female, presumably local English models whom Huxley enlisted for the purpose. The shock with which correspondents around the world received these images must have been considerable: little was left to the imagination. The photographs of the female model, young, buxom, and completely nude, would certainly have excited the attention of male officials to whom they were sent. There is some indication of this in the Huxley Archive itself, in which Huxley's copy of the head-on female sample is missing. A note is penciled in the margin next to its intended location. "Full face photo of accomp. woman *stolen* during Aug. '98. It is believed by a sapper who shortly afterwards went abroad and cannot be traced."[32] By using classical statuary, Louis Agassiz had at least avoided such difficulties.

Although Huxley's system was an improvement over the haphazard approach in *People of India*, his controls were not enough to produce the sort of measurable anthropological data he sought. The system was too fluid. As the anthropologist Frank Spencer has pointed out, Huxley failed to make adequate provision for accurately determining the subject's stature.[33] In the photographs it was unclear whether the subjects were standing completely straight, and even if they were, it could not be determined if their fully erect postures were equivalent. As Spencer has explained, the system was similar to one used more than a century earlier by the Flemish artist Jacob de Wit (1695–1754) in *Handbook of Proportion* (1747). The subject was directed to stand in the same plane as a measuring rod, with both rod and subject free-standing. This not only permitted the subject to hunch slightly, but also created some doubt as to precisely where the various parts of the sitter's body were in relation to the rod.

Huxley's project also presupposed the availability of a skilled international network of photographers, but few qualified photographers lived in the more remote areas of the Empire, and even the best might be prone to interpret his instructions slightly differently. Inconsistencies in the tilt of the head, the raising of the arm, and the thickness of the subject's hair rendered measurements unreliable. Differences in lens selection, focus, and manipulation of the camera bellows to adjust perspective rendered them virtually useless for scientific data.

As if that were not enough, reactions of the government bureaucrats whom Huxley and Granville approached ranged from enthusiasm to indignation, with apathy perhaps the most common response. The governors of Mauritius and Barbados replied that their colonies contained no populations of interest to the ethnologist,[34] while the governors of British Guyana, Tobago, and the Virgin Islands claimed they had no photographer qualified for the job.[35] The Central Board for Aboriginies [*sic*] in Melbourne considered it politically inadvisable:

> I am of the opinion that it would be unwise to ask the Aboriginies of this colony to submit themselves to the photographer in the manner described by [Professor Huxley]. In Victoria the Aboriginies, I am glad to say, are civilised as regards their habits, but they are not sufficiently enlightened to submit themselves in a state of nudity for portraiture in order to aid the advancement of science. Indeed they are very careful in the matter of clothing. If I empower a photographer to visit the stations and take photographs with professor Huxley's instructions in his hand, I would I am sure offend the Aboriginies. . . . I shall be glad to collect photographs of full length figures . . . but I can promise such only as have been gathered with the cheerful consent of the blacks.[36]

In Ceylon, the surveyor general in Colombo, Captain A.B. Fryers, sent photographs fulfilling Huxley's request, noting that he had "more difficulty than anticipated." He overcame resistance among the natives by using prisoners as subjects, producing just two sets of photographs: one of a tribal chief, "sentenced two years ago to 20 years imprisonment with hard labour for an alleged rape" and the other, a female Tamil, "21 years of age, a coffee picker by day, a prostitute by night."[37] In Grenada, the reaction was especially strong. The lieutenant governor reported that the House of Assembly had convened to debate the proposal and passed a resolution forbidding "the furnishing of photographs required in accordance with Professor Huxley's desire."[38] Although the bill tactfully cited pecuniary concerns, the serious and unequivocal reply conveyed a fundamental distaste for the undertaking.

Although Huxley gathered dozens of photographs, many of which are preserved in the Thomas Henry Huxley Archive at Imperial College London, his dream of documenting the peoples of the British Empire was never realized. The project was simply too ambitious.

Darwin and Photography: Early Indications

At the same time that Huxley was struggling to apply photography in his research, Darwin was also becoming interested in the scientific potential of the medium. His own involvement began with the exchange of portraits with colleagues, a practice that was becoming increasingly common in scientific circles. References to portrait photographs first appear in Darwin's correspondence in the late 1850s and were in keeping with the vogue for small carte-de-visite photographs that the photographer André-Adolphe-Eugène Disdéri (1819–89) had launched in 1854. Such personal use of photography was of particular value to Darwin, who was both a prolific letter writer and, due to ill health, a reluctant traveler. Many of his correspondents had never met him, and the exchange of photographs was an effective means of introduction. Darwin sent dozens of portraits to associates internationally and received many in return. Some of the exchanges were formal, as when an entire scientific society would send Darwin a souvenir album of its members. More often the exchange was casual, for example, with Asa Gray (1810–88), Darwin's strongest advocate in the United States (and who, like Louis and Alexander Agassiz, also served on the Harvard faculty). At Gray's request in 1861, Darwin sent him an unmounted crumpled portrait, suggesting apologetically that he might "iron it smooth."[39]

Darwin disliked having his picture taken. He was notoriously self-conscious about his appearance and unenthusiastic about photographs made of him.[40] Writing to the botanist Joseph Dalton Hooker (1817–1911) about Maull & Polyblank's portrait of 1855, he wrote, "You ask about my photograph. I have been done at the Club, but if I really have as bad an expression as my photograph gives me, how I can have one single friend is surprising."[41] Eight years later he sent Hooker an updated photograph, with a similarly wry description: "As regards my photograph, I believe I have very little expression. I have often remarked that I am not recognised except by those who know me tolerably well; that I have often to introduce myself—added to which all my photographs & portraits make me look either silly or stupid or affected."[42]

In his autobiography, Darwin described the alarming reaction his portrait had caused among the members of a German scientific society:

> A few years ago the Secretaries of a German psychological society asked me earnestly by letter for a photograph of myself; and some times afterwards I received the proceedings of one of the meetings in which it seemed that the shape of my head had been the subject of public discussion, and one of the speakers declared that I had the bump of Reverence developed enough for ten Priests.[43]

This was a more generous interpretation than that pronounced by Robert FitzRoy, the captain of the *Beagle*. He once confided that after their first meeting he had nearly decided against retaining the young Darwin as the ship's naturalist. A keen physiognomist, he had been troubled by the shape of his nose. According to Darwin, "he doubted whether anyone with such a nose could possess sufficient energy and determination for the voyage."[44]

Despite constant demand for his portrait, by 1863 Darwin had never had a professional carte-de-visite portrait done, relying instead on pictures made by friends and photographs made in larger formats. Writing in January of that year to the Swiss botanist Alphonse de Candolle (1806–93), he made this clear: "I have never had a proper 'carte photographique' taken of myself; but I enclose one done 2 or 3 years ago by my son, if worth your acceptance."[45]

Darwin's son William had taken up photography in July 1857. That summer, the twelve-year-old budding artist prepared, exposed, and developed photographs within the family home. An upstairs room had been converted into a temporary darkroom, and the boy proceeded to photograph members of the family and the grounds around the house. When he returned to boarding school, his father told him it was "odd not having you rushing up & down the house, with your Photographs & very dirty hands."[46] A tutor had evidently offered to adapt a room for him at school, but William had not brought his equipment. "By the way," Darwin admonished, "I think it not amiss that you did not take your Photography, as you wd. have no time for school work."[47]

The biographers Desmond and Moore describe William as "grasping glass plates" as he ran from room to room, implying that he used the wet-plate collodion process.[48] Wet-plate photography involved making photographic negatives by hand, using a sticky and noxious solution of silver nitrate, potassium iodide, and guncotton dissolved in ether. It is reasonable to suppose William used this technique, as it was becoming the most common form of photography at the time. However, Darwin's correspondence does not identify the materials his son actually used. For the boy to have worked successfully using any photographic process at such a young age would have been a precocious feat, because the medium was famously demanding. The intricacies of processing collodion wet plates posed challenges even for professional photographers, and amateur practitioners in the 1850s, though not rare, were still uncommon. It is unclear how William became interested in photography or who may have instructed him. Someone familiar with the process would likely have been required to train him, at least initially. Some parental supervision would have been necessary, too, given his tender age and the hazards associated with early photographic chemistry. Possibly Darwin joined his son in the enterprise occasionally, either out of curiosity or simply to check up on him. In any case, it is likely he was at least broadly familiar with the operation of a darkroom and the basic requirements of the materials and methods used.

The Darwin Family and the Invention of Photography

Ironically, it was around this time that members of Darwin's extended family were identified as contributors to the invention of photography. On 16 April 1863, the Birmingham *Daily Post* reported that it had uncovered the earliest known photographs, believed to be the result of experiments conducted about 1791 by James Watt, inventor of the steam engine.[49] The photographs he was said to have made, depicting eighteenth-century architectural features destroyed before the invention of any previously known photographic process, later proved to be forgeries. Nevertheless, the story propelled the members of the eighteenth-century Lunar Society, which included not only Watt but also Darwin's grandfathers Josiah Wedgwood and Erasmus Darwin, to the forefront of discussions about photography's invention.

Although the report was eventually disproved, it had challenged the accepted view that William Henry Fox Talbot (1800–77) and Louis-Jacques-Mandé Daguerre (1787–1851), the nearly simultaneous inventors of the earliest British and French processes, should share recognition for the invention. This, in turn, revived discussion of the contributions made by Charles Darwin's uncle, Thomas Wedgwood, who with pioneer chemist Humphry Davy had published the influential paper "An Account of a Method of Copying Paintings upon Glass and of Making Profiles by the Agency of Light upon Nitrate of Silver" in the *Journal of the Royal Institution* (1802). Some English writers argued, with patriotic zeal, that Wedgwood and Davy should be credited with the invention of the first truly photographic process, although the two did not devise a method for fixing the images they created. They did manage to produce evanescent images on paper and leather using silver nitrate and exposure to light. But without a means to preserve them, the importance of this advance was not immediately appreciated.

Although they were not friends, Darwin and Talbot were both alumni of the University of Cambridge and mixed in similar circles. Talbot, whose main occupations were linguistics and crystallography, was Darwin's senior by about nine years. However, by coincidence, Talbot's perfection of his photographic process and the initial conception of Darwin's theory of evolution by natural selection were nearly simultaneous. Darwin returned from his voyage on the *Beagle* in 1836, just as Talbot made his first successful photographs of his estate in Lacock, Somerset. In the late 1830s, Darwin and Talbot's audiences overlapped, as both men attempted to win support among London scientists. There are, however, no accounts of the two having met.

The most direct link between them was their mutual friend the astronomer Sir John Herschel (1792–1871), who was instrumental in perfecting Talbot's process. It was Herschel, for example, who suggested Talbot use a bath of sodium hyposulfite to fix light-sensitive silver nitrate, rather than the ordinary table salt first used. Talbot described Herschel's efforts as the "steps of a giant."[50] In addition to the invention of an effective fixer, Herschel's contributions included invention of the cyanotype process, the first practicable method of applying photographic materials to glass, the use of photography in astronomy, and the use of miniaturized photographs (an antecedent of microfilm) as a method for storing public records.

Even as he guided the progress of photography, Herschel also had a profound effect on Darwin's concept of natural history. Desmond and Moore have described Herschel as one of Darwin's heroes.[51] Darwin wrote that Herschel's *Introduction to the Study of Natural Philosophy* "stirred up in me a burning zeal to add even the most humble contribution to the noble structure of Natural Science."[52] He described the book as one of the two most influential he ever read. He met Herschel on several occasions and exchanged at least ten cycles of correspondence with him. In a long letter passed among Darwin's circle by the geologist Charles Lyell (1797-1875), Herschel expressed great interest in the scientific understanding of the origin of life, describing it as the "mystery of mysteries." The question of origins for Herschel was linked to the understanding of time itself. Lamenting that a literal interpretation of the Bible had hindered understanding of the true age of the Earth, he wrote emphatically, "We must interpret [time] in accordance with whatever shall appear on fair enquiry to be the truth for there cannot be two truths."[53]

Thus, Herschel joined the chorus of gradualists, who, like Lyell, argued that speciation can best be explained on a scale of millions, not thousands, of years. Dissatisfied with existing theories on the subject, Herschel offered no explanation for the origin of species himself, but maintained that a convincing scientific answer would soon be forthcoming. Darwin was, in effect, the most illustrious of those to pick up the gauntlet Herschel had thrown down.

It was in this context in the late 1860s that Darwin began to explore the use of photography in his research. Privately, through the exchange of portraits with other scientists, through his son's amateur enthusiasm, and through personal connections, he had become acquainted with its possibilities. Professionally, with the burgeoning use of photography in the research of Louis and Alexander Agassiz, Thomas Huxley, and others, he had witnessed its potential as a scientific tool. His repeated exposure to photography in these two spheres ultimately convinced Darwin that it could be useful in his own work.

His approach would be different from that of his friend Huxley. Darwin viewed photography as a potent source of information, as Huxley did, and as a means to illustrate his theories. But his understanding of the medium was more intuitive. Whereas Huxley attempted to create structured, measured records of carefully selected subjects, as Darwin began to study expression in humans he looked for spontaneous, candid pictures of ordinary people in various emotional states. There were to be no guidelines or measuring grids in his photographs. Instead, he began to gather pictures capturing natural behaviors, that is, pictures taken outside of a formal, experimental context. Herschel, in a prophetic article for the *Photographic News*, had described such photographs in 1860 as the "representation of scenes in action—the vivid and life-like reproduction and handing down to the latest posterity of any transaction in real life. . . . [T]he possibility of taking a photograph, as it were, by a snapshot—of securing a picture in a tenth of a second of time."[54]

Having spent most of his career considering the rise and fall of populations of animals over vast expanses of time, in the 1860s Darwin took a new tack. By obtaining photographs of emotional expressions taken in mere fractions of a second, he hoped to discover evidence of previously undocumented similarities between species. His interest lay not just in the anatomy of various organisms, but in the subtle and sometimes imperceptible ways they behaved. Martin Rudwick has described the established practice of portraying the "vanished reality of the past" through depictions of ancient animals and their habitats as creating "scenes from deep time."[55] Toward the end of his career, Darwin began to gather photographs that might be termed, in the spirit of Rudwick, "pictures from shallow time."

6

Method to Their Madness

How photography in mental hospitals influenced Darwin

The first photographs Darwin found that he considered suitable for publication had already been published in another groundbreaking scientific treatise. In 1862, the French neurologist Guillaume-Benjamin Duchenne de Boulogne published a photographically illustrated guide to facial expression, with the unwieldy title *The Mechanism of Human Facial Expression, or an Electro-physiological Analysis of the Expression of the Passions Applicable to the Practice of the Fine Arts*.[1] Darwin owned two copies of this magnificent book: a small, octavo version, and a rare, unbound, grand-in-quarto edition. Darwin's copy of this larger version may be the most complete in existence.[2]

Duchenne was a physician at the Salpêtrière, a hospital for the poor in the suburbs of Paris. The hospital treated patients with neurological problems, such as epilepsy, palsy, paralysis, and insanity. Believing these conditions to be the result of electrical dysfunction in the nervous system, Duchenne began to experiment with electrical current as a means of inducing neural action. He wanted to understand the way human bodies are wired, believing this could help him develop therapies to cure his patients. He soon learned that by using electric probes it was possible to cause isolated groups of muscles to contract. Applying this technique to groups of muscles in the face, he found that he could artificially induce a variety of recognizable facial expressions in his subjects.

By applying a gentle flow of electricity, Duchenne was able to hold the expressions he induced long enough to produce photographs of several seconds' exposure. Instead of accelerating the photographic process to produce instantaneous images, as others had tried to do, Duchenne devised a system for freezing the activity of his subjects long enough to accommodate the lengthy exposure times necessitated by photographic technology at the time. Although his photographs still required reasonably quick exposures to register, the electrical system allowed him to overcome the major limitation of photographic technology—its slowness. He merely held the subject in position as long as necessary for the camera to record the experiment.

Mechanism was produced with the assistance of the photographer Adrien Tournachon (1825–1903). Known as "Nadar jeune," after his brother, the great French photographer Félix Nadar (1820–1910), he achieved notoriety in the early 1850s for his photographs of horses and, in 1854–55, for a cycle of instantaneous portraits of the mime Debureau enacting various dramatic expressions.[3] Under Duchenne's supervision, Tournachon produced eighty-four large-format photographs depicting human subjects in various expressive poses. Published in *Mechanism*, these proved invaluable in Darwin's project to portray expression.[4]

Mechanism was organized in three sections. In the first part Duchenne introduced his experiments, providing an overview of the physiology of facial musculature. The heart of the volume lay in the second part, which Duchenne called the "scientific" section, and was of primary concern to Darwin. In it Duchenne described his electrical experiments and assessed the specific muscles that contribute to facial expressions. The section concluded with a criticism of anatomical errors made in several classical sculptures; in contrast to many of his contemporaries who accepted Greco-Roman statuary as ideal, Duchenne challenged their anatomical accuracy (figure 6-1). The third section, titled "Aesthetics," concluded the book. It included a sequence of photographs of a costumed blind woman enacting dramatic poses, with and without electrical stimulation. As a sightless person she afforded a distinctive view of how emotions manifest themselves. She had developed her expressions instinctively, without visual cues from others.

Six models participated in Duchenne's study: an elderly man, an opium addict, a child, a blind girl, a blind woman, and a sculptor. Only the latter was not a patient at the Salpêtrière, a man with an elaborate handlebar moustache whom Duchenne had met at the Academy of Arts. Duchenne was criticized for his choice of subjects, which was said to be unrepresentative of the general population. Indeed, aside from convenience, there was little reason to rely almost exclusively on patients. Healthy subjects would have served equally well for his photographs. The photographs are disquieting in part because of the state of the

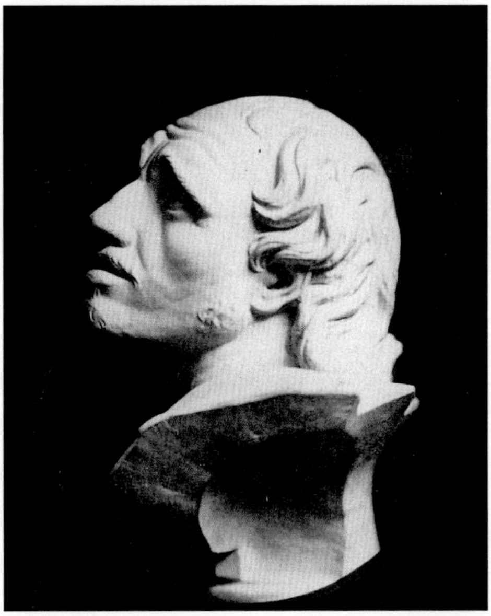

6-1 Guillaume-Benjamin Duchenne de Boulogne, Head of Arrotino, c. 1862. Albumen print. Plate 66 of *Mechanism*. Classical statuary was considered ideal; in this case Duchenne felt the forehead on the statue was depicted incorrectly.

participants' mental and physical condition. The electrode procedure may have been painless, but the expression of joy induced in a drug addict and the look of sadness inscribed on the face of a blind girl are provocative (figure 6-2). The subjects were vulnerable, and one cannot help but imagine, despite Duchenne's justifications, that they fell victim to his camera.

With Duchenne's permission, Darwin used eight illustrations from *Mechanism* in *Expression* (plate 2, figures 1 and 2; plate 3, figures 4–6; plate 7, figure 2; and the engraved figures 20 and 21). Initially, Darwin wrote asking permission to reproduce just four or five, all of which, he said, would be reproduced in wood engravings. He offered to pay for the right to publish them, but Duchenne replied that he was an admirer of his "magnificent" researches and declined compensation. He gallantly declared, "The question of money should not exist between men of science. Choose, in my album, all the figures you wish."[5] It was a generous offer, but Darwin had forced the issue somewhat with his initial request. While offering to pay a modest fee for the pictures, he hinted that he might lose interest if the demands were too steep. "I could manage very well without copying two first, if you objected, as I have photographs which would serve my purpose," he wrote, "but they are not so good as yours; nor are the others indispensable for my work."[6]

Six of the Duchenne plates published in *Expression* were of a single subject (figure 6-3), whom Duchenne described as "an old, toothless man, with a thin face, whose features, without being absolutely ugly,

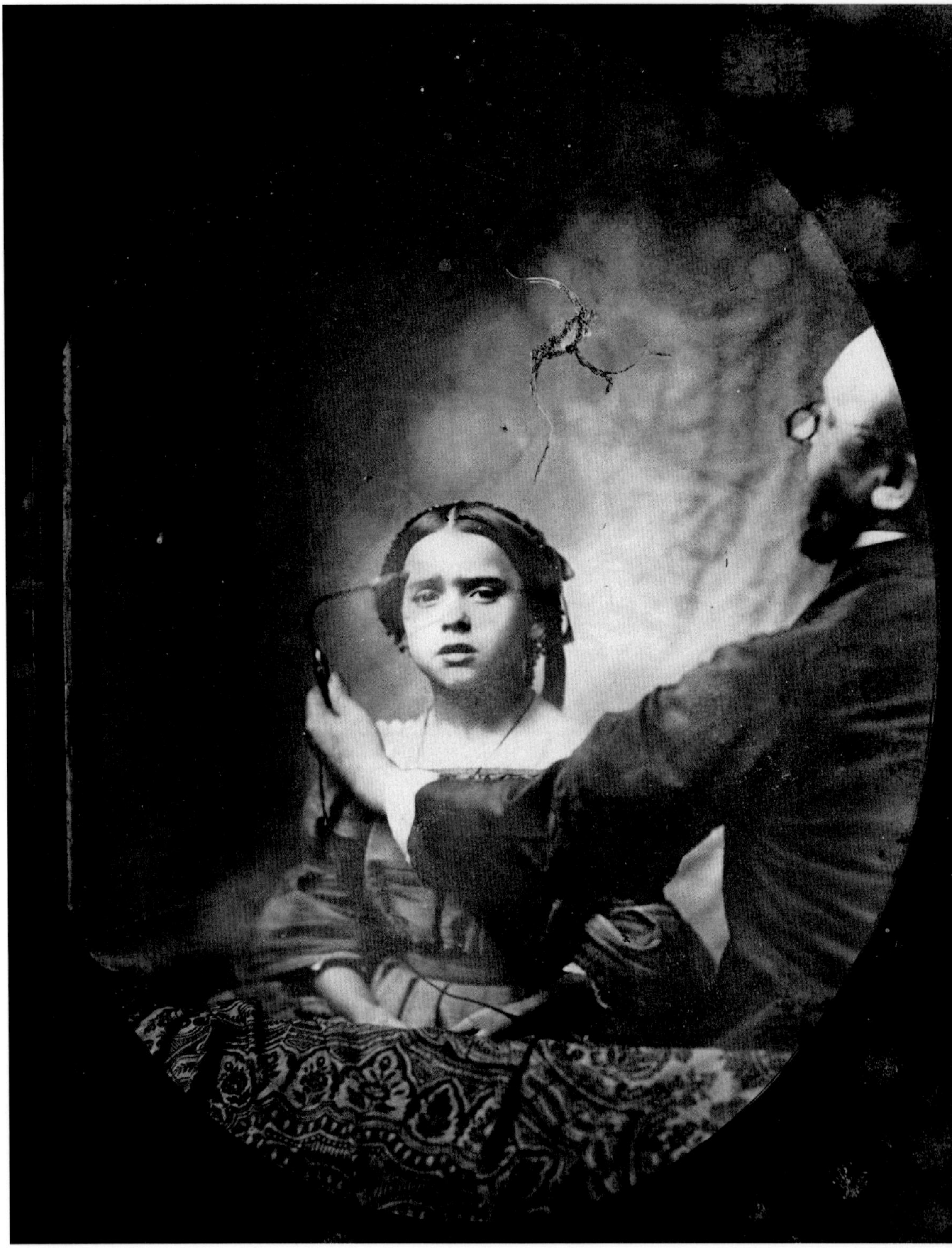

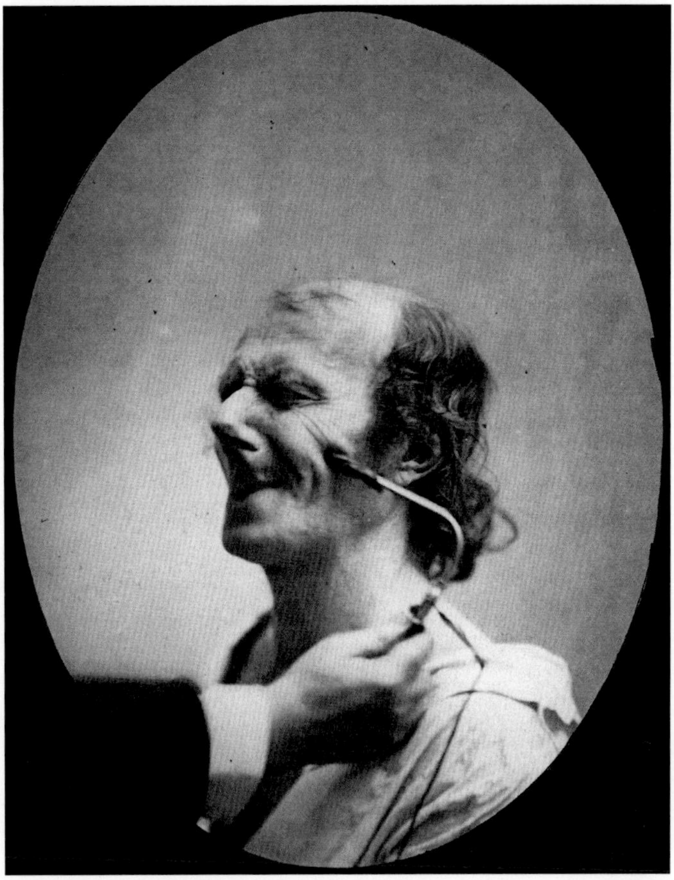

6-3 Guillaume-Benjamin Duchenne de Boulogne, Grimace, c. 1862. Albumen print. Plate 6 of *Mechanism*. An old man with a numb face was chosen for practical reasons.

approached ordinary triviality and whose facial expression was in perfect agreement with his inoffensive character and his restricted intelligence."[7] According to Duchenne, the subject, who appears in *Expression* plate 3, figures 4–6; plate 7, figure 2; and engraved figures 20 and 21, was chosen for practical reasons: he suffered from an anesthetic condition of the face, which made it possible to stimulate individual groups of his facial muscles without causing involuntary response among others. It was as if, as Duchenne remarked, he were "working with a still irritable cadaver."[8]

Darwin made changes to some of Duchenne's plates before publication. Two were copied into wood engravings (figures 20 and 21 in *Expression*); the others were reproduced photographically and appear in plates 3 and 7 of Darwin's book (figures 6-4 through 6-7). Duchenne's toothless man is the only subject portrayed in the book both in photographs and in engravings. Although the photographic plates are accurate copies of Duchenne's originals, the engraved reproductions are substantially different from their predecessors.

Opposite:
6-2 Guillaume-Benjamin Duchenne de Boulogne, Expression of Pain in a Young Girl, c. 1862. Albumen print. Plate 28 of *Mechanism*.

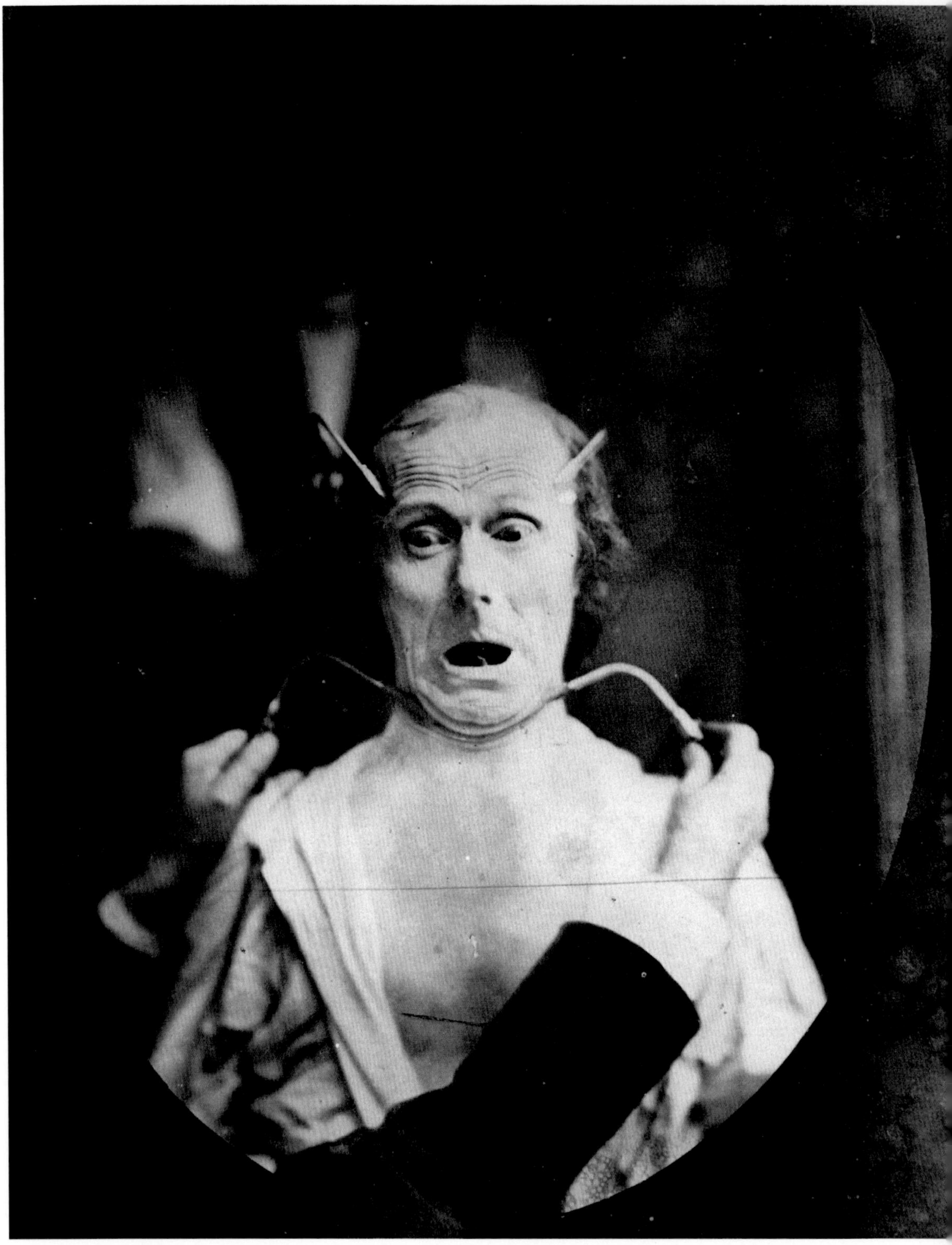

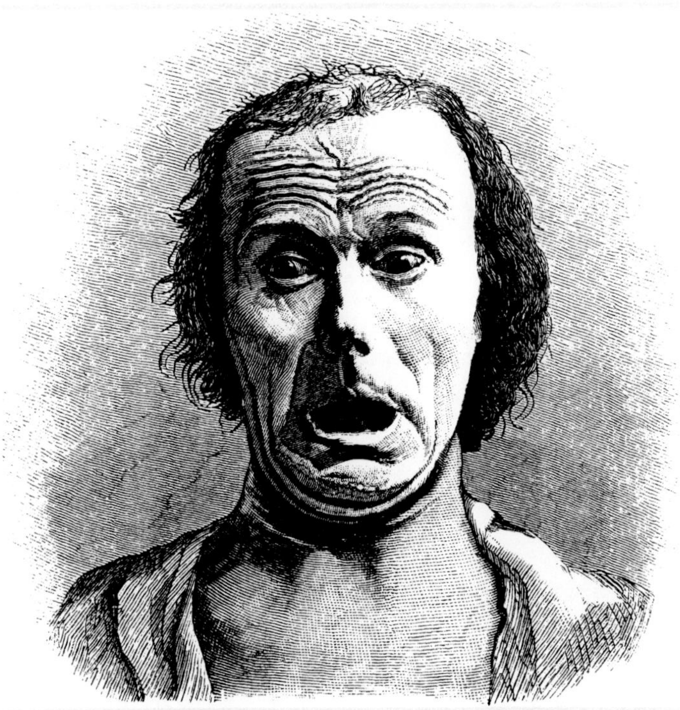

6-5 James Davis Cooper after Guillaume-Benjamin Duchenne de Boulogne, *Terror*, 1872. Wood engraving. Darwin instructed Cooper to remove the electrical apparatus.

In the engraved figures, Darwin instructed the engraver, James Davis Cooper, to "Attend to the wrinkles on neck & forehead—form of mouth. Omit galvanic instruments and hands of operator."[9] Thus, the finished wood engravings (figures 6-5, 6-7) show the subject completely out of context—there is no indication of the electrical probes that caused the expression registered on the man's face. The metallic rods used as stimulus looked more like torture devices than the instruments of a more or less benign test. Duchenne's experiments had been performed with the consent of his subjects and with minimal discomfort to them. Nevertheless, the results were chilling. The subjects had been stimulated to assume false expressions, and the wires and probes were clearly visible in the original photographs. Darwin thought it best to remove them; they were distracting and unnecessary.

This decision gave the final engraved versions new strength. They immediately grab the reader's attention; there is a revelatory quality about them. Without probes to ground the images, and without text to explain them, readers were confronted with iconic pictures of unmitigated terror, horror, and agony. Duchenne's two engraved figures in *Expression* (figures 6-5, 6-7) are bold, commanding images. As Darwin well understood, emotional expressions do not merely emanate from the person who displays them; they are also received in the mind of the person who witnesses them. They veritably jump from the page, bursting with energy and import.

Opposite:
6-4 Guillaume-Benjamin Duchenne de Boulogne, Terror, c. 1862. Albumen print. Plate 61 of *Mechanism*. Duchenne's original photographs revealed a great deal about the circumstances of their creation.

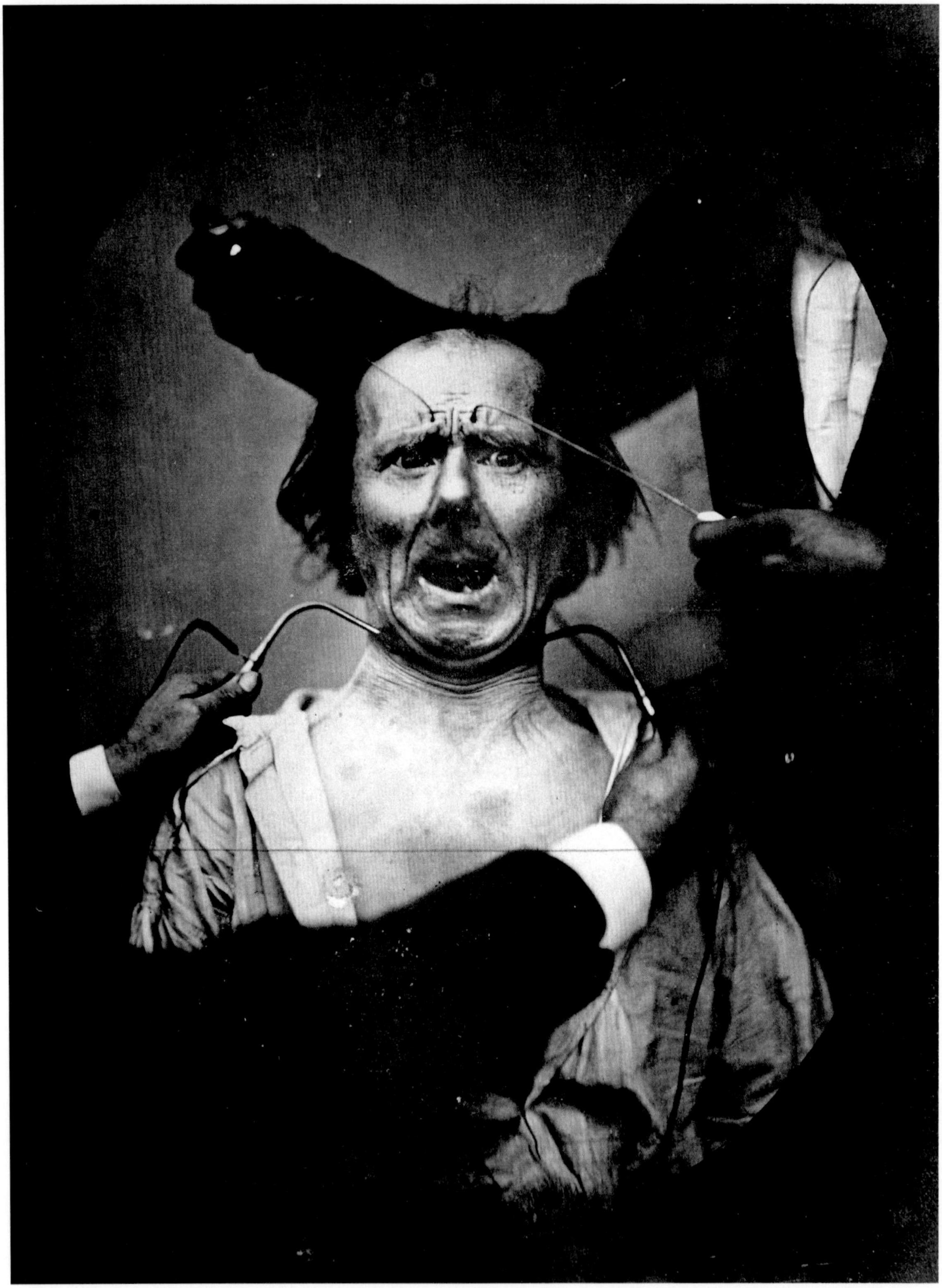

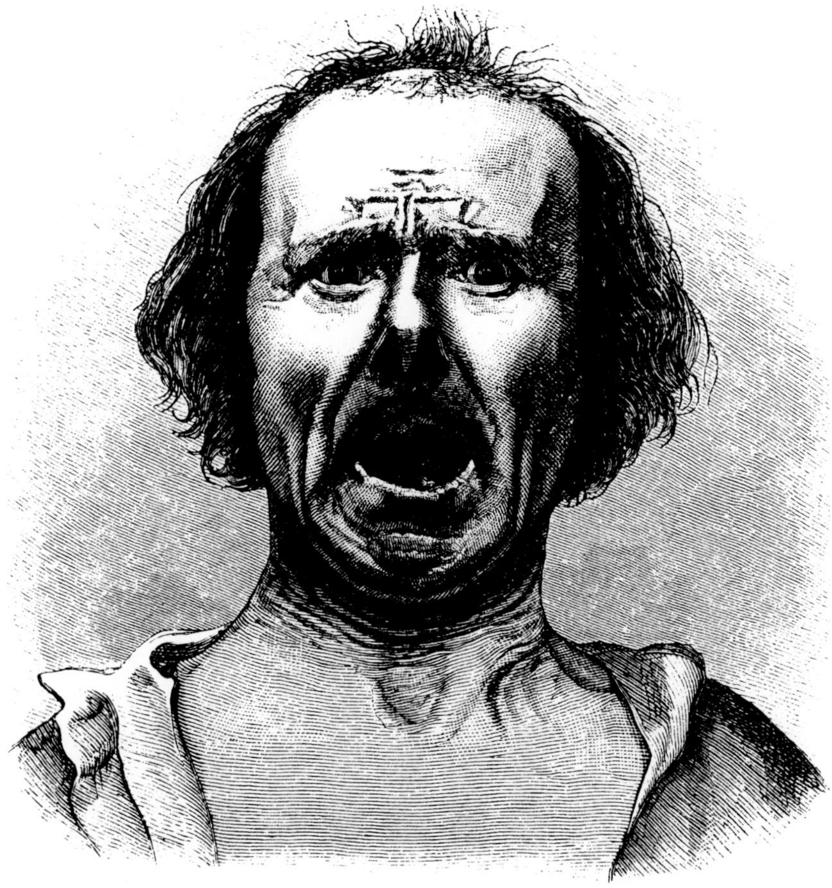

6-7 James Davis Cooper after Guillaume-Benjamin Duchenne de Boulogne, *Horror and Agony*, 1872. Wood engraving. The engraved version was striking.

Not only did Cooper remove the probes and hands of the operator, but he also removed any solid background. In the engraved figures, the old man is shown within an oval vignette, built up of tightly stacked horizontal hatch marks. Faint jagged lines animate the area surrounding the figure, suggesting motion, energy, or perhaps clouds. The man is rendered with varying densities of the same style of marks that make up the background, giving the figure an ethereal, otherworldly quality. He is given more hair in the engraved version, but the gape of his shirt is preserved. This, combined with his direct gaze, form an incomplete narrative. In "Terror" (figure 6-5), the subject looks down and away as if seeing something dreadful beyond the frame of the picture. In "Horror and Agony" (figure 6-7), the last illustration in *Expression*, the subject looks directly at the reader.

Opposite:
6-6 Guillaume-Benjamin Duchenne de Boulogne, Horror and Agony, c. 1862. Albumen print, plate 65 of *Mechanism*. The version Darwin owned.

In *Expression*, Darwin praised Duchenne's photographs. However, he did not embrace them fully. Some of the pictures, such as those of horror and agony, could not be mistaken for anything else. Others were more ambiguous. Mindful that these might be misinterpreted, Darwin showed them to friends and colleagues and asked them to judge the expression portrayed. In *Expression* he explained:

> It fortunately occurred to me to show several of [Duchenne's] best plates, without a word of explanation, to above twenty educated persons of various ages and sexes, asking them, in each case, by what emotion or feeling the old man was supposed to be agitated; and I recorded their answers in the words which they used . . . the most widely different judgements were pronounced in regard to some of them. This exhibition was of use in another way, by convincing me how easily we may be misguided by our imagination; for when I first looked through Dr. Duchenne's photographs, reading at the same time the text, and thus learning what was intended, I was struck with admiration at the truthfulness of all, with only a few exceptions.[10]

Duchenne had photographed people in such unusual circumstances that viewers had difficulty resolving what they were meant to communicate. Their appearance was abstracted, with codifying elements that could be used to deduce meaning removed. In Duchenne's pictures, perhaps the abstraction had gone too far. While Darwin marveled at the "truthfulness" of the images, those less schooled in the intricacies of expressive musculature had trouble interpreting them. Their indeterminacy undermined their effectiveness, and they required text to make them truly meaningful. Perhaps this is why Darwin considered Duchenne's images insufficient to illustrate expression. They were anatomical, but they left readers without a clear idea of how to place them. He needed them to reinforce and illustrate his text, not to distract and confuse readers.

Duchenne's electrical experiments gave the human face the status of an appliance; expressions were just so many bundles of muscles to be tweaked and stimulated at the will of an experimenter. He was principally an anatomist, concerned with the physiology of facial mechanisms, and did not attempt to explain the origins of expression as Darwin did. Duchenne was also concerned with beauty; having dissected expression on a mechanical level, he proceeded to ask what makes certain expressions appealing. In a largely neglected section of *Mechanism*, he attempted to reconcile his discoveries regarding the structure of the human face with contemporary notions of aesthetics.

Apologizing for having used the face of a common, ugly man in the preliminary or "scientific" section of the book, in his "aesthetic" section Duchenne employed an attractive young adult woman for his experiments. His stated goal was to explore the intersection between beauty and expression, or "the set of conditions that constitute beauty from an aesthetic point of view."[11]

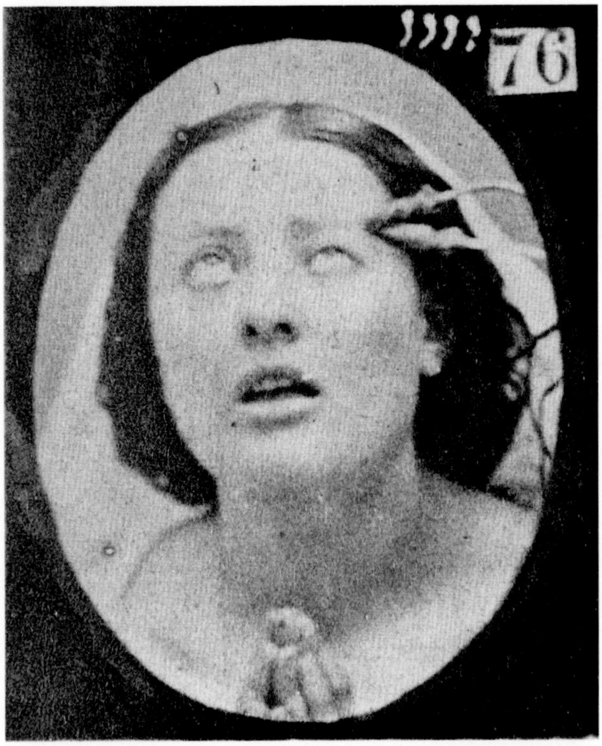

6-8 Guillaume-Benjamin Duchenne de Boulogne, *Woman's Spirit Being Exalted by Her Ardent Faith*, c. 1862. Albumen print. Duchenne apologized for using an ugly man in his experiments and in a later section introduced a beautiful woman instead.

Using his electrical procedure to isolate fine gradations of expression, and taking advantage of advancements in photographic technology since the completion of his so-called scientific pictures in 1856,[12] Duchenne caused his subject to display degrees of expression, differentiating between full and partial contractions, and experimenting with symmetry in the face. Relying only on "my judgement and on my artistic feeling,"[13] Duchenne attempted to create archetypal depictions of such narrow classes of expression as "remembrance" or a "young woman's spirit being exalted by her ardent faith" (figure 6-8). He even dressed the subject in theatrical garb. The depth of faith Duchenne sought to convey was expressed in the text:

> [D]oes not her white veil and homespun dress signify that she is doing something great, that she is going to renounce this world? If you cover the eye and forehead of the left side, the sadness of her features (due to the slight lowering of the angle of the mouth) makes you feel that she is not leaving her dearest loved ones without some regrets. And if you cover the eye and forehead of the other side, you see that her face is sorrowful; you feel that the heart of the nun, who is perhaps leaving her dear mother and family, has not yet been withered through the exaltation of religious feelings.[14]

Duchenne came to believe that his photographs were too mechanical to engage his readers without supporting prose. Too far removed from daily life, his photographs could be criticized as inaccurate. By attempting to reinject beauty into his work, Duchenne attempted to legitimize his methods. He wished to demonstrate that highly complex expressions, even beautiful ones, could be attained through electrical stimulation.

James Crichton-Browne

References to *Mechanism* first appear in correspondence between Darwin and the psychiatrist and amateur photographer James Crichton-Browne in 1869. Crichton-Browne was another contributor to *Expression*. As director of the West Riding Lunatic Asylum in Yorkshire commencing in 1866, he had adopted photography as a tool in the treatment of those in his care. He was a progressive administrator, treating inmates as patients rather than charges. He arranged potted plants in the halls, invited musical performances, and even allowed the patients to drink beer.[15] It was a far cry from the methods used in many contemporary asylums, in which residents were belittled and confined. Crichton-Browne also created a lecture series on the grounds of the asylum, in which medical problems were discussed, and published a journal, *West Riding Asylum Medical Reports* (1871–76), which for a time was the preeminent publication in the field.

As a forward-thinking physician, Crichton-Browne embraced research developments that promised to improve the health of his patients. Photography was one such development. In fact, psychiatry was one of the few scientific disciplines in which the use of photography had become established at this time. In the 1850s, the psychiatrist Hugh Welch Diamond (1809–86) reported success in using photographs of patients as a therapeutic tool. In an oft-quoted paper read before the Photographic Society of London in 1856, he reported: "The photographer catches in a moment the permanent cloud, or the passing storm or sunshine of the soul, and thus enables the metaphysician to witness and trace out the connexion between the visible and the invisible in one important branch of his researches into the philosophy of the human mind."[16] Diamond claimed that he had been able to cure some of his patients by showing them their photographic portraits. The shock of seeing oneself depicted in a diminished state, combined with the psychological distance afforded by using a static picture, was said to be enough to bring some patients to their senses.

Although not widely reproduced, these findings attracted publicity, and the usefulness of photography as a remedy was widely touted. Photography also offered a second, indirect benefit. In addition to being useful to patients as catalysts for mental transformation, photographs enabled physicians to catalogue behavioral disorders. Crichton-Browne was one of numerous Victorian psychiatrists who followed in Diamond's footsteps by assembling a photographic record of patients organized according to ailment. One goal of this effort was to document and analyze the physiognomy of the mentally ill.

Darwin found pictures of the insane especially revealing. He believed that one of the things that characterized the mentally ill was their inability to conform to social norms; often they flouted conventional rules of behavior. This was seen by many, including Darwin, as reversion to a primitive state. According to this idea, their expressions are uninhibited and unconstrained because they have lost the ability to control their emotions. By studying their behaviors in photographs, Darwin hoped to see raw emotional expressions sometimes suppressed in adults. In this regard, the insane were seen as comparable to children, of whom he also strived to obtain good photographs.

Darwin had a long-standing interest in the physiology of the insane. As a student at the University of Edinburgh, he had been nominated for membership in the student-run Plinian Society by his fellow student William Browne, who had devoted himself to studying the nature of madness. Browne wrote a thesis in 1826 focusing on inmates of the Montrose Lunatic Asylum in Angus, Scotland. It was his contention that religious fervor was a kind of insanity and that the charismatics who had been canonized by the church in prior centuries would have been committed to asylums in nineteenth-century Britain.[17] He believed this demonstrated the hopeless inconsistencies of church doctrine, and criticized the orthodox church. It was Browne who gave the lecture refuting Charles Bell's pious *Anatomy and Philosophy of Expression as Connected with the Fine Arts* on the night of Darwin's petition for admission into the society.

Among those who followed Diamond's example was Henry Hering (active 1857–74), who in the 1860s documented patients of Bethlem Royal Hospital in Beckenham, Kent. Little is known about the circumstances under which Hering produced pictures of the insane, because he is usually eclipsed by Diamond, his contemporary. Stylistically, their works are very similar and so are easily confused. Hering is better known as a portraitist; his 1857 image of Florence Nightingale is perhaps his best-remembered work. Darwin owned a photograph Hering took in Beckenham: on the back it is annotated, "Insane Woman from Bethlem Hospital," and "bought at Hering, Nov. 1866"[18] (figure 6-9).

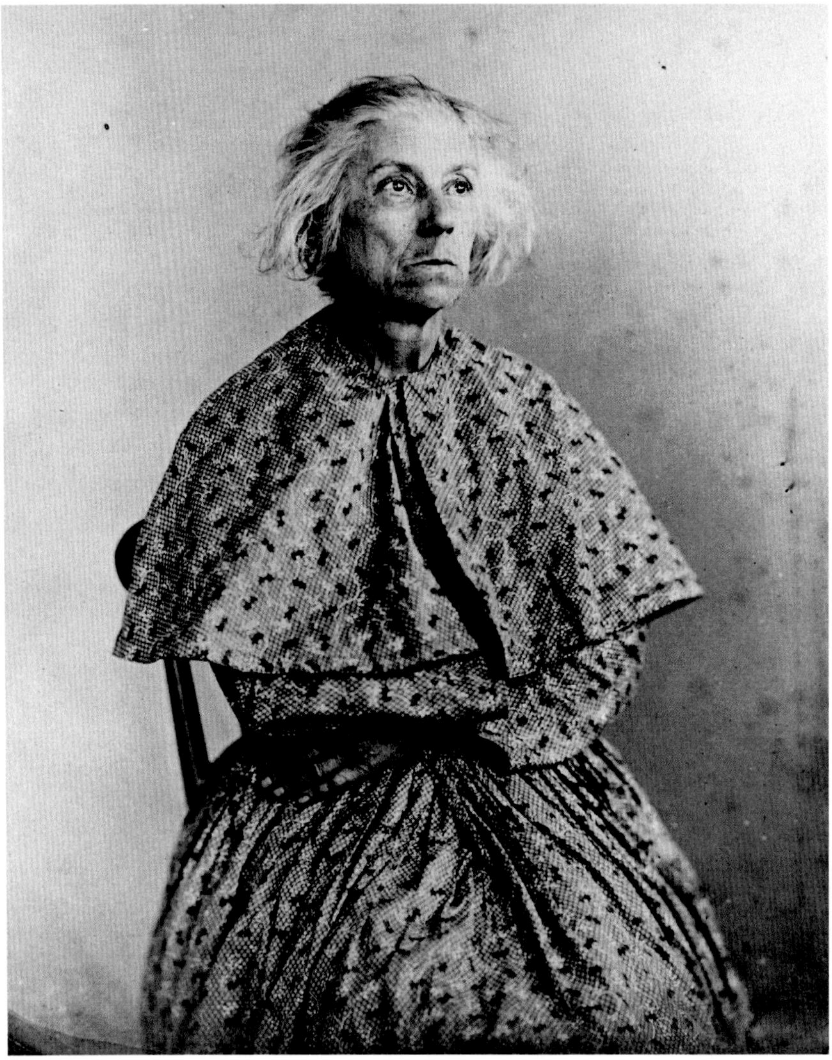

6-9 Henry Hering, *Insane Woman—from Bethlem Hospital*, c. 1866. Salt print. Darwin purchased this photograph of an inmate at Bethlem Hospital in 1866.

Hering's family maintained a studio at 137 Regent Street in London from 1857 to 1874, along the route Darwin is known to have taken in his search for pictures.[19]

A duplicate of Darwin's photograph is held in the archives of the Bethlem Royal Hospital, Beckenham, Kent.[20] It is a salt print, rich in tone and in pristine condition, suggesting it was little used. The date on the back indicates that it was one of the first photographs Darwin acquired and is unusual in the Darwin Archive for its exquisite quality. Unusually, Darwin did not make any annotations about the diagnosis of the sitter or her name, and there is little about her expression that would recommend its inclusion in *Expression*. It is possible that he was drawn

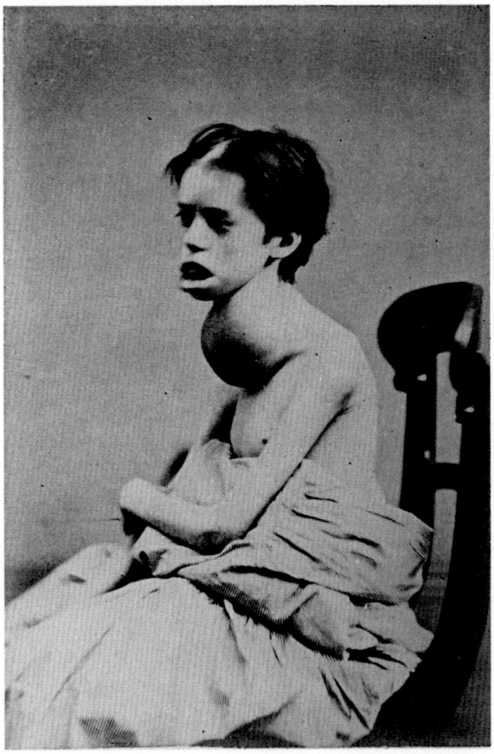

6-10 G. & J. Hall, *Male Subject with a Disfiguring Growth on His Neck*, c. 1865. Albumen print. Darwin could find little use for some of the photographs Crichton-Browne sent.

to the woman's tousled hair. Later, he obtained another photograph of a woman with more obviously disheveled hair from Crichton-Browne, which he included in the book.

Darwin and Crichton-Browne corresponded regularly between 1869 and the publication of *Expression* in 1872. Crichton-Browne's letters to Darwin were often accompanied by portraits of patients in the asylum. Among them were images of people with various physical abnormalities or suffering from mental illnesses from "melancholia" to "homicidal impulses."[21] For example, he sent five portraits of a man with a disfiguring tumor on his neck by the Wakefield firm of G. & J. Hall (figure 6-10). They were arresting, but Darwin bundled them together with a cover slip labeled "Idiots—of no use."[22] The photograph of an "Insane Ear" was more noteworthy, because Darwin had expressed an interest in the evolution of ears in his *Descent of Man* and revised the illustrations in the second edition (figure 6-11). However, the photograph Crichton-Browne sent, a close-up of the deformed ear of a male patient, did not mesh with Darwin's analysis and was left out. The image was titled "The Insane Ear," which suggested only one kind of ear was ever seen among the mentally ill. Dar-

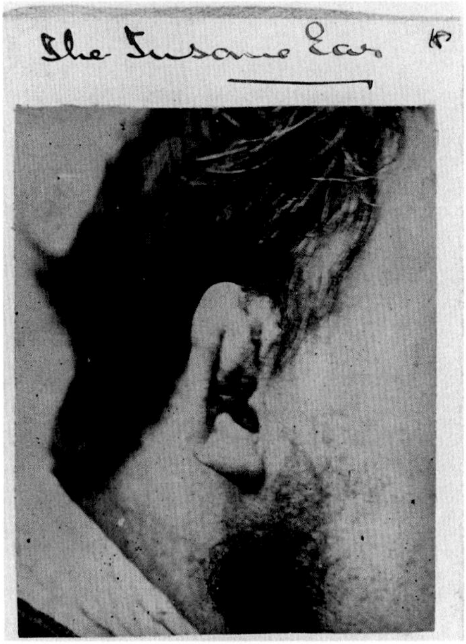

6-11 Studio of James Crichton-Browne, *The Insane Ear*, c. 1873. Albumen print. Crichton-Browne did not prove there is such a thing as a typical insane ear.

win was not sure what to do with it. He put it in an envelope labeled "Photograph of the insane from J. Crichton Browne, March 4, 1873. Perhaps of no good use—the insane ear."[23]

Another example of this type is an image of a family of mentally ill patients in Scotland. Photographed by a firm called Bryson Studios, it was labeled "Brother and Sister 'Imbeciles' All Inmates of One Asylum in Scotland" (figure 6-12). On close inspection it is clear this image was made from seven separately photographed portraits, cut out and collaged to make the image. The print that Darwin received was a photograph of this collage. Crichton-Browne meant well, but the expressions on the subject's faces, which ranged from stupefaction to absence, were not in the slightest bit revealing. Evidently, he could not resist sharing photographs he found interesting, even if they did not relate to Darwin's investigations. From a clinical point of view, they may have been highly compelling, but they were useless to Darwin.

Crichton-Browne also sent photographs of people whose physical appearance belied their mental disquiet. Crichton-Browne had personally participated in the making of many of these images. In total, he sent Darwin at least thirty-seven photographs of his patients, many accompanied by lengthy descriptions of their symptoms. One wonders what transgressions caused the woman diagnosed as suffering from "Erotomania" to be confined in Crichton-Browne's ward (figure 6-13). She glares at the camera with eyes half shut, half-knowing, contemptuous, and indifferent to the forces that sent her there.

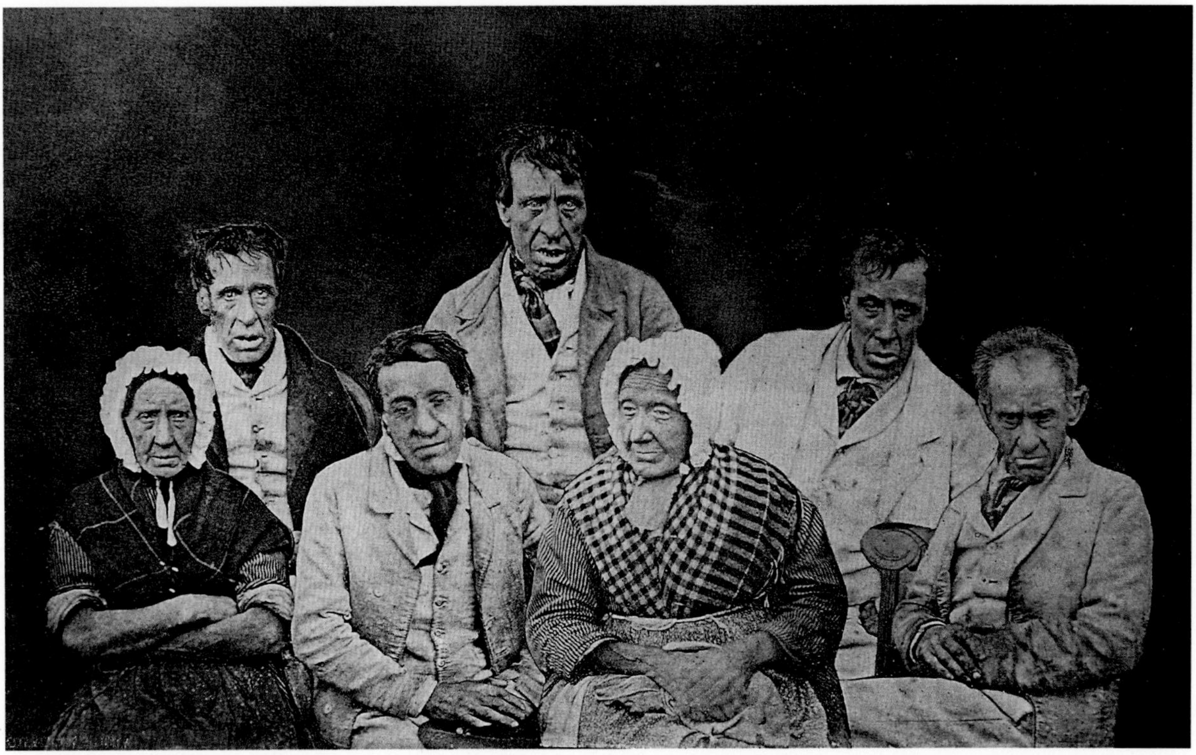

6-12 Bryson Studio, *Brother and Sister "Imbeciles" All Inmates of One Asylum in Scotland*, c. 1870. Albumen print. The expressions were not particularly revealing and were useless to Darwin.

A similar portrait of an "Obstinate Imbecile" also invites reflection (figure 6-14). Stout, round-faced, and with square jaw, he looked the part. But how did his physician define obstinacy? Crichton-Browne's pictures represent people categorized according to the mores of the time. Moreover, they represent a typological approach to personality that Darwin strived to avoid. To the extent that Crichton-Browne captured expression, his successes were limited to "abnormal" individuals, and they were of minimal use to Darwin as illustrations. In fact, he chose only one, the head of a "lunatic" woman with frizzy hair, to reproduce in *Expression* (figure 6-15). Darwin wrote on the back of it in pencil, "Insane woman with horror prevalent—taken in paroxysms—hair like wire." As with several other illustrations in *Expression*, Darwin had it copied as a wood engraving (figure 6-16). He credited the image to his colleague in the text.

Ironically, it seems that Crichton-Browne may not have made this particular photograph but had rather collected it, even though it is signed by Crichton-Browne on the mount. An identical print in the University of Edinburgh Library is marked "Dumfries" on the back, suggesting it was made by another photographer.[24] Psychologist Paul Ekman has argued that the engraved version is actually incorrect, be-

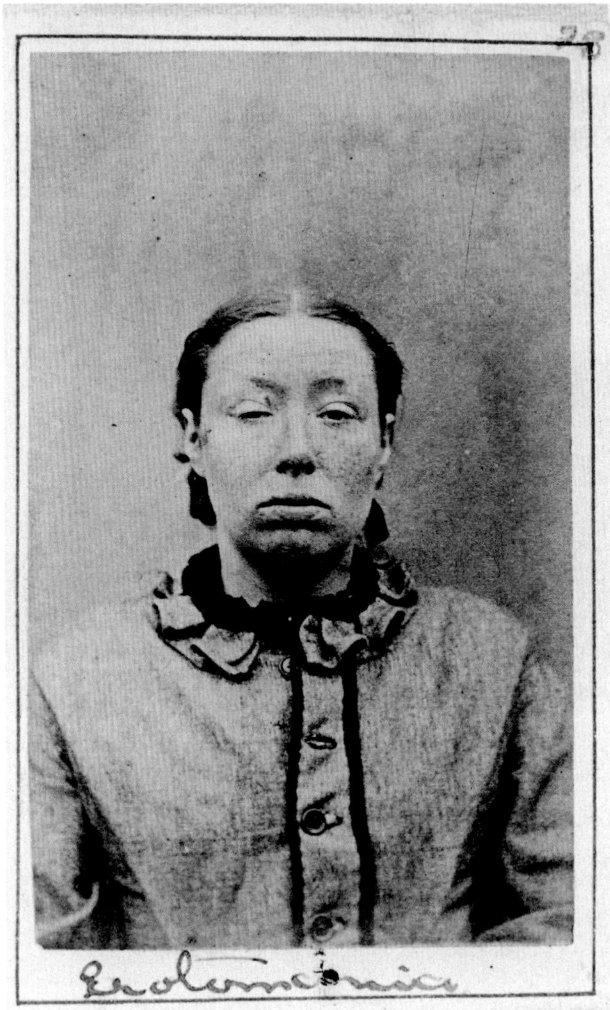

6-13 Studio of James Crichton-Browne, *Erotomania*, c. 1870. Albumen print.

cause the wrinkle on the subject's forehead is anatomically impossible. Because the wrinkle is not present on the original photograph from which the copy was made, Ekman believes the engraver added it, possibly on his own initiative.[25] Because both extant copies of this image have faded with time, it is impossible to know what would have been visible in the photograph at the time of its creation.

Crichton-Browne's main contribution to *Expression*, then, was conceptual. Darwin had asked his opinion on anatomical matters relating to expression, including his interpretation of Duchenne's *Mechanism of Human Facial Expression*. Darwin's large copy of *Mechanism* is full of marginalia, inscribed by Crichton-Browne and Darwin as they passed the book back and forth. Darwin incorporated so many of Crichton-

Opposite:
6-14 Studio of James Crichton-Browne, *Obstinate Imbecile*, c. 1870. Albumen print.

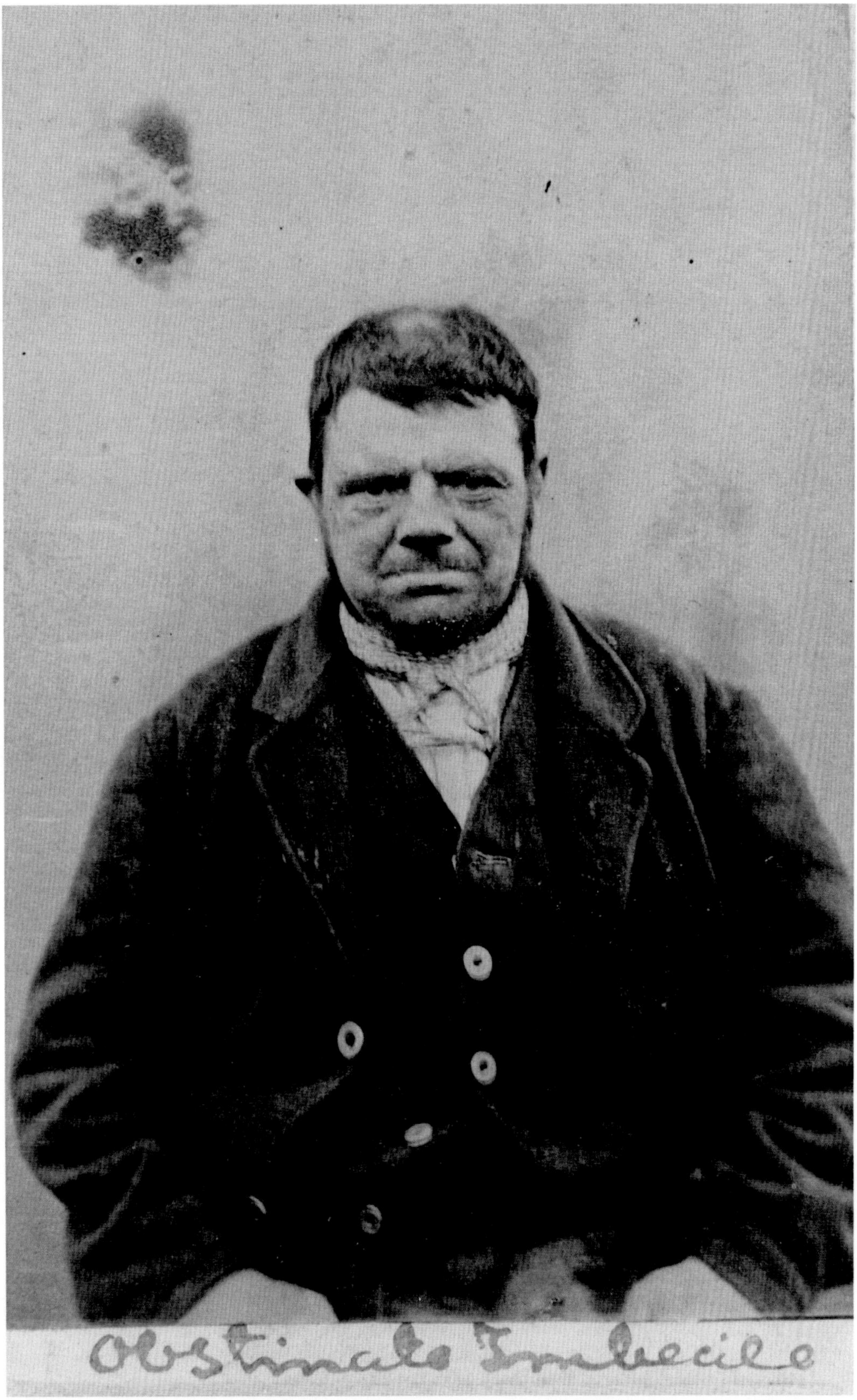
Obstinate Imbecile

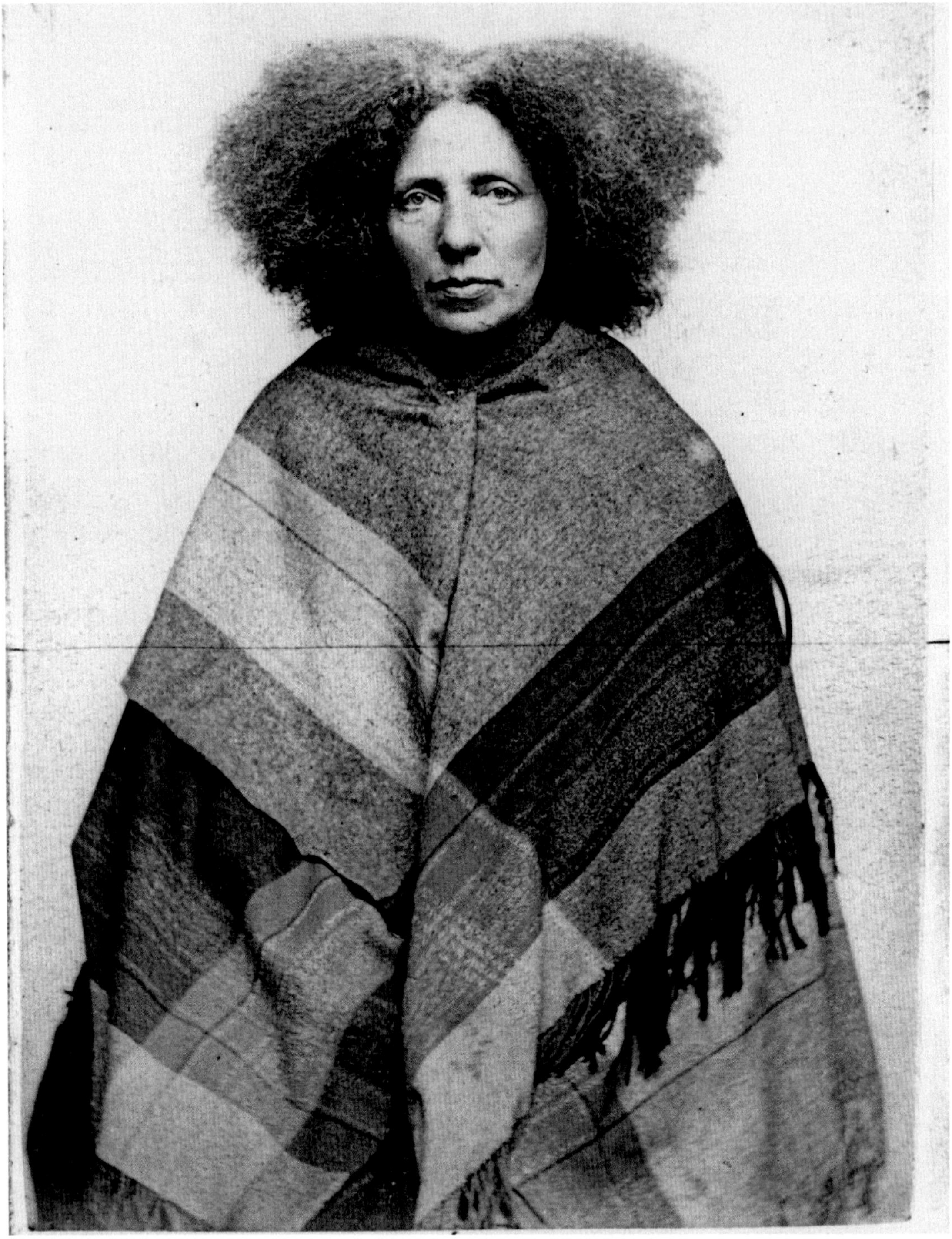

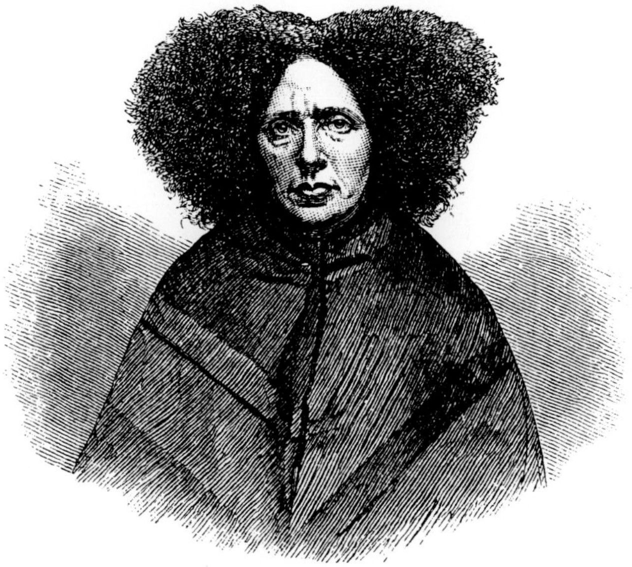

6-16 James Davis Cooper after a photographer in the Circle of James Crichton-Browne, *Insane Woman, Showing the Condition of Her Hair,* 1871–72. Wood engraving. Figure 19 of *Expression.*

Browne's observations in *Expression* that in March 1871 he wrote suggesting that he should be credited as coauthor of the book.[26] Crichton-Browne declined politely. It was, perhaps, an exaggeration of his role. However, Darwin's gratitude was genuine. His contribution to the project, practical or otherwise, was considerable.

Opposite:
6-15 Photographer in the Circle of James Crichton-Browne, *Insane woman, Showing the Condition of Her Hair,* c. 1870. Albumen print. The original contained little detail in the forehead.

7
Laughing and Crying

Darwin's quest for pictures of expressive babies

Virtually everything in the life of Charles Darwin was the subject of research and experiment, so it was not long after he married and began to raise a family that his wife and children became prime specimens in his personal laboratory. His home, Down House, was by all accounts a warm and loving place to grow up, but it was also a place where the residents knew that even the smallest quirks might be jotted down for future analysis.

The progression of infancy was a matter Darwin took seriously, so when Darwin's first son, William, was born in 1839, his development immediately came under scrutiny. Notes on the the mental and physical growth of William and other Darwin children form a substantial part of Darwin's notebooks "M" and "N," two volumes he devoted to observations on expression. These notes, in turn, became crucial evidence in the formulation of his ideas about evolution; many subsequently found their way into *Expression*. Together with a diary centering on the developmental phases of William himself, they also formed the basis for an article he published in the journal *Mind* in 1877, "Biographical Sketch of an Infant." In it Darwin described what amounted to his son's decidedly secular private Genesis: "During the first seven days various reflex actions, namely sneezing, hiccupping, yawning and of course sucking and screaming were all well performed by my infant."[1]

Such parental pride as may be detected in this account of his son's accomplishments was offset by more clinical reports of the savage qualities of his offspring.Darwin saw children as primitive, uncontaminated versions of adults, plainly displaying the ancestral characteristics of their animal progenitors. His notebooks are full of observations likening children to various animals, such as his speculations on the origins of hide-and-seek: "Children have an uncommon pleasure in hiding themselves and skulking about in shrubbery, when other people are about: this is analogous to young pigs hiding themselves."[2] Similarly, viewing animals at the London Zoological Gardens he drew comparisons with human children: "The orang outang [when touched] threw itself down on its back & kicked and cried like a naughty child. Do monkeys cry? (They whine like children.)"[3] He was fascinated by infant expressions of irritation, jotting in his Notebook "M": "whine of children, puppies do so, dogs nearly silent, so with men."[4] In his article in *Mind*, Darwin even likened the unintentionally violent arm movements of young children to the snapping of jaws in newborn crocodiles.[5] With fish, however, he drew a mental line, remarking blithely, "What a contrast does the mind of an infant present to that of a pike."[6]

Darwin always acknowledged a developmental dimension to his evolutionary theories. In *Origin of Species* he devoted a subchapter to embryology, explaining that embryos of vastly different species can appear indistinguishable in their primordial state. Citing the experience of his rival, Harvard scientist Louis Agassiz, he wrote that once when Agassiz forgot to label a number of embryonic specimens, Agassiz could not tell "if it be that of a mammal, bird, or reptile."[7] Darwin concluded:

> Thus, as it seems to me, the leading facts of embryology, which are second in importance to none in natural history, are explained on the principle of slight modifications not appearing, in the many descendants from some one ancient progenitor, at a very early period in the life of each, though perhaps caused at the earliest, and being inherited at a corresponding not early period. Embryology rises greatly in interest, when we thus look at the embryo as a picture, more or less obscured, of the common parent-form of each great class of animals.[8]

By the time he wrote *Expression*, however, his attention had shifted slightly. Darwin studied only humans from a stage after which their organs had differentiated; there could be no mistaking the body of a one- or two-year-old child for an equivalently developed monkey, for example. Whereas in *Origin* he had pointed out the similarities between the fetal states of different species, in *Expression* he studied children as fully formed human beings with untrained animal minds.

Darwin regarded all people, young and old, as governed by the same biological forces. However, the expressions of children are more transparent, as their moods are less controlled than in adults, and because children are less likely than their elders to suppress their true feelings; they have not yet internalized social constraints. As a result, he reasoned, their expressions are honestly expressed. Infants were especially useful, because they are too young to have learned to manipulate their expressions in any way. He viewed them as ideal for studying the similarities between human and animal behaviors, revealing human instincts at their least contaminated. Infants, therefore, became a focus of *Expression*.

Children and Animals

Darwin treated children as experimental blank slates on which to study the expression of emotions.[9] Though their bodies contain all of the muscles used in the expressions of adults, they lack adult control and conditioning and therefore unconsciously reveal instinctive tendencies. Darwin believed that such tendencies are common to all humans, regardless of culture. He claimed that the universality of most expressions could be determined, especially by observing "very young children, those born blind and the most widely distinct races of man."[10] He valued observations of the insane for the same reason, because he considered their actions unguarded compared to ordinary persons.

Of the five behaviors Darwin identified in the first seven days of his son William's life, only one was an emotional expression: screaming, which he associated with the experience of pain, suffering, grief, dejection, despair, fear, anger, or terror.[11] He considered it pivotal to his investigations, because it is one of the most common expressions and one of the first exhibited. In "Biographical Sketch," Darwin described circumstances in which William exhibited the expression:

> [W]hen a little over seven months old, and perhaps much earlier, there could be no doubt, from the manner in which the blood gushed into his whole face and scalp, that he easily got into a violent passion. A small cause sufficed; thus, when he was a little over seven months old, he screamed with rage because a lemon slipped away from his hands.[12]

In *Expression* Darwin elaborated, quoting novelist Elizabeth Gaskell's (1810–65) *Mary Barton* (1848). "An excellent observer," he wrote, "in describing a baby crying whilst being fed, says, 'it made its mouth like a square, and let the porridge run out all four corners.' "[13]

Fourteen of the thirty photographic illustrations in *Expression* are of children. Five of them, the first photographs in the book, and the only pictures of babies, appear on plate 1. An older child appears in another figure on plate 1, and other children appear on plates 2–4. Of these, ten were made by Oscar Rejlander, who was well known for his seemingly spontaneous photographs of children. The oceanographer and photographer George Charles Wallich provided one example (plate 3, figure 2), and another three were provided by the ironically named Hamburg photographer and portrait artist Adolph Diedrich Kindermann (1823–1892) (plate 1, figures 2 and 5).

The order of illustrations forms a pseudoevolutionary progression, from cats, dogs, and birds to monkeys, and then from infants to older children and adults. The photographs of infants on plate 1, illustrating the chapter "Special Expressions of Man: Crying and Weeping," mark a narrative transition from animals to humans.

Ginx's Baby

The first of these photographs, referred to as "Mental Distress" in the text but known to the public as *Ginx's Baby*, is the most storied of the illustrations in *Expression*[14] (figure 7-1). It was one of the photographs Darwin commissioned from Rejlander, but unlike any of the other illustrations in the book, it grew to have a commercial life of its own. Hailed as one of the first "momentary" images ever made, Rejlander sold thousands of copies as individual photographic prints. The picture was considered a technical *tour de force* because Rejlander supposedly had captured body and facial muscles in vigorous movement. The picture

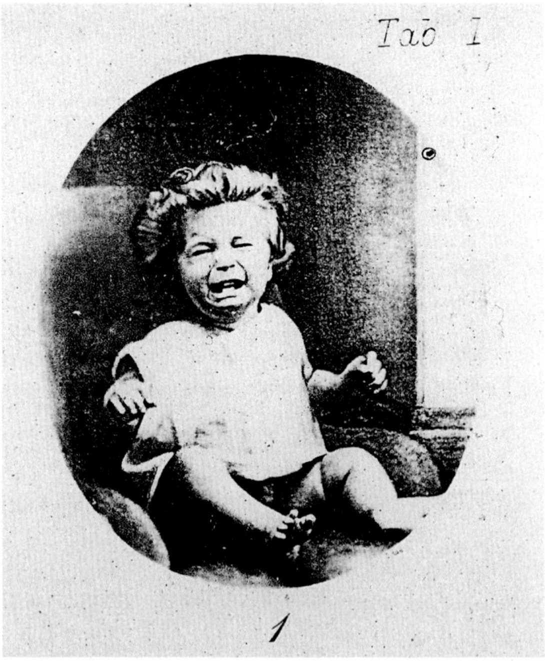

7-1 Oscar Rejlander, Mental Distress, or Ginx's Baby, 1871–72. Heliotype. Plate 1, figure 1 of *Expression*. The version that appeared in most copies of Darwin's book.

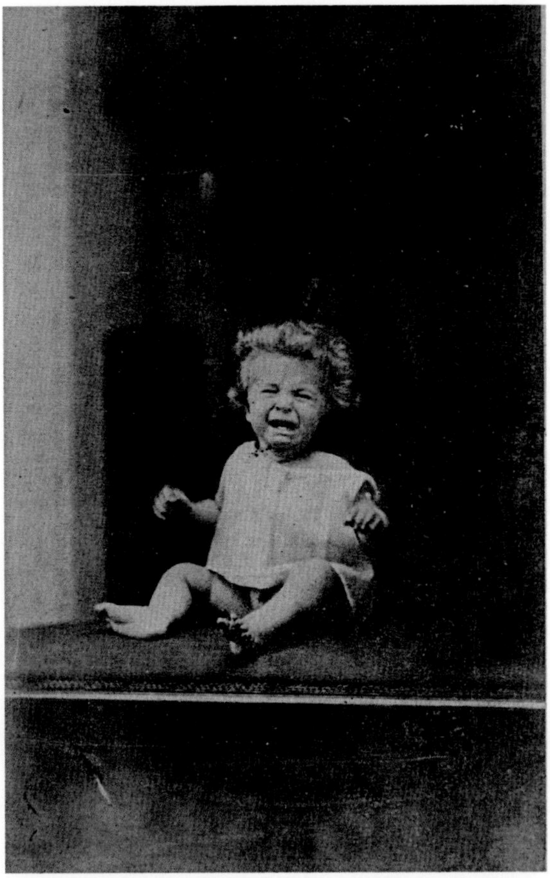

7-2 Oscar Rejlander, *Mental Distress*, or *Ginx's Baby*, 1871–72. Albumen print. Rejlander's original, small photograph.

has been described as exemplary of the collaboration of Darwin and Rejlander.[15]

"Mental Distress" was intended to demonstrate the actions of a child in a fit of rage:

> Infants, when suffering even slight pain, moderate hunger, or discomfort, utter violent and prolonged screams. Whilst thus screaming their eyes are firmly closed, so that the skin round them is wrinkled, and the forehead contracted into a frown. The mouth is widely opened with the lips retracted in a peculiar manner, which causes it to assume a squarish form [much as Gaskell had described]; the gums or teeth being more or less exposed.[16]

But the picture was deceptive. Although widely praised for its instantaneity, it is now apparent that the picture was not a true photograph, but a drawing made to look like an original photograph.

Although Rejlander had succeeded in producing a modest photograph of a crying child for Darwin, it was too small and unclear to be used in *Expression* (figure 7-2). To get around this problem, he made a large chalk drawing of the original, then photographed it. The negative, when printed, looked like an original. Darwin knew that the image had not been made through conventional photography, as readers were led to believe. But neither admitted the sleight of hand, preserving its reputation as a significant photographic achievement.

Ginx's Baby is the only photograph in *Expression* that changed between editions. The rephotographed drawing is more common, appearing in most known versions of the book. Strangely, the unaltered photograph does appear in some early editions, too, particularly in certain issues of the first edition.[17] Darwin never commented on this variation, so it is impossible to know whether it was intentional. One can only speculate that when the actual photograph was tried, it looked inferior to the drawn version and was quickly replaced.

Heliotype Printing

The two variations might also be traced to the printer, the newly formed Heliotype Company, which struggled to produce the plates Darwin needed on time. More than 7,000 copies of the first edition of the book were printed in Britain alone, putting enormous pressure on the firm. The plates seem to have been the company's first large commercial order, and the company had not yet perfected mass production. There was also labor unrest at the company, and a strike was called just six days before publication. Darwin wrote to the offices of his publisher, John Murray, expressing his concern:

> It is a very bad job about the Heliotype plates and will I fear cause us loss; but it is obvious that the Co. must be at least as anxious as we are, and therefore I honestly hope the difficulty will soon be overcome. It will be necessary to have a look at the plates, to see that they do not palm off poor copies owing to the employment of incompetent workmen.[18]

Produced in a rush and possibly even with replacement workers, the plates were of uneven quality. Darwin wrote to his publisher again, suggesting they lay down the law: "One of my sons saw a copy of 'Expressions' the other day with very poor Heliotypes. Would it not be well to caution the Co. that they will lose credit if they distribute poor copies?"[19]

Heliotype had been Darwin's idea. Normally, photographic illustrations had to be glued, or "tipped," into the pages of books. This was labor intensive, and the quality of the finished product was difficult to control. In an effort to reduce costs, he had canvassed photographic printers in London for suggestions of alternative ways of reproducing photographs. Heliotype was a new photomechanical method of reproduction invented by the photographer Ernest Edwards (1837–1903), for whom Darwin had sat for a portrait in 1868.[20] Although he had no experience in photographic publishing, Darwin suggested this new technique to John Murray. In April 1872 he wrote to Murray's assistant, Robert Cooke, telling him what he had learned: "I think it would be worthwhile to get estimates," he suggested, from "Edwards & Kidd, 22 Henrietta Street. For reproduction of photographs the size of paper of Origin. By this process common paper is printed on & there is no need for mounting. I have seen many very good illustrations of this process."[21]

The use of heliotype for photomechanical reproduction was innovative. Unlike ordinary photographs, which must be printed individually, heliotypes are made using printing plates and may be mass-produced. The heliotype process involves coating a printing plate with light-sensitive gelatin emulsion, which is exposed photographically using an ordinary negative. When exposed to light, the gelatin emulsion reticulates, or develops tiny fissures, in a pattern corresponding to the negative used. These fissures form a relief copy of the photographic image on the plate, which is then inked and run through a printing press using ordinary paper. The finished prints may be bound together directly with pages of text, eliminating the need for expensive, tipped-in photographs. Although low in detail and high in contrast compared to alternatives such as the rival Woodburytype (a photomechanical process involving lead intaglio plates), heliotype reduced the costs of production considerably, enabling Darwin to afford the unprecedented number of photographs appearing in *Expression*.

Darwin was not without his doubts about the suitability of heliotype. "I am very much perplexed what to decide about printed photographs & woodcuts," he confessed in May.

> I believe the former would certainly be best. I'm equally perplexed whether to try the Woodbury or the Heliotype process. You [Cooke] think the latter best; Mr. Rejlander [the main photographer on the project] (who wishes that his photos should be well given) feels confident that the Woodbury process is best. Dr. Wallich has just been here & he considers the autotype [a photomechanical technique using carbon pigments] process superior to either of the above.[22]

Two weeks later, he was leaning toward heliotype, and a trial run was proposed. "The Heliotype seems consistently the cheapest, & I think I could not do better than try one set of plates. What do you think?"[23]

Eventually, Darwin and Murray settled on heliotype. Using the process, it was possible to publish photographs without relying on copyists to make facsimile engravings. This allowed Darwin to claim that the details of his originals had been reproduced with absolute precision. Once the plates were printed, however, Murray complained about escalating production costs. Darwin defended himself, saying, "I could not possibly have given many of the expressions by wood-blocks," but conceded, "I would not have given so many photos had I foreseen the cost."[24]

Confusion about how best to illustrate *Expression* is evident in the mixture of photographs and wood engravings it contains. By the time Darwin and Murray had agreed to use photographic illustrations, Darwin had already paid the engraver James Davis Cooper to prepare wood-engraved versions of three photographs.[25] Rather than waste them, he appended them, somewhat incongruously, alongside the photographic plates. These are figures 19–21 in the book, the illustrations of a woman with wiry hair, and photographs from Duchenne illustrating looks of horror and agony. Darwin was pleased, but remained committed to true photographic illustrations. "They are excellently done," he admitted, "but some of my friends tell me they think photos would be better than even facsimile engravings, in as much as the former are a guarantee of absolute accuracy."[26]

Production turmoil may have caused the actual photograph of *Ginx's Baby* to be substituted for the drawn one. Considering the uncertainties at the factory, it is also possible the job was subcontracted to another firm and a fresh set of plates created. Heliotype plates are made of fragile gelatin and, unlike more robust printing methods like wood engraving, must be replaced frequently. Plates could easily have been mixed up and the correct placement confused. In all of the seven plates, apparently arbitrary reversals may be found from one edition to the next, as new plates were created and images mistakenly flipped. In fact, few of the photographs in *Expression* initially appeared with the correct left-right orientation. Only recently was this problem corrected, in the psychologist Paul Ekman's 1999 edition of the book.

Rejlander to the Rescue

But such printing difficulties occurred after *Ginx's Baby* had been made; the biggest obstacle to the publication of the image was its creation in the first place. The first written reference to the difficulties Darwin and Rejlander anticipated in making this and other commissioned photographs appears in a letter of November 1871. Responding to Darwin's request, Rejlander wrote, "In spring-time one might get up such scenes with the chance of taking it photographically, instantaneously, but not now—not now. . . . You ask more than I can do—at this time of the year—at least."[27]

This marked the second time Rejlander had written to Darwin to discuss his commission. An illustration of a screaming child was the picture he most prized. He wanted an image of an infant uttering "violent and prolonged screams,"[28] caught on camera at the height of distress. The photograph would have to be taken so quickly that the minutiae of muscle movement could later be analyzed and studied. Rejlander knew such a photograph would be a challenge to produce. The best available technique, collodion photography (which involved coating glass plates with a sticky photosensitized solution of guncotton mixed with ether), was on the face of it too slow for the task, as it usually involved exposures of several seconds or more. The time of year exacerbated the problem, because the pale, angled light of an English winter meant low natural light levels. Yet he managed to deliver the photographs Darwin requested in the spring of 1872, including the challenging crying-baby image. This picture, and the obstacles that had to be overcome to make it, became a centerpiece of their collaboration.

Although *Expression* sold well, the popularity of *Ginx's Baby* owed more to Rejlander's manufacture of carte-de-visite and cabinet card (6-½ x 4-½ in) versions of the image than to its appearance in the book. In the three years between the publication of *Expression* and his death in 1875, Rejlander is reported to have sold more than 60,000 nine- by twelve-inch prints and a quarter of a million cartes-de-visite of *Ginx's Baby*[29] (figure 7-3). These were sold independently, as mounted prints, with no commercial connection to *Expression*. Separated from the book, *Ginx's Baby* developed a popular status removed from its scholarly origins. Most who bought the photograph were probably not even aware that it had once served the cause of science.

There are two reasons for the extraordinary reception of *Ginx's Baby*. One was the innovative way it portrayed a child in action. The second was its timely association with Edward Jenkins's satirical novel, *Ginx's Baby, His Birth and Other Misfortunes* (1870), the story of a child born into a poor London family.[30] In the book, the baby's father, Mr. Ginx, grief-stricken at his inability to provide for his thirteen children, attempts to drown his newest child at birth, sparing him the misery of his family's meager existence. Caught in the act by a policeman, he is forced to surrender the child to a nunnery. According to the novel, because the mother was not Catholic, an evangelical group files suit to save the child from papist hands. They succeed, but the money they raise to provide for his care is instead spent on the lawyers who free him. As a result, the boy is returned to his father, who abandons him on the front steps of a gentlemen's club. The club raises him for a while, employing him as a page, but soon he is fired for stealing silver spoons. Broken, the hopeless child throws himself off Vauxhall Bridge and drowns. A potent blend of Dickensian storytell-

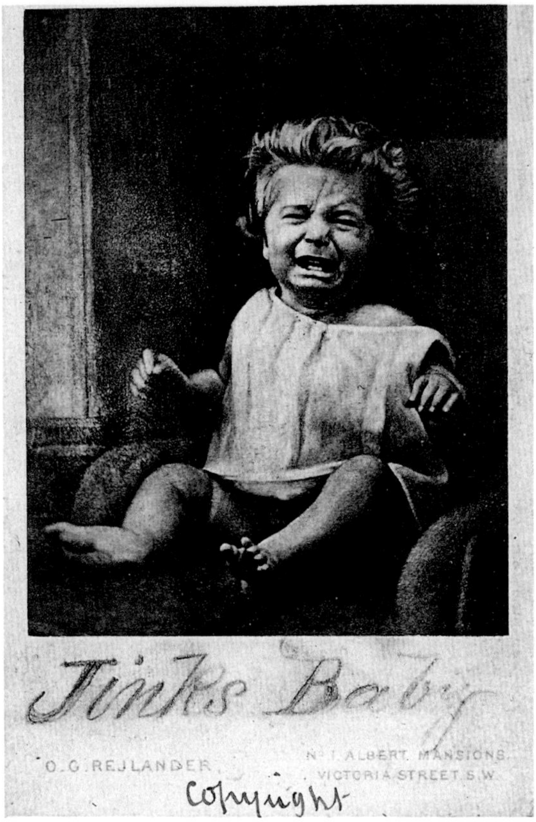

7-3 Oscar Rejlander, *Mental Distress*, or *Ginx's Baby*, 1871–72. Carte-de-visite albumen print. Rejlander sold thousands of copies of the image in collectible formats.

ing and biting political satire, the prose version of *Ginx's Baby* enthralled readers of the early 1870s. Thirty-six editions of the book were printed in Britain in the first five years of its publication, and a similar number were produced and distributed in the United States.

The screaming baby Rejlander photographed for Darwin, with his preternaturally furrowed brow, became identified with Jenkins's hero. It is not clear who first identified the photograph with the character. Readers may have imagined the picture illustrated the child's chaotic baptism in the Catholic faith:

> I have no hesitation about describing the impious behaviour of little Ginx. Whatever swaddled infant could do in the way of opposition, with hands, and legs, and voice, was done by that embryo saint. The incense made him cough and sputter; the lights and singing raised the very devil within him. His cries drowned the prayers. He frightened the conductress by the redness of his face. He ruined the red cross with ejected matter. You would have taken him for an infant demoniac.[31]

Inspired by such vivid descriptions, quantities of Ginx-themed goods entered the marketplace. The songwriter J.C. Drane composed a *"Ginx's Baby* Polka," which was published as sheet music with a litho-

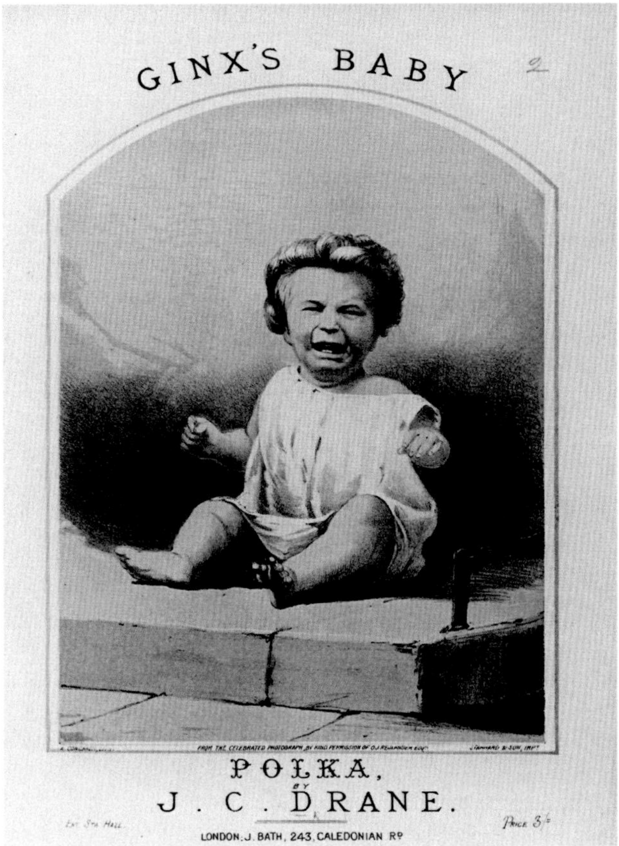

7-4 Alfred Concanen after Oscar Rejlander, *Ginx's Baby*, c. 1874. Lithograph. Sheet music cover for the "*Ginx's Baby* Polka" by J.C. Drane.

graphic reproduction of Rejlander's photograph printed prominently on the cover[32] (figure 7-4). Later editions of the novel were illustrated by former Charles Dickens illustrator Frederick Barnard (1846-1896), with etchings loosely based on Rejlander's image (figure 7-5). The photograph soon became indistinguishable from Jenkins's character. Although Darwin—and, at least publicly, Rejlander—objected, they were powerless against the might of the marketplace.

The Birth of Ginx

Darwin was determined that his illustration of a crying child should be a photograph.[33] Babies' features are tiny, and the subtle movements that make up their expressions can change too quickly to be observed accurately with the unassisted eye. As with adults, photography promised a relatively objective tool for recording infant expression. As Darwin explained: "It is easy to observe infants while screaming; but I have found photographs made by the instantaneous process the best means for observation, as allowing more deliberation."[34]

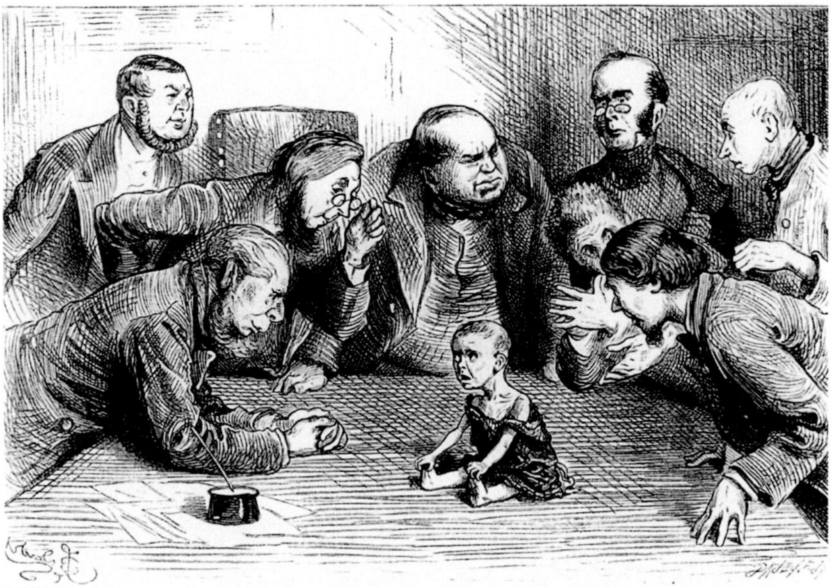

7-5 Frederick Barnard, The Board of Governors Took a Good Look at Him, c. 1874. Wood engraving, from *Ginx's Baby, His Life and Other Misfortunes.* Ghosts of Rejlander's baby even turned up in later editions of Jenkins's novel.

Rejlander was not the first photographer Darwin approached to produce his crying baby image. In 1870, he had contacted another photographer, unidentified by name but probably George Wallich, with the same request.[35] Wallich was a medical doctor by training, best known for discoveries he made while serving as oceanographer on the voyage of the *Bulldog* in 1860. Retiring from marine biology, he opened a professional photographic studio. Darwin must have known him from his natural history work, and he was one of Wallich's first customers. Darwin lamented, "I have been trying to get a London photographer to make one of a young baby screaming or crying badly, but I fear he will not succeed."[36]

His pessimism seems to have been justified, because Wallich's efforts resulted in only two photographs: a picture of a nun in prayer (figure 7-6) and a charming but somewhat conventional portrait of Wallich's daughter Beatrice smiling while wearing a fancy bonnet. The praying nun was useful for comparison with José de Ribera's *Mary in Egypt* (see figure 8-6), but it was not reproduction quality. The picture of his daughter served to illustrate moderate laughter in an older child, so Darwin included it as plate 3, figure 2 of *Expression*.[37] In the text Darwin remarked only that her expression had been a "genuine one."[38]

A German contact, presumably Darwin's former governess at Down House from 1859 to 1865, Camilla Ludwig, put him in touch with Adolph Kindermann. (The contact is identified only as "Miss

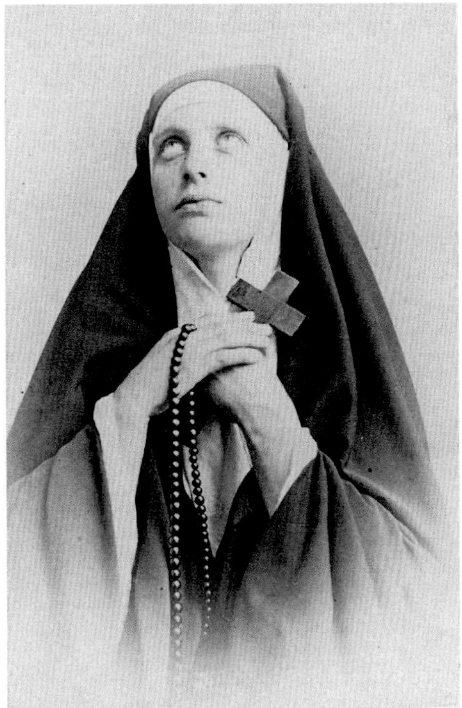

7-6 George Wallich, *Nun in Prayer*, c. 1870–72. Albumen print. Not quite what Darwin was looking for.

Ludwig" in Darwin's correspondence.) In March 1871, Darwin wrote to Kindermann:

> I am writing an essay on the Ex. of the emotions in man & the lower animals. Miss Ludwig forwarded to me several of your photographs of young children, which are quite excellent. I write now to know whether you will grant me the favour to allow me to have 2 or 3 copied & engraved for the sake of illustrating my book.—I should of course have printed beneath each, "copied from a photograph by M. A. Kindermann of Hamburgh."
>
> I sent a message telling Miss L that I shd. be very glad if you could make for me a photograph of an infant, only a month or two old, which weeps with the eyes firmly closed, adequately to [illegible] for the eyes, so as to show the wrinkles round them, & permit me to have it engraved.[39]

In reply, Kindermann sent Darwin more than a dozen lively portraits of children he had photographed in his studio[40] (figure 7-7). In the text Darwin praised Kindermann's pictures as some of "the best photographs in my collection," and reproduced three of them on plate 1, figures 2 and 5, and plate 2, figure 2.[41] Better than the images Wallich had provided, they succeeded in depicting children and infants actually crying,

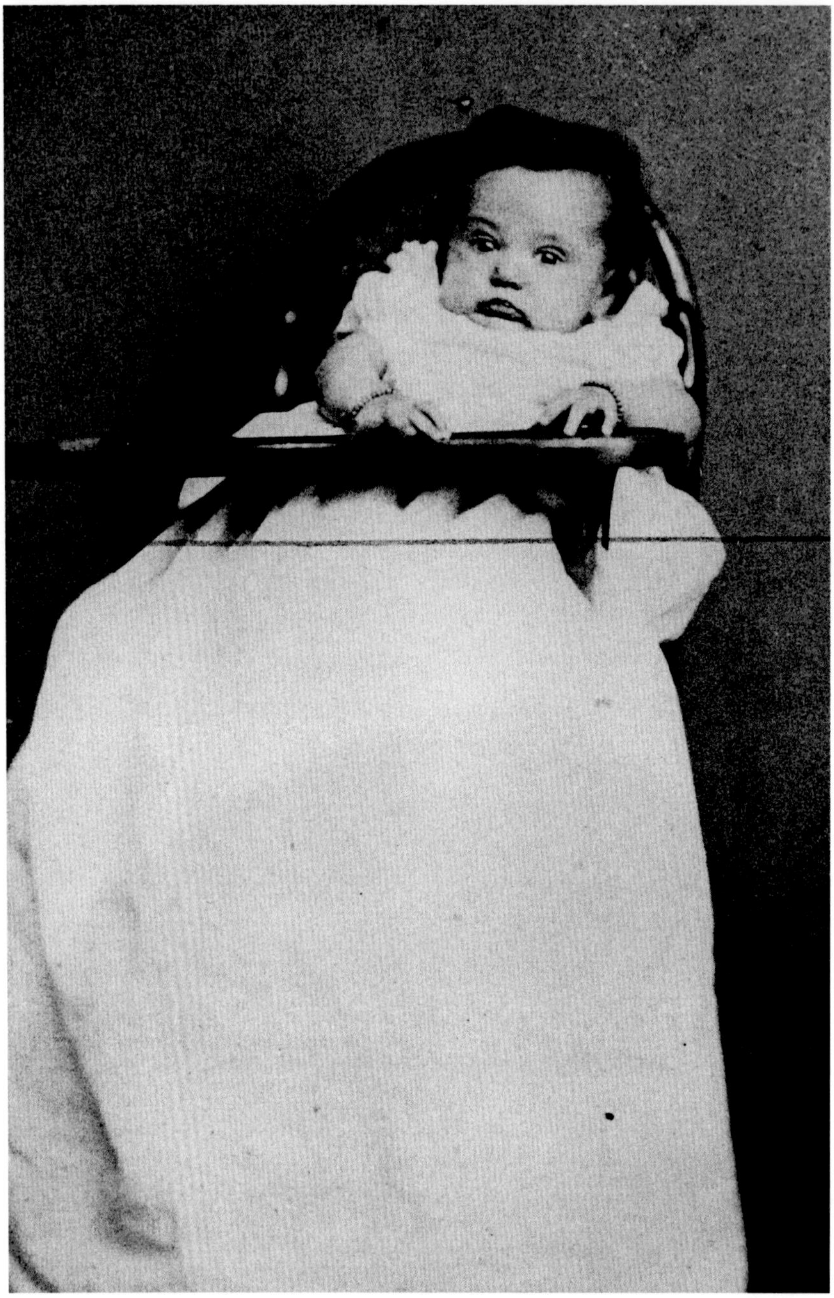

7-7 Adolph Kindermann, *Astonished Baby*, c. 1865–70. Albumen print. Kindermann's pictures were "some of the best" in Darwin's collection. The horizontal line is a cropping mark for publication.

open-mouthed with eyes clenched shut. The children themselves were oblivious to being photographed, slumping and leaning awkwardly in a way any parent would quickly recognize. The natural feeling of these photographs satisfied Darwin's demand for realistic portrayals of expression in children. Lacking in detail, however, they were ineffective in conveying the contractions of distinct muscle groups.

There is evidence in the Darwin Archive that Darwin bought photographs of crying children during his London picture-shopping excursions. The dates of these purchases are not clear, but he may have begun as early as 1869, when the first known reference to collecting photographs appears in his correspondence.[42]

Darwin owned a copy of American portraitist James M. Landy's celebrated *Expressive Pets*, a montage of a dozen photographic portraits of human babies compared with dogs in sympathetic poses. Presented as an eleven- by fourteen-inch grid of carte-de-visite–sized portraits, the dogs appeared to be mimicking the expressions of the children, and vice versa. Viewers enjoyed the strange parallels between the expressions of babies and dogs. One reviewer simply described Landy's crying babies as "very amusing."[43] Another, writing in the *Photographic Times*, declared that Landy's "precocious" babies should be classed among the photographic wonders of the era:

> If you want to insure calm repose on the part of your restless sitters, hang one of Mr. Landy's large pictures upon your studio wall. Your customers will laugh so violently, that they will be so exhausted when they go to sit for their picture, that there will be no move in them. We guarantee a pleasant expression, however. Every gallery ought to have a set of Landy's *Pets*. Twelve laughing, and crying, and screaming babies. . . . Never was so much fun concentrated on so small a sheet of albumen paper before.[44]

Expressive Pets was a commercial success. Landy sold hundreds of copies of the image set for $1.50 each, then reissued the composition in a deluxe version with 28 plates.[45] The piece earned him a gold medal for portraiture at the Cincinnati Exhibition of 1872.[46] The idea of uniting portraits of dogs and babies, both beloved subjects of nineteenth-century consumers, was inspired. Yet it was the execution of Landy's *Pets* that ensured its success, because it had captured the fleeting expressions of its subjects with clarity. A reviewer in *Anthony's Photographic Bulletin* concluded in June 1874, "This group of infants is exceedingly interesting," for it "represents every stage of infantile infelicity and satisfaction."[47]

Darwin was not deterred from his investigations by the work's humorous theme. He analyzed the expressions it portrayed with scientific detachment, dissecting the work from a biological perspective. Darwin cut the image into eight strips to form his own comparisons, and taped the strips back together (figure 7-8). The result was three new, smaller combinations, each of which contained four portraits from the original photograph.

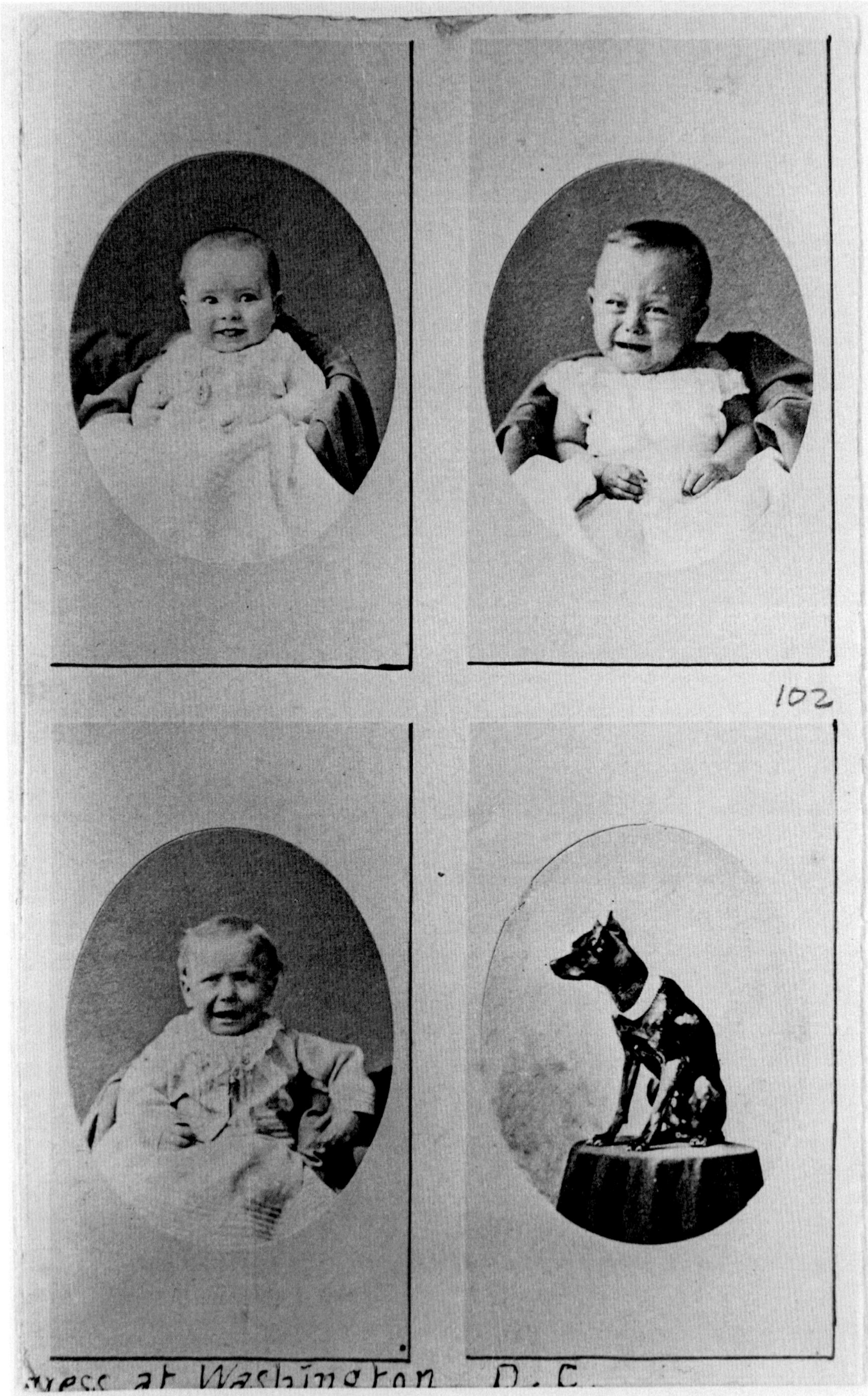
102
ress at Washington D.C.

It may seem surprising that Darwin would have considered *Expressive Pets* worthy of study; it was designed only to entertain. However, he did not believe commercial intent negated a picture's empirical value. Darwin believed that interspecies comparisons are basic to understanding the evolutionary origins of expressive behaviors. He also held that observations made of other animals could shed light on aspects of human expression, noting in *Expression* that "facts observed both with man and with lower animals will here be made use of; but the latter facts are preferable, as less likely to deceive us."[48] *Expressive Pets* may have been comical, but that alone did not disqualify it as evidence.

Charles Bell Returns

Even as he studied contemporary images obtained on his London shopping trips, Darwin returned to imagery familiar from earlier in his career. In April 1871, the same month that the first reference to Rejlander appeared in Darwin's correspondence, Darwin wrote to John Murray asking to borrow a first edition of Charles Bell's 1806 treatise, *The Anatomy and Philosophy of Expression as Connected with the Fine Arts*.[49] Darwin owned a copy of the posthumous third edition of 1844, given to him by his brother Erasmus in 1866,[50] but the first edition, published in full quarto size and illustrated with handsome stipple engravings, was more visually effective. In the later edition, Bell's engravings had been reduced to a cheaper octavo format, and they were not as detailed or as fresh as they had been in the first edition.

The clarity of illustrations in Bell's first edition and their successful integration into the text also provided a useful model for the design of *Expression*. Darwin described the book as "in every way deeply interesting; it includes graphic depictions of the various emotions, and is admirably illustrated."[51] The loan from Murray came just three weeks after Darwin had written to James Crichton-Browne noting that he had met Rejlander, which, in turn, was just a few months before the creation of *Ginx's Baby*. The final version of Rejlander's image bears a striking resemblance to Bell's drawn illustration of a crying child, the only illustration of an infant expression appearing in that book[52] (figure 7-9).

Bell included his crying baby in a section devoted to weeping but did not comment on the illustration itself. Referring obliquely to the manifestation of grief in adults and children, he remarked, "I have thrown the expression of weeping, from pain, into the face of a Faun; for such expression is inexpressibly mean and ludicrous in the face of a man."[53] The animal parallel may have intrigued Darwin; however, in the margins of his personal copy of the book, he wrote that Bell's observation that the lachrymal or tear glands are the first organs affected in crying

Opposite:
7-8 James Landy, Portion of the montage *Expressive Pets*, c. 1872. Albumen print. Darwin cut out the portraits to make interspecies comparisons.

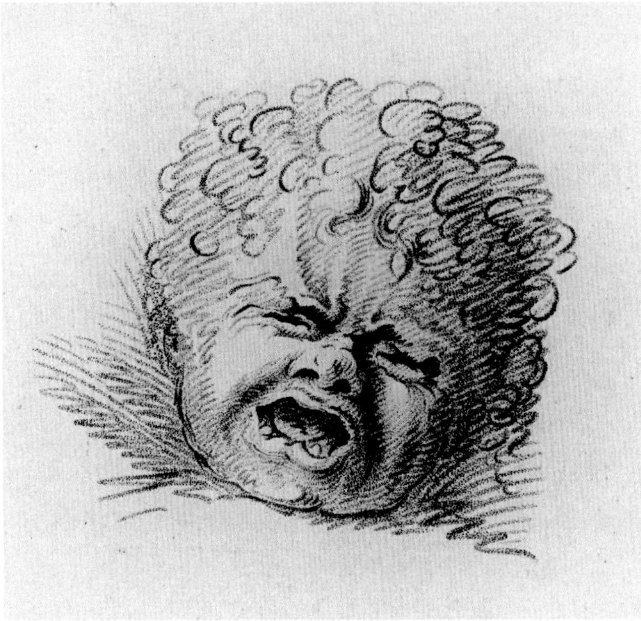

7-9 Charles Bell, Crying Baby, c. 1806. Stipple engraving. From *The Anatomy and Philosophy of Expression as Connected with the Fine Arts*. Bell's screaming baby may have been the model for Ginx.

does not apply to babies,[54] a point that he stressed in *Expression*. Evidently, Darwin repeated the experiment himself, also noting his results in the margin, "upturned corners give look of silly complacency."[55]

The remarkable similarities between Bell's crying child and *Ginx's Baby* suggest that the former may have been used as a model for Rejlander's commission. The child has the same curly hair, is approximately the same age, and adopts a similar posture, with his head tilted slightly and mouth equally agape. Unlike *Ginx's Baby*, however, Bell's drawing illustrates only the child's head, whereas *Ginx's Baby* is full-length. Darwin may have passed Bell's book, perhaps even Murray's first edition, to Rejlander to copy or reinterpret photographically. Such reinterpretations were not unusual. Rejlander's friend Henry Peach Robinson (1830–1901), for example, is known to have sketched elaborate preparatory drawings, down to the location of individual sheep in a landscape, before taking up his camera to execute the same composition.[56] Photographs made in this way were consistent with the concept of conscious composition, which Robinson advocated, and which formed the basis of Rejlander's claim that photography should be considered a fine art.

Back to the Drawing Board

The original drawing of *Ginx's Baby* is in the collection of the National Museum of Photography, Film and Television in Bradford, England[57] (figure 7-10). Mounted on a yellowing piece of cardboard, it is a large

Opposite:

7-10 Oscar Rejlander, *Mental Distress*, or *Ginx's baby*, 1871–72. Albumen print after a chalk drawing. Darwin's copy of the drawing.

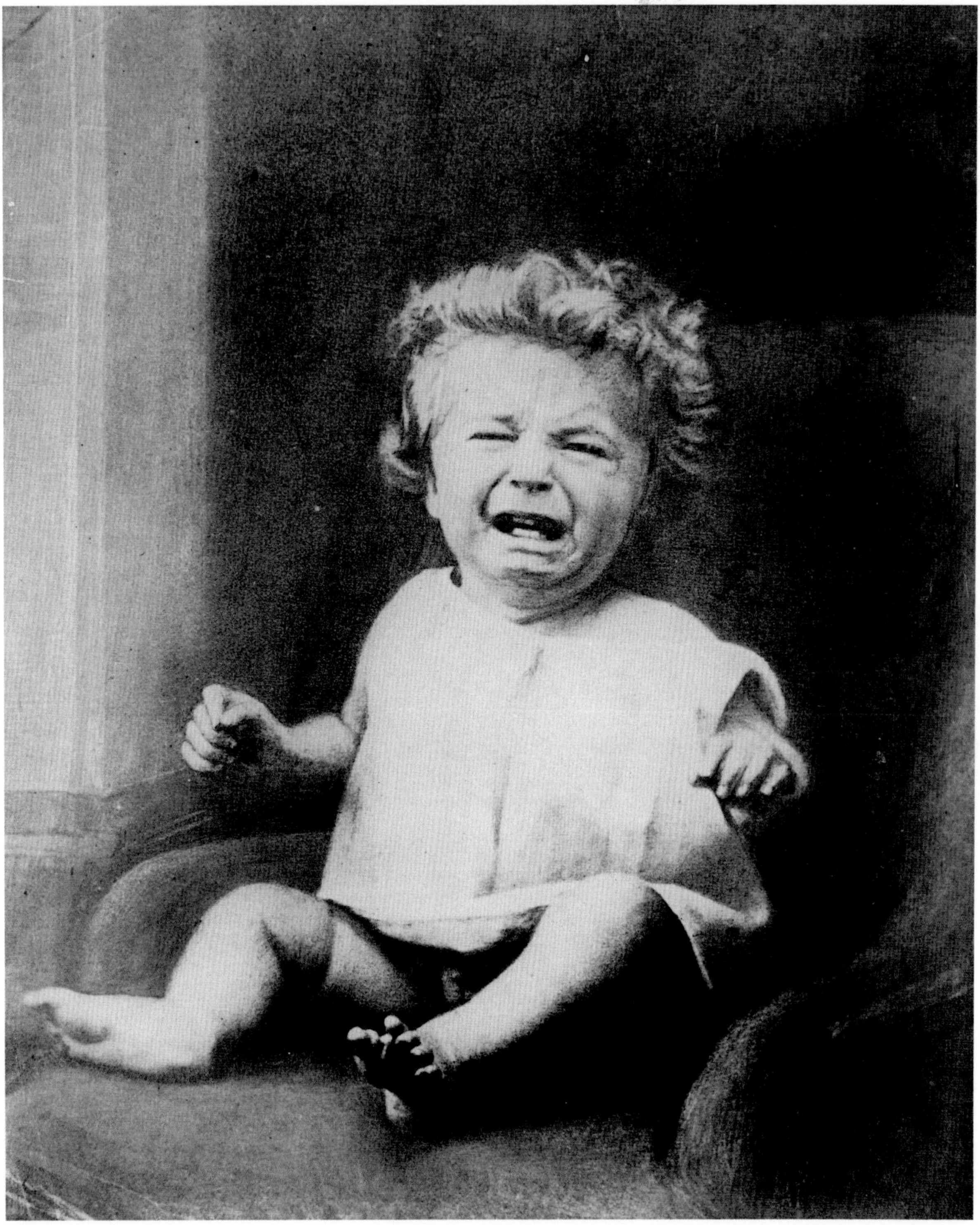

black and sepia chalk drawing, heightened with white. It is skillfully drawn and, apart from its unusual scale and color, is a convincing replica of the original albumen photograph. In fact, when it belonged to the Royal Photographic Society, it was catalogued as a hand-colored photograph. The three shades used to define the image were probably chosen to make copying easier. Through a quirk of chemistry, early photographic materials were relatively insensitive to low-frequency light; that is, colors in the blue range of the optical spectrum. Areas colored in reddish sepia, however, recorded easily, appearing in photographic copies as a subtle, light gray tone. This would have been difficult to achieve using black and white chalks alone. Large swathes of the drawing were rendered in this color, including portions of the face, hands, and legs, as well as the folds of the child's clothing and selected parts of the background.

Rejlander's first successful attempt to produce Darwin's crying-baby image is preserved as a small, faded photograph in the Darwin archive (see figure 7-2). It is a carte-de-visite albumen print, made from a wet-plate collodion negative and glued to a Rejlander studio card mount, measuring approximately four by two and one-quarter inches. At first, it is indistinguishable from the published version. The crying baby closely resembles his published counterpart; his bearing, clothes, and expression are nearly identical to those that appear in the book.

Because the final image was to be reproduced using the newly invented heliotype process, resulting in high-contrast prints of limited detail, any photographs submitted to the printer had to be as clear and detailed as possible. The negative was too small to reproduce successfully in heliotype, and the child fills just a small part of the frame. To achieve better results, Rejlander would have had to photograph the child at close range, yet it seems he was unable to take the photograph any closer than he did in this instance, perhaps as a result of the lens he used.

To make an instantaneous image of this sort, Rejlander would have had to have used every method at his disposal to speed up exposure. His fastest lens would have been his shortest one; that is, the one in which the front and rear objectives were closest together. Such a lens has the additional advantage of producing a final photograph with maximal depth of field. Unfortunately, it would also have the effect of widening the field of view, much as a wide-angle lens does today. He might have been able to get away with such an arrangement photographing a subject the size of an adult, but not with an infant.

The original also contained chemical blotches that would have been unappealing in reproduction. Rejlander had fallen prey to pitfalls he had warned his colleagues about in the *British Journal of Photography* in

1864, that they should contain "[n]o draining lines, nor crape, nor 'stars and stripes,' nor comets!"[58] It is hard to separate flaws intrinsic to the original negative from those that have developed over time. However, it is likely that many of the defects visible in the photograph would have existed at the time it was made. Wet-plate collodion photography was an awkward process, but the pressures of capturing rapid action made it additionally difficult. It is likely Rejlander rushed to produce the negative, which may have caused the quality of development, and washing, to suffer. In addition, he may have overdeveloped, or "pushed" the negative to compensate for underexposure. This also could have caused some of the chemical discoloration visible on the print.

Although Rejlander's crying baby was imperfect, it did capture a crying child at the height of agitation, just as Darwin had requested. The picture was unique, and Rejlander was unlikely to replicate its success in a separate sitting, even if the same model were still available and could be coaxed to act in the same way. It was a point of pride with Rejlander that his infant sitters should develop their poses naturally. As Rejlander's biographer A.H. Wall explained:

> One of his favourite plans was to get up morsels of mimicry or little plays, if I may so call them, into the spirit of which the children were sure to enter, often with a wonderful degree of histrionic cleverness. But then, having inspired them with his idea and feeling—the studio ringing, perhaps, with childish merriment—he generally left them to carry it out themselves, suggesting, perhaps, in a merry way, never gravely commanding, alterations. . . . He appealed to the imagination. . . . As the result, some of the most child-like portraits of children ever produced emanated from Rejlander's studio.[59]

It is clear Darwin knew the final image was drawn, because his copy of the new version is inscribed "Photograph of a chalk enlargement of action photograph,"[60] while another image in which the drawing appears is inscribed in Rejlander's hand, "Child, copy of my large drawing."[61] To make the drawing, Rejlander would almost certainly have used the procedure he outlined in his article "Photography as the Handmaid of Art," published in the *British Journal of Photography* in 1866. In the article he described a method for projecting photographs to provide drawing templates. A contact transparency of a photograph, such as a lantern slide, would have been hung in a window against the sky, with the rest of the window darkened around it. A view camera with its film plate removed was aimed at the transparency, transforming the camera into a makeshift projector. The artist projected the image onto a canvas or sheet of drawing paper of the requisite size. It could then be traced, ensuring fidelity to the original photograph. Rejlander assured his procedure could "save for you the most precious thing you have, your time—no mean service in these electric-torrent-living times."[62]

There is substantial evidence in the Darwin Archive that Rejlander wrestled with the problem of enlarging and reducing his photographs for Darwin. The archive contains multiple printings of a number of the photographs used in the book, some with inscriptions identifying them as "copies." These are resized versions of photographs that were originally produced for sale as cabinet cards or larger commercial prints, or which happened to have been made earlier using large-format negatives. For example, the photograph published as plate 3, figure 3 of *Expression*, a young girl with her hand to her mouth, was sold in the early 1860s in cabinet-card size[63] (see figure 9-8). The archive contains two reduced versions of the same image, made using a stereo camera, in which the original print was propped on an easel for reproduction[64] (figure 7-11).

Because a number of the prints Darwin took from Rejlander were chosen from stock that Rejlander had on hand, they existed in odd sizes.

7-11 Oscar Rejlander, *Young Girl Standing with Hand to Mouth*, 1862. Albumen print, 1871–72. Rejlander tried reducing some oversized pictures using his stereo camera.

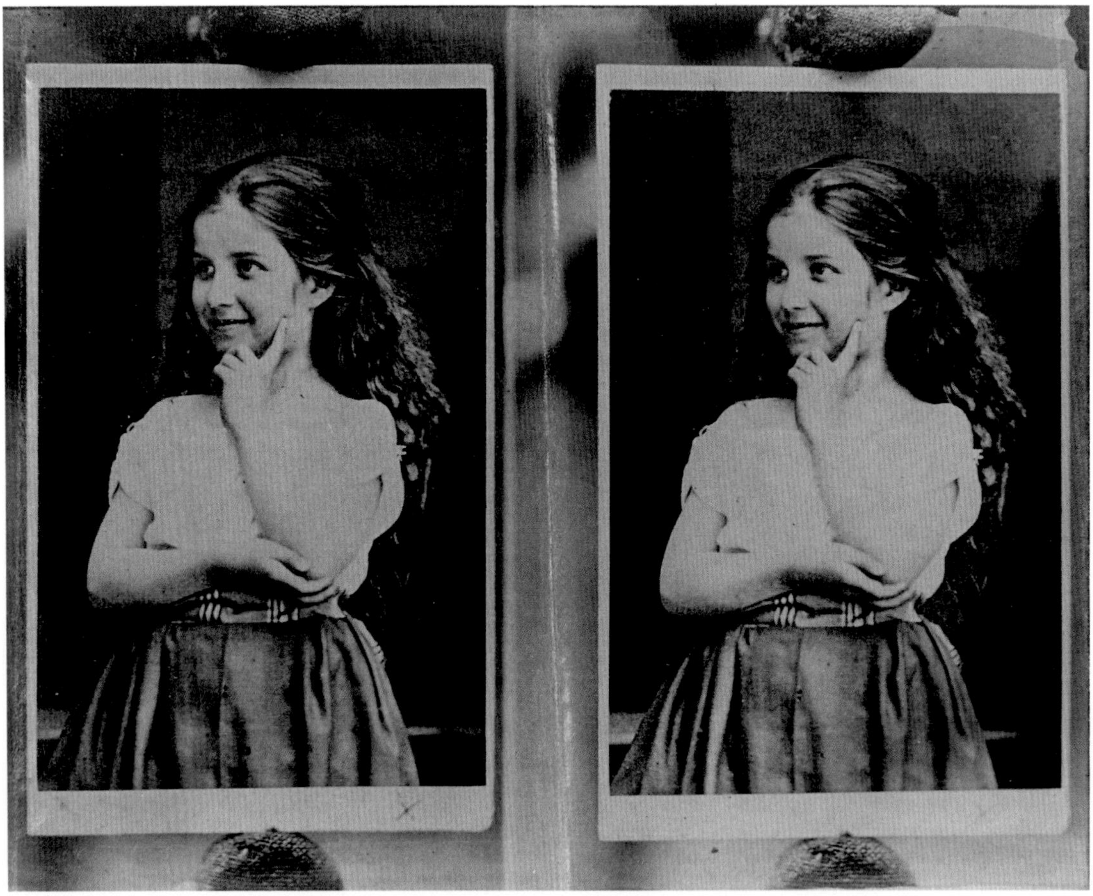

Other than those that had been made as cartes-de-visite, many were too large to be published in heliotype, because the process required negatives of the same size desired in the finished print. Routine enlargement and reduction of negatives was not possible until the introduction of gelatin dry plates and gelatin paper in the mid-1870s and early 1880s. Although commercial enlargers, known as solar cameras, were available as early as the 1850s, these were cumbersome and expensive to use. Furthermore, the quality of the enlargements that could be made using solar cameras was generally poor, and they were capable only of simple enlargements.[65] Given the substantial investment required, and the indifferent results that could be obtained, it is clear that Rejlander did not use one. However, he did put considerable effort into making photographs of an appropriate size and sharpness for publication. For example, in two photographs in the Darwin archive, he even tried using cyanotype (a process similar to blueprinting) to make corrected negatives. Although they are now badly faded, one is inscribed on the reverse in Rejlander's hand, "Better photographs of these. I have used for making better negatives."[66]

The most obvious difference between the large drawing and the photograph on which it is based is the addition of a large padded chair to support the crying baby. In the original photograph, he sits on a table, without support. The chair in the drawn version is relatively small, giving him a slightly larger-than-life appearance. Rejlander may have altered the scale of the chair to give the baby a more commanding presence. The perspective of the chair is also slightly incorrect, suggesting it was drawn freehand. If Rejlander had used his projection technique, the chair would have been perspectivally perfect—it was a prop Rejlander kept in his studio, because it appears in at least one other carte-de-visite produced at approximately the same time as *Ginx's Baby*[67] (figure 7-12). In the drawing, the chair serves to ground the child in a domestic setting, giving it an illusion of domesticity lacking in the original.

With Rejlander's corrections, the drawing conveyed a tangible moment in everyday life readers could mentally verify, and with which they could sympathize. In the drawn version of *Ginx's Baby*, the wall behind the child is more fully delineated than in the photograph, and the edge of the wall to the right is moved into the child's plane. The drawing reproduces the limited depth of focus evident in the original photograph, however, leaving the background blurred and giving the drawing a distinctly photographic appearance. The child is only slightly changed. His left eye is enlarged a little, better revealing his pupil, and his hair is enlivened and brushed over the top of his head, exposing the wrinkles of his forehead, which seem prematurely well defined. The groin of the child is darkened, deemphasizing his exposed genitals.

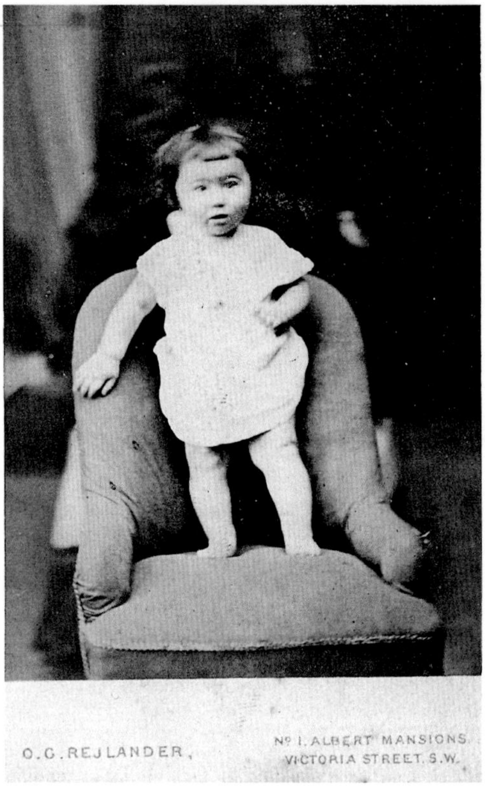

7-12 Oscar Rejlander, *Daisy at One*, 1869–75. Albumen print. The chair shown in the photograph was a prop in Rejlander's studio.

Rejlander's naturalization of the sitter contrasts with the working methods of Lewis Carroll (1832–1898) and Julia Margaret Cameron, two other celebrated photographers of children with whom Darwin communicated. As the art historian Marcia Pointon has observed, they had an opposite tendency:

> There are instances in the photography of [Carroll and Cameron], and in some of Millais's child portraits, of repetitions and thematic recurrences that suggest an overdetermined relationship of artist to subject. Making the child twice other by locating him or her in another time and/or another culture safeguarded the desirability of the object of the gaze, requiring the adult spectator's knowing engagement with a past that is neither his or her own nor that of the portrait subject.[68]

Role Play

Rejlander made his drawing of *Ginx's Baby* approximately life-size, which made it a suitable prop for role-playing and experimentation. The drawing appears in two stereo photographs in the Darwin Archive, in which Rejlander himself mimicked the child's expression. Edgar Yoxall

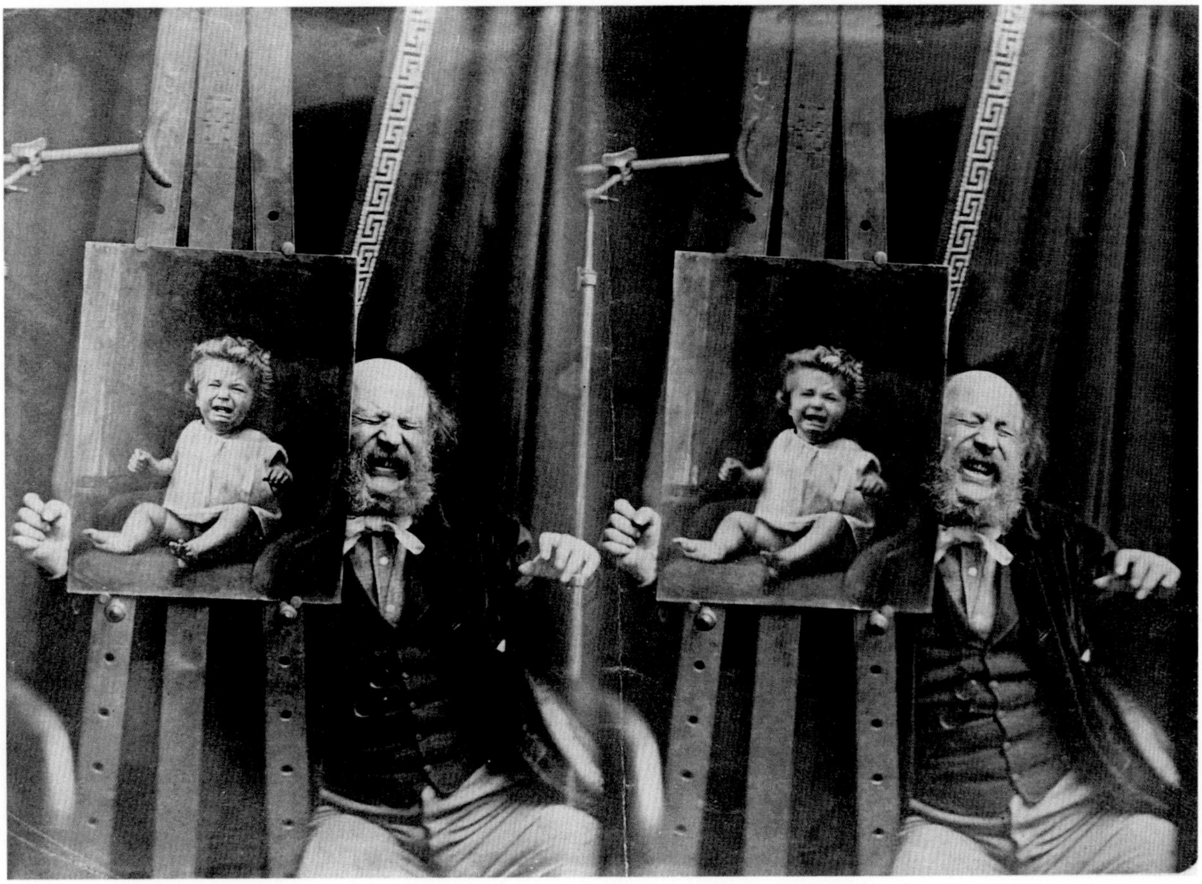

7-13 Oscar Rejlander, *Laughing/Crying* (Rejlander mimicking *Ginx's Baby*), 1871–72. Albumen print. Comparing adult and infant expression, including the hands.

Jones described one of these as Rejlander "laughing at himself," musing "who would have dreamed that *Ginx's Baby* would have made him solvent?"[69] Indeed, Rejlander annotated on the back of one, "Fun, only. There I laughed! Ha! Ha! Ha! Violently—In the other I cried—e, e, e, e. Yet how similar the expression"[70] (figure 7-13).

It is unlikely Darwin would have dismissed the images as a good-natured prank; he was keenly interested in examining physiological differences between infants and adults. In each set of stereos, Rejlander assumed the same posture as the infant in *Ginx's Baby*, and in one he attempted to replicate the position of his hands. On one side he pretended to be crying, on the other he pretended to laugh. This constituted a further innovative use of the stereo camera in Rejlander's work for Darwin. Had Darwin chosen to view the images through a stereopticon (a stereo viewer), he would have seen Rejlander's laughing and crying faces superimposed, enabling him to study the divergences between the muscle contractions associated with each expression. Each set of images was useful for different reasons. The close-up version (figure 7-14) af-

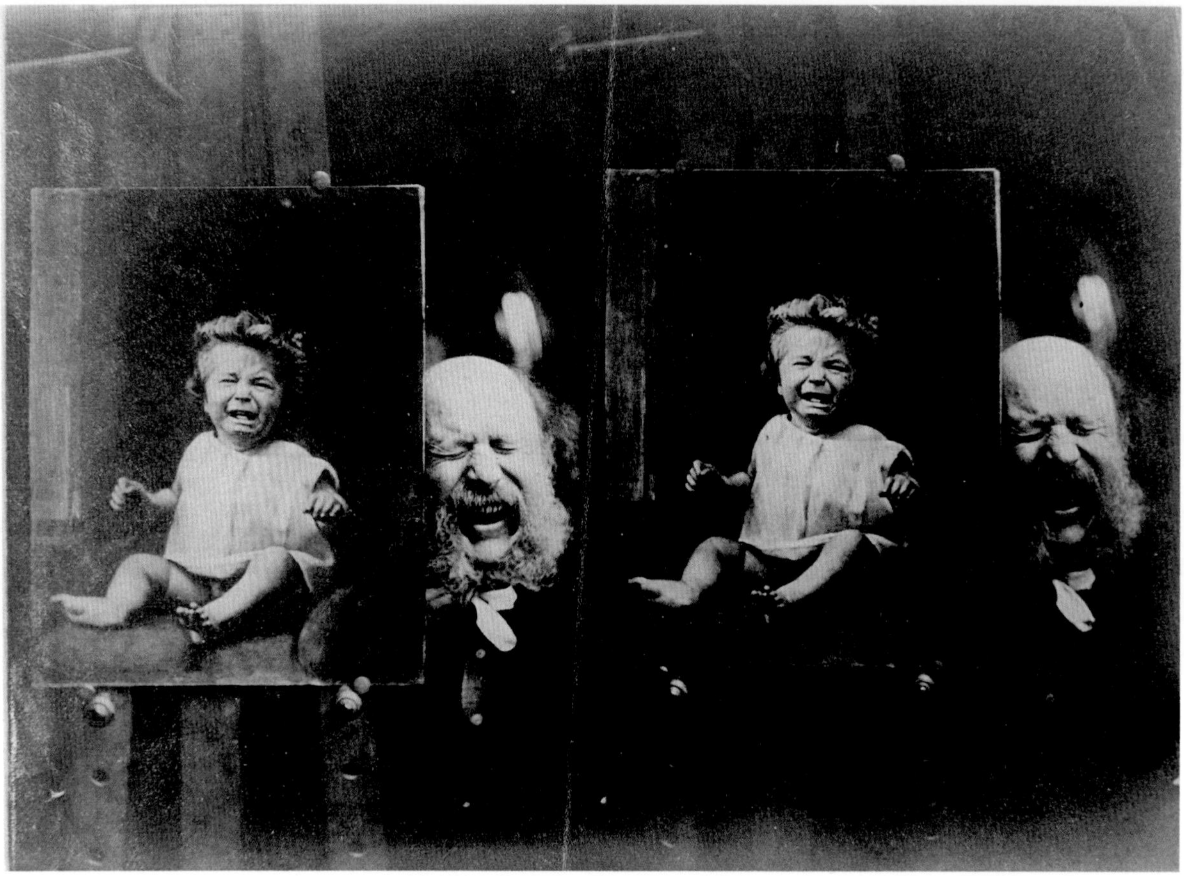

fords a better view of facial musculature, whereas in the other one the hands are better seen. The differences between childhood and adult expressions were not a trivial matter for Darwin. He had vested a great deal in the idea that children use the same expressive muscles as adults. Rejlander's mimicry tested this assumption.

7-14 Oscar Rejlander, *Laughing/Crying* (Rejlander mimicking *Ginx's Baby*), 1871–72. Albumen print. Another role-play, this time close-up.

Although Rejlander described his experiments as "fun, only," Darwin was intrigued by the physiological parallels between laughing and crying. For example, he owned a photograph by the London Stereoscopic Company depicting a pair of laughing and crying sculptures by the French artist Adolphe Itasse (1830–93).[71] The Itasse stereo print, made in 1871, employs the same technique of superimposition that Rejlander used in his poses with *Ginx's Baby*. A different image was placed on either side of the stereo card, so that the viewer could see Itasse's laughing and crying sculptures appearing one on top of the other.

While playful, Rejlander's description of the exclamations "Ha! Ha! Ha!" and "e, e, e, e" also had serious implications. The German physiologist Hermann von Helmholtz (1821–94), whom Darwin admired, studied the effects of mouth shape on vocal pitch. With regard to young infants, Darwin wrote,

> When [they] cry they open their mouths widely, and this, no doubt, is necessary for pouring forth a full volume of sound; but the mouth then assumes, from a quite distinct cause, an almost quadrangular shape, depending, as will hereafter be explained, on the firm closing of the eyelids, and consequent drawing up of the upper lip. How far this square shape of the mouth modifies the wailing or crying sound, I am not prepared to say; but we know from the researches of Helmholtz and others that the form of the cavity of the mouth and lips determines the nature and pitch of the vowel sounds which are produced.[72]

In 1872 Rejlander sent Darwin a final photograph featuring the *Ginx's Baby* drawing. Titled *O. G. R. the Artist Introduces O. G. R. the Volunteer*, or *Introducing . . . Mr. Rejlander*, it is a double self-portrait made from two negatives posed in the Rejlander studio (figure 7-15).

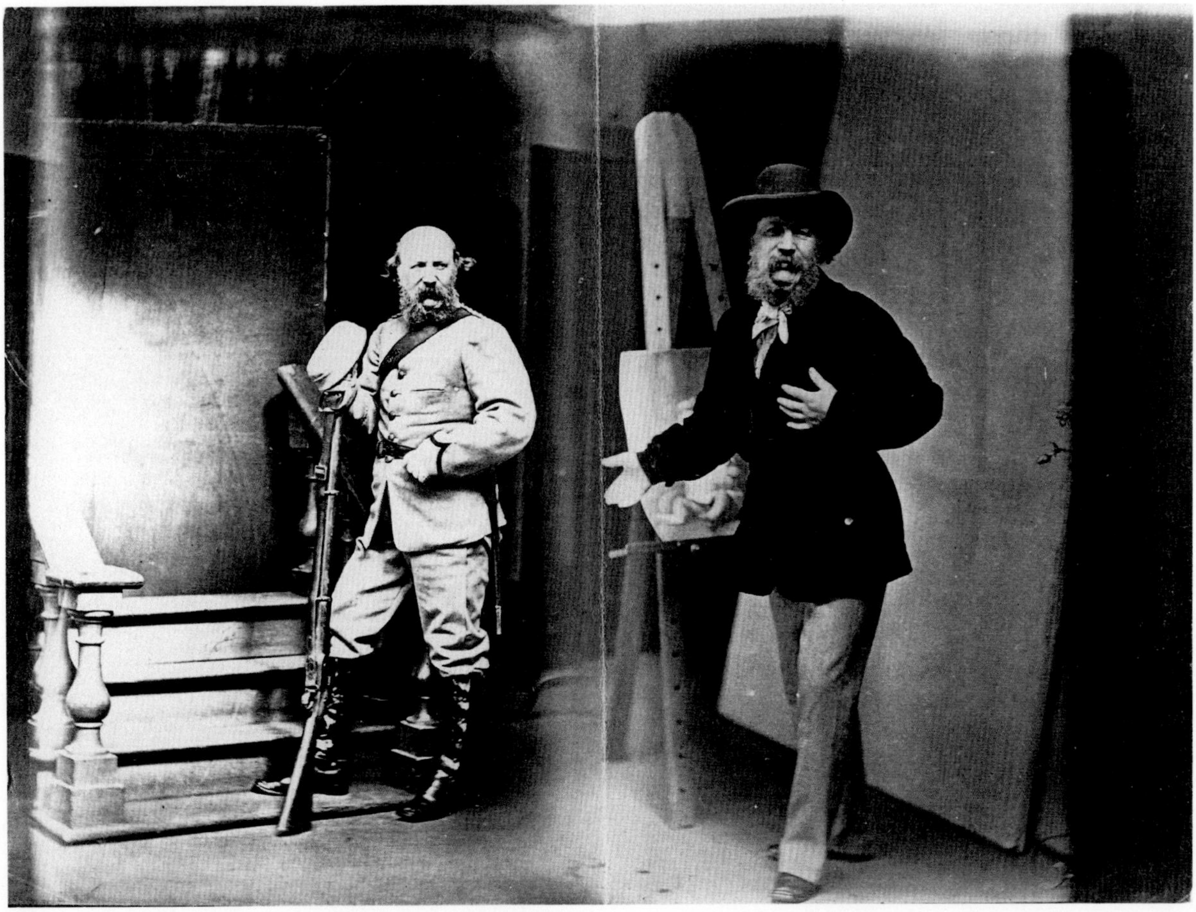

7-15 *Oscar Rejlander, Introducing . . . Mr. Rejlander,* 1871–72. Albumen print. Possibly this was a veiled commentary on Rejlander's life's work.

On the right stands Rejlander dressed as he was in *Indignation* (plate 6, figure 3) and *Surprise* (plate 7, figure 1), illustrations taken from the first batch of photographs he made for Darwin. On the left Rejlander appears again, wearing the uniform of the "Artist's Rifles," 38th Middlesex Regiment, his gun by his side. The artist Rejlander motions with his hands in the direction of his soldier twin, as if to introduce him to the viewer. The soldier replies with a suspicious stare. The soldier Rejlander leans on the banister of a set of stairs, while the artist Rejlander stands in front of the enlarged *Ginx's Baby*, propped on an easel.

Historians have traditionally treated *Introducing . . . Mr. Rejlander* as little more than a curiosity, a late-period print in a checkered career. Rejlander was notorious for making composite photographs of this type, in which two separate negatives were combined in a single print to create the illusion of a scene that never existed. The photohistorians Heinz Henisch and Bridget Henisch, for example, described the image as a last attempt at composite photography, remarking on its military aspect.[73] However, it was almost certainly designed as a serious and complex self-portrait. Contained within it are references to some of the most significant accomplishments of Rejlander's career.

Humor was certainly one aspect of its message; in sending a copy to Darwin he wrote, "My wife insists upon me sending this for you that your ladies may see that I can put on a more amiable expression."[74] Rejlander cherished art's ability to entertain, and it is fitting that he would have chosen a light-hearted tone for his self-portrait for Darwin. His double appearance in the image also conveyed his two strongest professional loyalties, as an artist and a volunteer soldier. The image was also a commentary on the medium and its possibilities. As the photohistorian Robert Sobieszek noted, "Gesturing, grimacing and doubling himself, Rejlander clearly anticipated more contemporary artistic conventions of feigning passions and self-referentiality."[75]

Given the financial difficulties Rejlander experienced as an artist, he may have intended to equate his military voluntarism with his inability to prosper in the studio. The stairs on which the volunteer Rejlander stand are the same as those used in his renowned photograph *Poor Jo* (1860), a portrait of an abandoned orphan boy, bent in despair, his head pressed miserably in his hands (figure 7-16). Was Rejlander's portrait meant to invoke this melancholy image? Before *Ginx's Baby*, *Poor Jo* was his most commercially successful picture. Like *Ginx's Baby*, it depicted a child in circumstances similar to those described in Jenkins's novel. The inclusion of these very recognizable stairs thus united Rejlander's two most illustrious photographic children.

Produced just three years before his death, *Introducing . . . Mr. Rejlander* foreshadows his reported last words: "I shall die a volunteer, let

Opposite:

7-16 Oscar Rejlander, *Poor Jo*, or *Night in Town*, c. 1860. Albumen print. One of Rejlander's most celebrated images; the stairs reappeared in the photograph he made for Darwin.

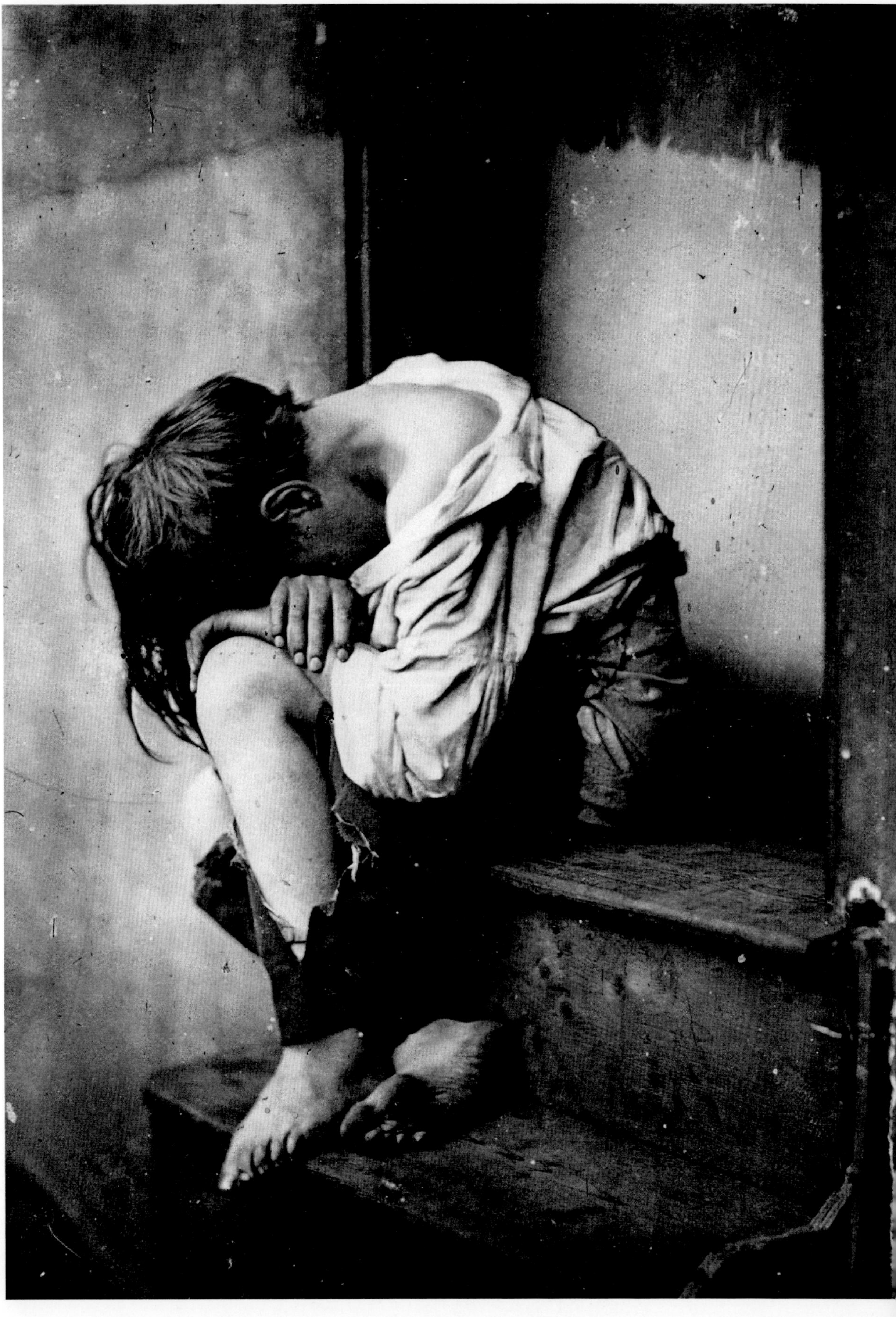

me be buried as one."[76] Viewed as a reflection on his career, it commemorates his devotion to his adopted country and to his art. It is also a tribute to his remarkable collaboration with Darwin. Like a magician performing his best trick in slow motion, the image shows the artist standing next to an easel holding his most enduring illusion, *Ginx's Baby*. His double appearance hints at the trickery embedded in *Ginx*, hinting at his facility with composite printing. The true nature of *Ginx* remained undiscovered in his lifetime. Perhaps Rejlander left *Introducing* as a time capsule, to be decoded and appreciated by future audiences. The artifice involved in the production of *Ginx's Baby* was not easily detected. Professionally, it was a final triumph—its authenticity was never challenged.

8

Darwin's Eyes and Ears

The artists who guided Darwin's search for pictures

The Pre-Raphaelite sculptor and poet Thomas Woolner left England for Australia in 1852, lured, as the *Beagle* voyagers had been some twenty years earlier, by the continent's distinctive natural history. Woolner, however, did not go seeking pure knowledge—he hoped to profit from the gold rush that gripped Australia in the 1850s. Nor was he planning to return, at least not soon. Disappointed by the response his art had received in England and perhaps sensing the imminent breakup of the original group of seven Pre-Raphaelite brothers, Woolner abandoned his artistic career and planned to settle in the young colony, trading his hammer and chisel for a pickaxe and prospecting pan.

The venture lasted just over two years. Mining was more difficult than he had imagined, and while there was gold to be had, extracting it was arduous and the returns unpredictable. Woolner did not find gold himself, but he did discover a rich vein of colonists who had, many of whom were eager to be immortalized by a former student of the Royal Academy. Embracing this unexpected market, he quit the mining profession and returned to making artwork, accepting commissions for portrait medallions. In two years in Australia, he produced twenty-four portrait sculptures, more than he had made in England in the preceding decade and breathing life into what had been a flagging career.[1] In 1855 Woolner returned to England, rededicated to his art and bolstered by increased critical respectability.

Woolner's Australian expedition marked a new phase in his career. Previously, he had specialized in idealized renderings of fictitious and mythological scenes. He did produce a handful of portrait medallions during these early years, including plasters of the essayist Thomas Carlyle and the poet Alfred Tennyson in 1852, but it was only in Australia that he received enough orders to prosper in such work.[2] Upon returning to England, he continued to work principally as a portrait sculptor. This, in turn, provided the financial security he needed to be able to pursue narrative subjects occasionally.

Woolner as Darwin's Guide

Although their stories were different, both Darwin and Woolner had embarked on youthful journeys to the far reaches of the globe, returning to England focused and invigorated. This similarity, though coincidental, helps to explain the friendship the two men developed after their introduction in the early 1870s. They had visited many of the same places, and Darwin came to regard Woolner as a seasoned observer. Darwin routinely tapped friends and acquaintances for information, valuing especially those who had been exposed to extraordinary situations. He viewed artists not as subordinates but as colleagues and included them in the network of contacts with whom he corresponded. Sometimes he found artists that simply corroborated his views, but often they enriched or challenged his understanding. As his own drawing skills were limited, he relied on the talents of draftsmen to help illustrate his ideas. Yet he also appreciated the insight of seasoned artists.

By the late 1860s, when Darwin first became acquainted with Woolner, his clients ranked among the most celebrated sitters of his day. Among them were a number of Darwin's colleagues, including the natural historians Adam Sedgwick (1860), William Jackson Hooker (1860, 1866), John Stevens Henslow (1861), and William Spottiswoode (1864). Not all of the portraits Woolner made of these figures were considered successful. In a letter to Woolner in 1860, the painter and poet Edward Lear (1812–88) wrote commiserating with him about the criticism his bust of Sedgwick had received during its exhibition at the Royal Academy.[3]

In 1869, Woolner completed a portrait medallion of Darwin for the Wedgwood company.[4] It is unclear who initiated this commission. In June, Darwin's son William wrote to Woolner thanking him for a cast of the medallion and inquiring whether his version, or the original in Woolner's possession, should be sent to the firm to be copied.[5] The following year Woolner completed a bust of Darwin, more than five years after the project was conceived.[6] The scientist John D. Hooker had commissioned the bust in 1864, but Darwin declined to pose at that time on grounds of ill health. With Hooker's encouragement, Darwin eventually agreed to sit for Woolner. The first sitting was to have occurred shortly after Darwin recovered, but no sooner did he feel better than Woolner also excused himself due to illness.

When Woolner finally began Darwin's portrait, Darwin wasted little time probing the artist's knowledge of expression. Coming from a lesser authority, Darwin's request for information in 1871 might have seemed indecent:

> I daresay you often meet & know well painters. Could you persuade some *trustworthy* men to observe young & inexperienced girls who serve as models, & who at first blush much, how low down the blush extends. . . . Moreau says a celebrated French painter once saw a new model blushing all over her body. So that I want much to hear what the experience is of cautious & careful English artists: I always distrust memory—can you aid me?[7]

Woolner's response is not preserved, but Darwin concluded in *Expression* that "as a general rule, with English women, blushing does not extend beneath the neck and upper part of the chest."[8] The French editor Jacques-Louis Moreau de la Sarthe (1771–1826) had cited the authority of a "celebrated painter" regarding the French propensity to blush in his edition of Johann Caspar Lavater's *Fragments of Physiognomy* in 1820; Darwin cited this observation in *Expression*. Darwin had also sought advice on the subject of blushing from several physicians, including James Paget (1814–99), Charles Langstaff (n.d.), and James Crichton-Browne.[9] All confirmed the view that blushing was generally confined to the face and neck, noting that in some exceptional cases it had been observed on lower extremities. This approach to fact-finding was typical of Darwin. When searching for information beyond his experience, he routinely consulted friends and colleagues.

Woolner worked according to the Pre-Raphaelite creed, which called for exactitude and attention to detail. He has been described as one of the Pre-Raphaelite brothers most committed to artistic naturalism.[10] Like many artists of his generation, it is likely he used photographic studies to make his renderings more accurate.[11] Ironically, Woolner's friend and housemate, the art historian and poet Francis Turner Palgrave (1824–97), cautioned against the increasing trend toward using photographic studies in late Victorian sculpture. Palgrave, the nephew of renowned Kew Gardens botanist William Jackson Hooker (1785–1865), admonished those who embraced the new medium in his *Essays on Art* (1867):

> And let us here remark . . . on the ludicrous impossibility of effecting anything in sculpture that can in the very least deserve the name, by any mere mechanical process, be the means ever so ingenious. We would apply this criticism to what is named "Photo-sculpture." Those, indeed, who "patronise the invention," as the phrase goes, from the petty passion for novelty, deserve no better fate than to throw away their cash. We cannot pretend to pity them, if the ugly stare of the photograph—which it is happily hopeless to secure from fading—be perpetuated in distorted and lifeless plaster.[12]

Even though Woolner evidently used photographs as studies, Palgrave enthusiastically praised his sculpture in his anonymous handbook to the South Kensington International Exhibition of 1862.[13] In 1871, he commissioned Woolner to provide the frontispiece for his *Golden Treasury of the Best Songs and Lyrical Poems of the English Language* (1871), an illustration of a piping shepherd. Woolner dabbled in poetry himself; the same character returned in his epic poem of 1884, *Silenus*, as the god Pan.[14]

However, it was not Pan but Puck who became the focus of discussions between Darwin and Woolner from 1869 to 1871. The Celtic hobgoblin, an appropriate subject given the Pre-Raphaelite preference for British rather than classical Greek and Roman mythological characters, was featured in Woolner's first well-known sculpture (figure 8-1). Woolner had exhibited a plaster of the subject at the Royal Institution in 1847.[15] Traditionally, Puck was believed to be a sprite who played tricks on unsuspecting persons. He was the mischievous protagonist of *A Midsummer Night's Dream*:

> Thou speak'st aright;
> I am that merry wanderer of the night.
> I jest to Oberon and make him smile
> When I a fat and bean-fed horse beguile,
> Neighing in likeness of a filly foal:
> And sometime lurk I in a gossip's bowl,
> In very likeness of a roasted crab
> And when she drinks, against her lips I bob
> And on her wither'd dewlap pour the ale.
> The wisest aunt, telling the saddest tale,
> Sometime for three-foot stool mistaketh me;
> Then slip I from her bum, down topples she.[16]

The sculpture, which Woolner exhibited in bronze at the Royal Academy in 1866, portrayed a bald imp with pointed ears, partly clad in a cape and loincloth, standing on a giant toadstool. Unlike most of

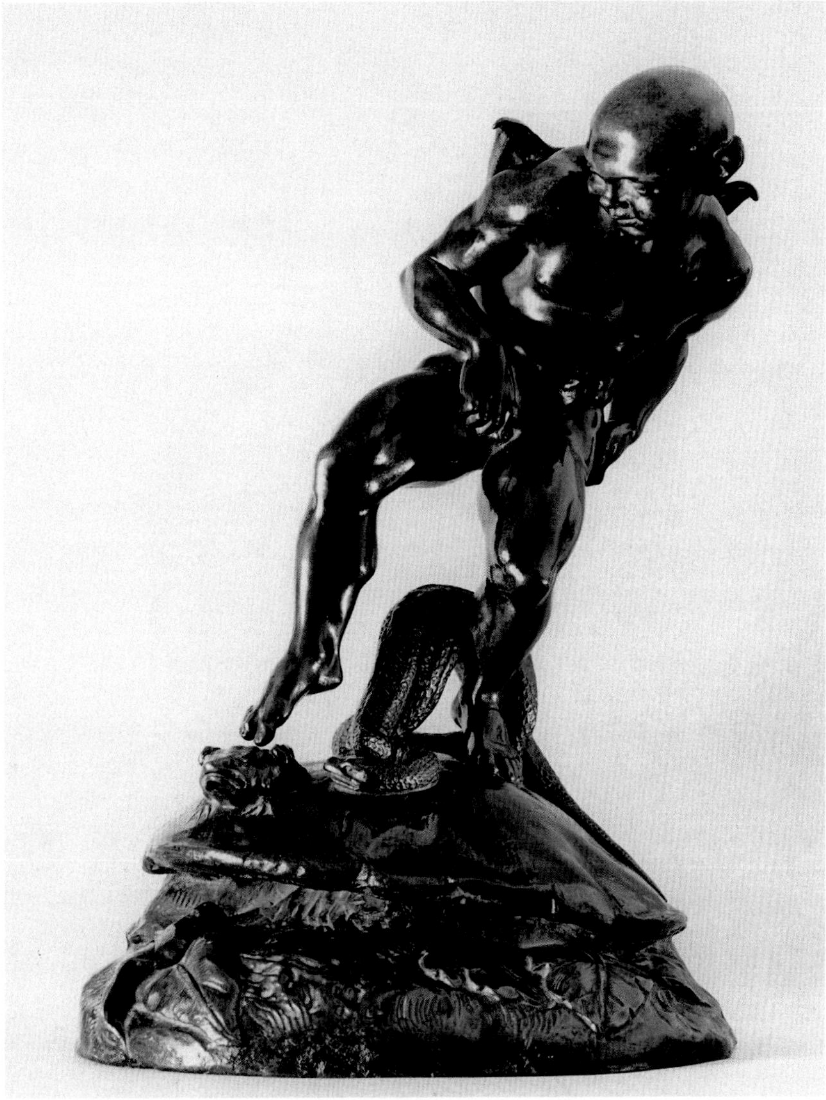

8-1 Thomas Woolner, *Puck*. Bronze, after the plaster exhibited 1847. More than twenty years after it was made, the ear of this statue became the subject of scientific scrutiny.

his works, which depict immobile subjects, *Puck* is an action scene. The central figure rests heavily on his left foot and twists his body to the right, leaning back and extending the toes of his right foot to touch the back of a crouching toad. A snake coils over Puck's planted foot, preparing to attack. In her 1917 biography of the artist, Amy Woolner reprinted an explanation of the work first published at the time of its exhibition: "A snake is stealthily creeping towards an unconscious toad. Puck is about to touch the toad with his foot, that thus warned it may escape the jaws of the enemy. A smile of half mischievous satisfaction is on his face."[17]

Darwin, however, was intrigued by the structure of Puck's ears. The structure and use of ears as expressive organs were of considerable concern to Darwin as he began to prepare the *Expression* manuscript. In *Descent of Man*, Darwin argued that the human body exhibits residual features of animal progenitors. Among these was a slight inward indentation of the outer ear, "a little blunt point, projecting from the inwardly folded margin, or helix,"[18] which is visible in certain individuals. This, Darwin believed, was a vestigial form of the pointed ears of apes and other simians. Since he proposed that humans have evolved from apes, he reasoned that human ears must have evolved from ape ears. The slight inward protrusion of some human ears is all that remains of the pronounced point evident in apes.[19]

The example was effective as it enabled individuals to appraise their relation to apes by scrutinizing the shape of their own ears. In *Descent*, Darwin credited this discovery to Woolner:

> The celebrated sculptor, Mr. Woolner, informs me of one little peculiarity in the external ear, which he has often observed both in men and women, and of which he has perceived the full signification. His attention was first called to the subject whilst at work on his figure of Puck, to which he had given pointed ears. He was thus led to examine the ears of various monkeys, and subsequently more closely those of man.[20]

In 1869 Woolner sent Darwin a drawing after the sculpture to be included in *Descent*. Darwin replied gratefully, saying "it does excellently—the 'Woolnerian tip' is worth anything to me."[21] The drawing appeared in engraved form as figure 2 of *Descent* (figure 8-2). Two years later, Darwin wrote to Woolner again: "The tips to the ears have become quite celebrated. One reviewer (Nature) says they ought to be called, as I suggested in joke, Angulus Woolnerianus."[22]

The example of *Puck*'s ears demonstrates the utilitarian approach Darwin took to his illustrations. He accepted imagery because it conveyed what he wanted to show. To him, it was sufficient that the depiction of Puck had been carefully researched by comparison with living models and came from a reputable source. The origin of the drawing, which he acknowledged in his text, does not seem to have concerned his readers, either.

Although he remained convinced of Woolner's observations, Darwin was dissatisfied with the published engraving of *Puck*. In 1871, he wrote to the zoologist Edwin Ray Lankester (1847–1929), asking him

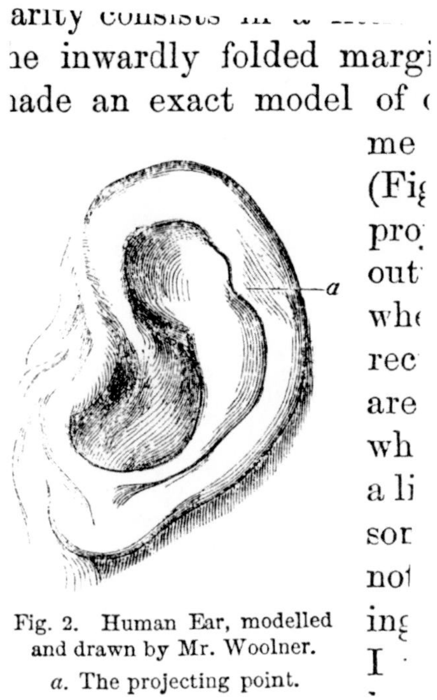

8-2 After Thomas Woolner, Human Ear (Puck), 1871. Wood engraving. Figure 2 of *Descent of Man*. The model for the illustration was Woolner's sculpture *Puck*.

to procure a photograph of a friend whom Lankester had noticed had the "Woolnerian tip." Photographic evidence, he reasoned, would be more convincing than Woolner's drawing, which he confided, though "no doubt accurate does not seem to me good."[23] The friend, a German zoologist named Hinrich Nitsche (1845–1902), sent Darwin photographs of both his ears (figure 8-3), together with a photograph of an orangutan fetus, in April 1871. Nitsche, wary that his photographs might cause some people to deduce that he was a close relative to a simian, provided them on the condition that they remain anonymous:

> I send you the life sized photographs of both my ears. I must however remark that only the photograph of the right ear can be considered as quite successful, and only on this one is the projection of the helix (made prominent by you) strongly marked and even as you will easily perceive, much more strongly developed than in the engraving given by you. Both ears however are on their upper ends really pointed, from which my friends maintain that I have Satan ears. I add a photograph of a fetus of an Orang, life size; it is in the Zoological University Collection of this place. Its ears are, as you will easily perceive, much further removed from human ears, than one usually finds in the ears of full grown anthropoid monkeys. I shall be very glad if the photographs are really of use to you. You can dispose perfectly freely of them, I should

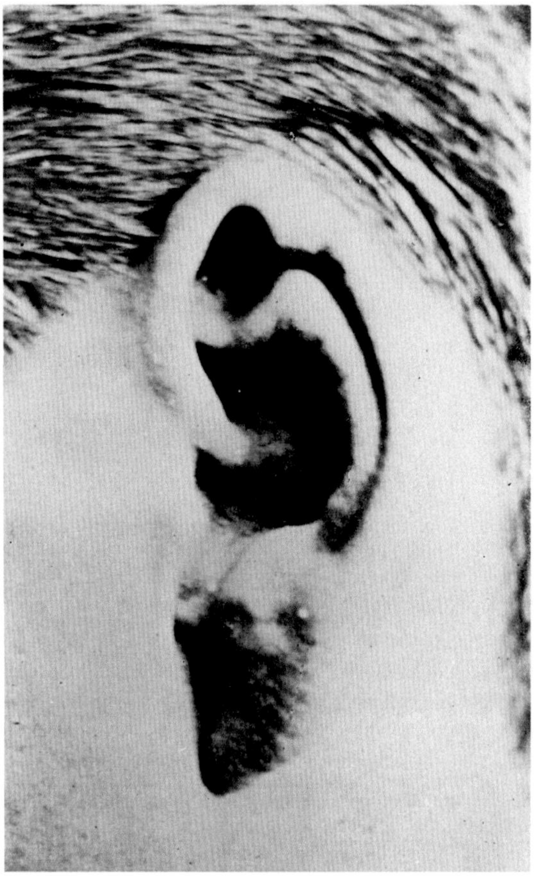

8-3 Circle of Hinrich Nitsche, *Hinrich Nitsche's Left Ear*, c. 1871. Albumen print. Nitsche sent Darwin pictures of his own ears but asked that he not tell anyone where he got them.

only beg you not to mention my name, in case the photograph of my ears should be incorporated into a new edition of your book.[24]

Darwin decided against publishing the photographs of Nitsche's ears and retained the *Puck* engraving in the second edition of *Descent*. However, he did use Nitsche's photograph of an orangutan fetus (figure 3 in the second edition) as evidence of the phenomenon in apes. The original photograph was engraved, showing a profile view of the head, "an exact copy of a photograph."[25] The Darwin Archive contains the original photograph from it was made[26] (figure 8-4). While the engraving is accurate, it is a substantially cut-down version of the photographic original, which shows the full length of the animal.

Darwin continued to ask colleagues for information about interesting ears. The physiologist William Preyer (1841–97) offered Darwin information about paintings he had seen in German museums, noting peculiarities in the renderings of the ears of Egyptian mummies. "It is situated higher up than in Europeans" Preyer explained. "Now I have never seen in civilised company anything like [it]. But apes shew the

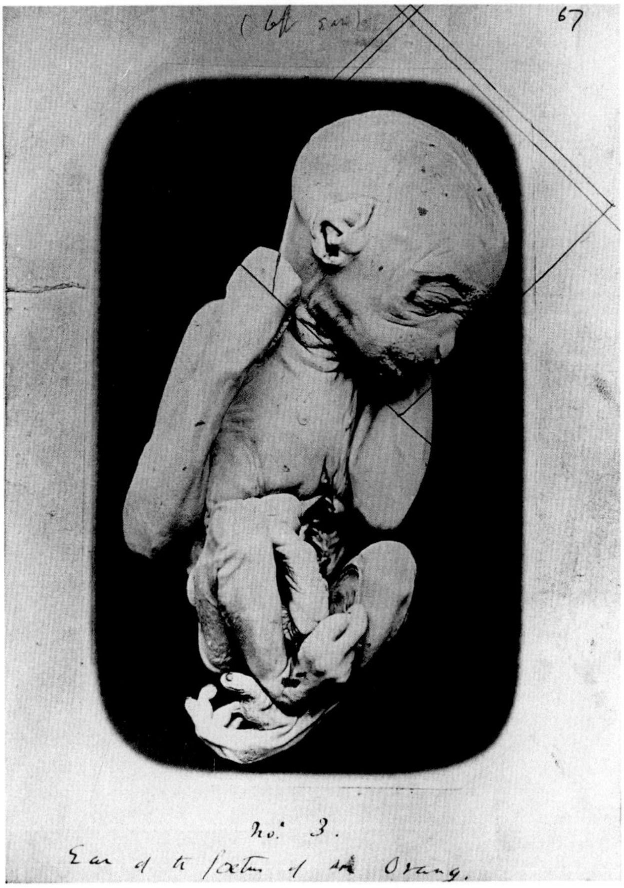

8-4 Circle of Hinrich Nitsche, *Ear of the Fetus of an Orang*, c. 1871. Albumen print. Darwin published a wood engraving of the section outlined in ink.

same position of the ear. Now I should very much like to know whether any existing race has this peculiarity and whether you think it might be considered as a connecting link." Preyer continued: "It was interesting to me to see that Rubens in his beautiful picture in the Dresden gallery gives the satyr pointed ears. The ear is very large but human with the only difference that it is pointed."[27]

The painting in question was almost certainly Peter Paul Rubens's *The Drunkenness of Hercules*, also known as *The Effects of Wine* (c. 1612–14), depicting the inebriated god slumped in a daze, his right arm supported by an adoring woman, and his left arm held by a mischievous satyr[28] (figure 8-5). Recent scholarship has revealed that this painting is not an original composition by Rubens, but a copy of an ancient Roman relief in the Mattei Collection in Rome.[29] The ear of the satyr to which Preyer referred, exaggerated beyond the limits of ordinary human development, is also evident in its Roman predecessor.

After *Expression* was published, Darwin also discussed facial asymmetry in Rubens's portraits with his colleague the physician Thomas

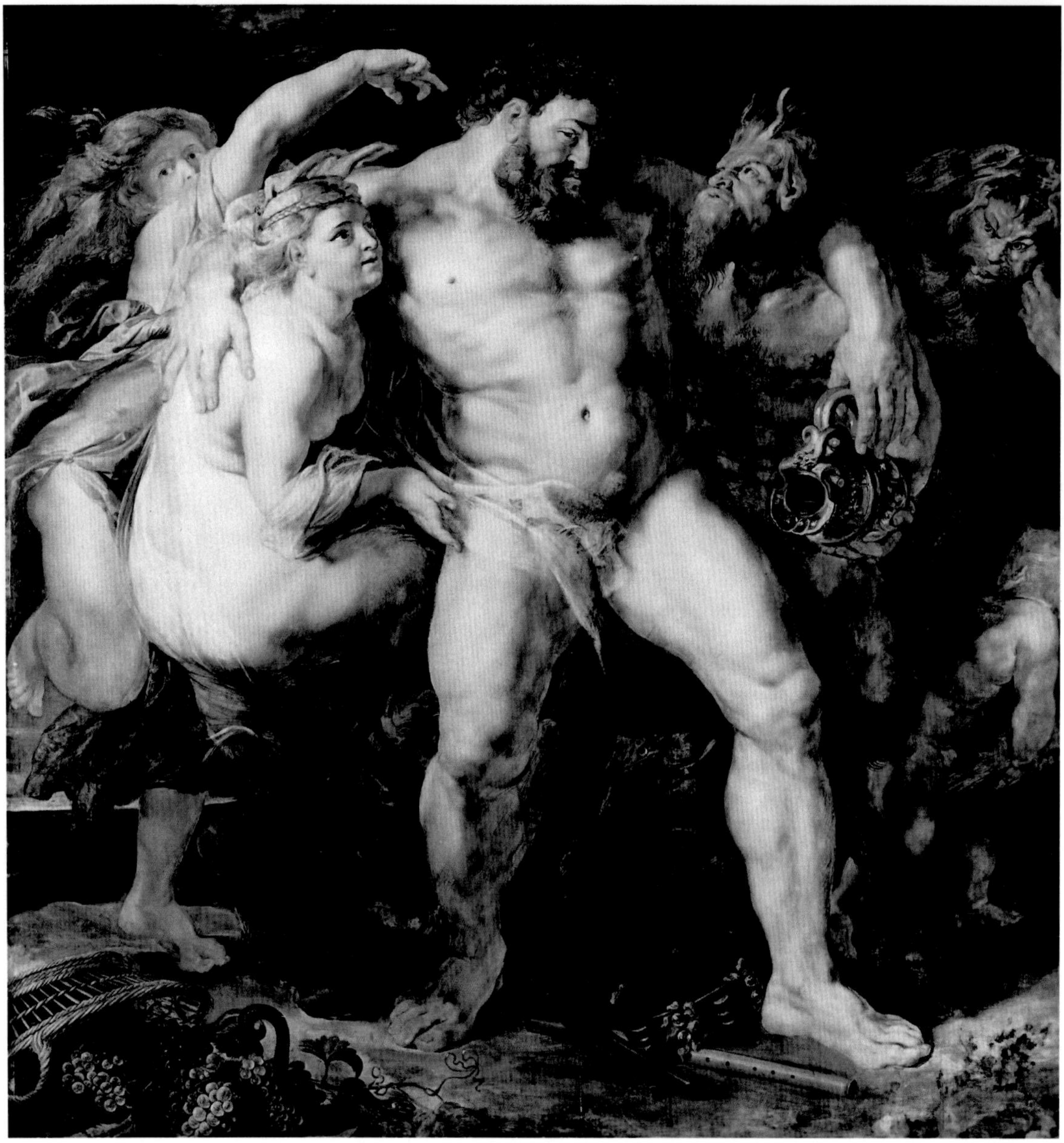

Lauder Brunton (1844–1916).[30] Darwin offered to send a photograph of a Rubens painting to Brunton in exchange for a cabinet card then identified as José de Ribera's (1591–1652) *Mary in Egypt* (1641), but now known to represent Saint Agnes[31] (figure 8-6). Brunton claimed that one side of the central figure's face expressed repentance, while the other expressed devout joy. Darwin received the Ribera, which he showed to his family for comments. On the mount Darwin wrote, "Hensleigh W [Wedgwood] thinks one side more seraphic than the other. F [Francis Darwin] thinks one side more hopeful, the other sadder."[32]

8-5 Peter Paul Rubens, *The Drunkenness of Hercules*, or *The Effects of Wine*, c. 1612–14. Oil on panel. The German physiologist William Preyer thought Darwin might find the satyr's pointed ears interesting.

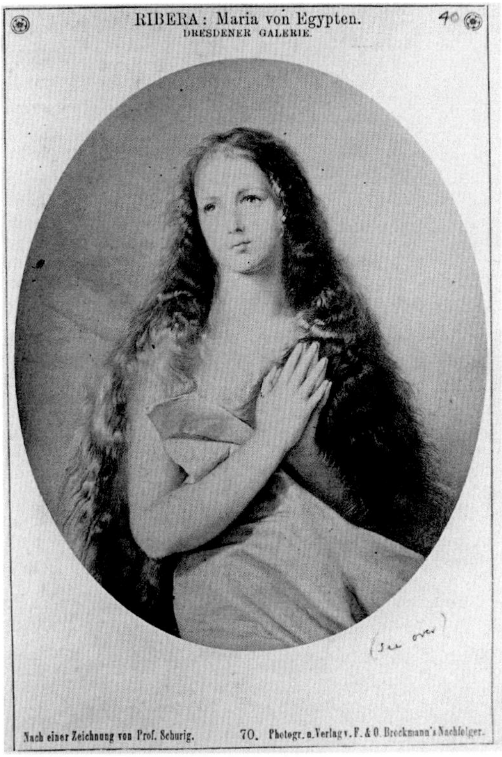

8-6 F. & O. Brockmann Studio, *José de Ribera's Mary in Egypt*, n.d. Albumen print, after a drawing by an artist called Schurig. Darwin asked friends if they could see differences between the two sides of Mary's face.

Joseph Wolf: As Truthful as a Photograph

The two engravings of the black ape *Cynopithecus niger* (figures 16 and 17 of *Expression*) were made by the artist Joseph Wolf (1820–99), whom his biographer, A.H. Palmer, described as a friend of Thomas Woolner[33] (figure 8-7). The original conté drawing for these figures, preserved in the Darwin Archive, contains numerous annotations in Darwin's hand, directing Wolf to "attend to the ears in both figures"[34] (figure 8-8). Wolf prepared his drawings from specimens in the London Zoological Gardens, but Darwin did not believe they represented the ears correctly, and asked him to lay the ears back further than he had. Wolf evidently complied, as the drawings show evidence of extensive reworking around the tips of the ears.[35] Darwin had additional instructions for the engraver, James Davis Cooper. "The point of the ear is decidedly too sharp—Please look at the ear in the other drawing, & round off the point a little—The outline of upper part of ear is perhaps a little too black & thick."[36] These annotations appear in pencil on a proof of Cooper's plate (figure 8-9).

Wolf's account of the creation of *Cynopithecus niger, Pleased by Being Caressed,* is related in Palmer's 1895 biography:

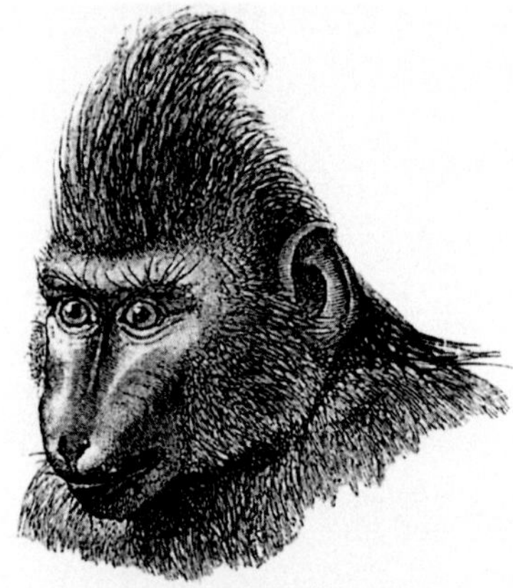

Fig. 16. *Cynopithecus niger*, in a placid condition. Drawn from life by Mr. Wolf.

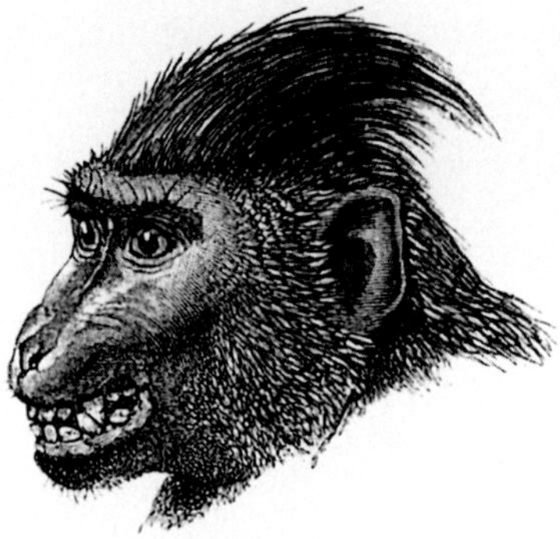

Fig. 17. The same, when pleased by being caressed.

8-7 After Joseph Wolf, *Cynopithecus niger*, in a Placid Condition, and The Same, When Pleased by Being Caressed, 1871–72. Wood engravings. Figures 16 and 17 of *Expression*.

Early in 1871 Darwin was preparing the materials for his *Expression of the Emotions in Man and Animals* (published in the following year), and mentioned to Mr. Bartlett his wish to have some work done at the [Zoological] Gardens which required unusual care. The Superintendent spoke of Wolf's accuracy and closeness of observation in high terms, and in due course introduced the two men to each other. Darwin, with a view to that section of his fifth chapter dealing with "Astonishment" and "Terror" in Monkeys, caused a living fresh-water Turtle to be placed in one of the cages. Wolf's account of the incident is this:—"One of the Turtles was put into a covered basket, and the keeper was asked to place it carefully under a heap of straw which was in the cage. Whilst that was being done, the Monkeys suspected something and kept looking down from on high. Clever fellows! I shall never forget that. The keeper then retired, and presently the heap of straw began to move. The Turtle came out, and instead of showing fear, the Monkeys crept nearer. The Black Crested Ape [*Cynopithecus niger*], came and looked at it, and walked in front of the Turtle as it

Opposite:
8-8 Joseph Wolf, *Cynopithecus niger*, 1871–72. Conté crayon, heightened with white. Wolf's drawings were altered for publication.

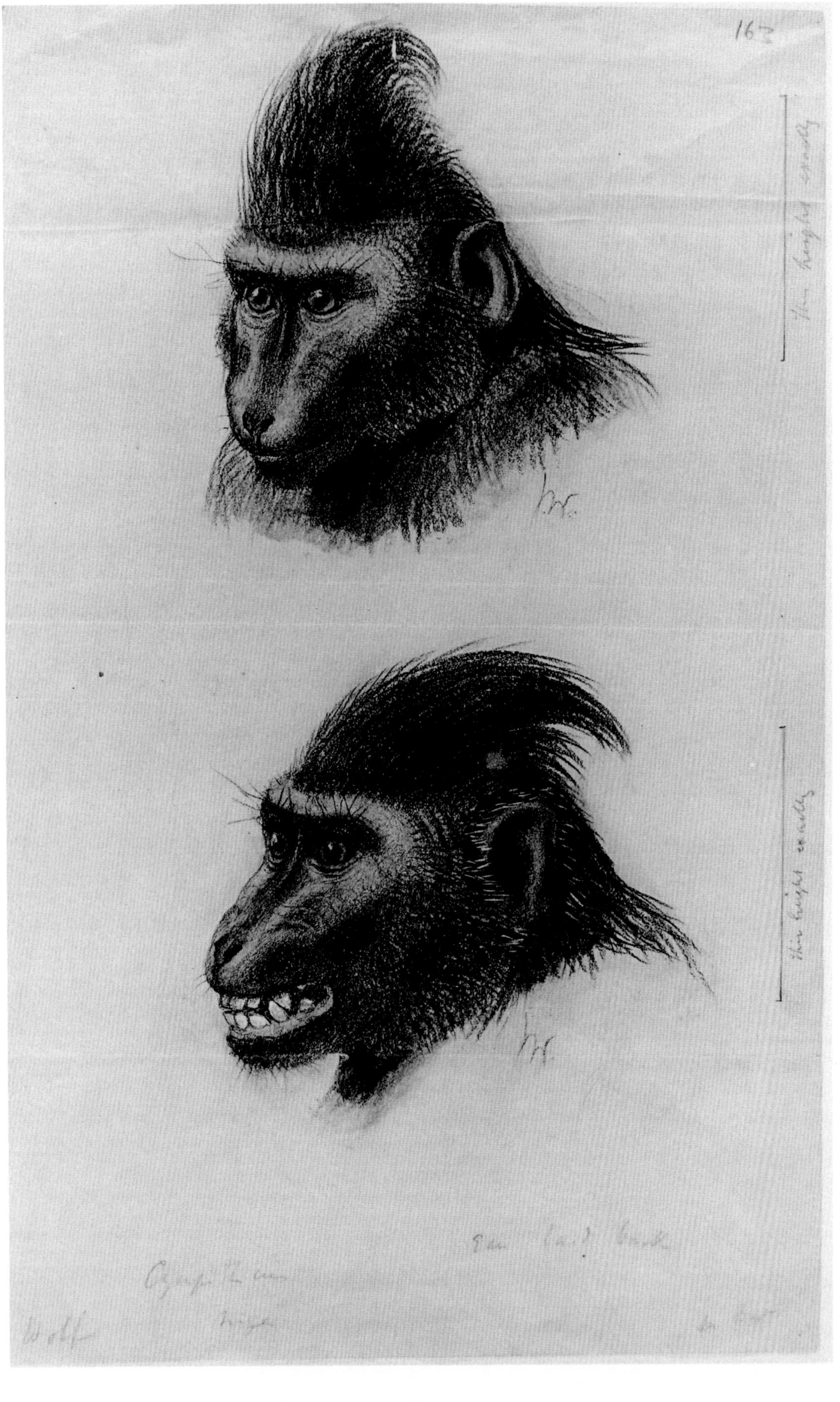
163
this height exactly
this height exactly
Wolf

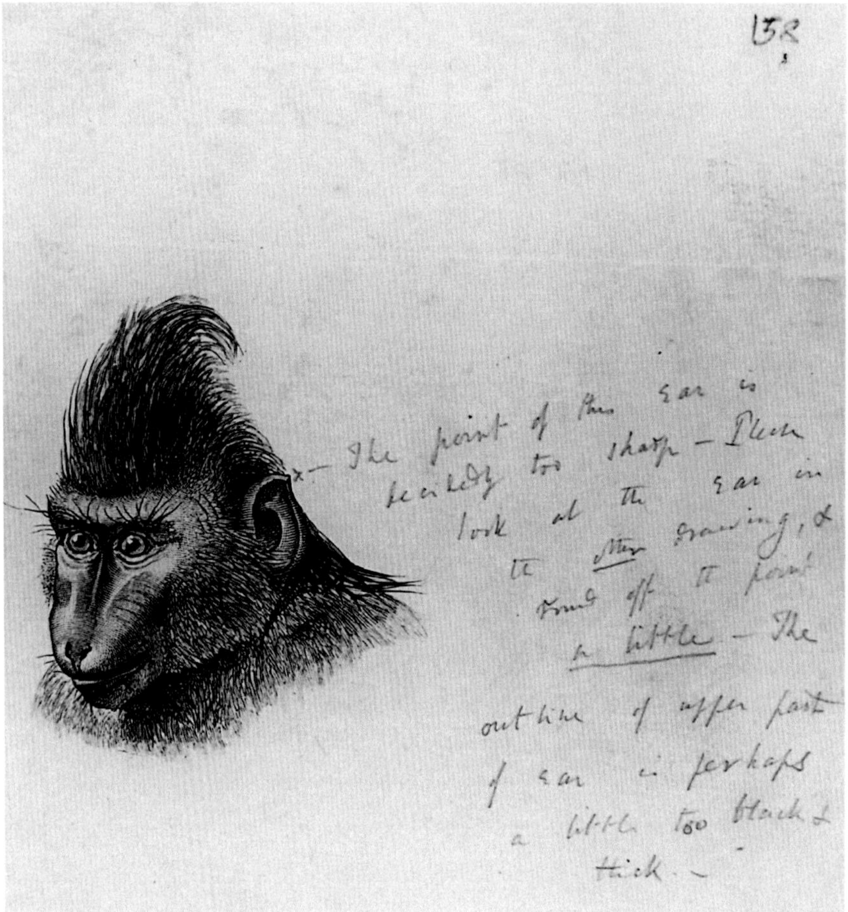

8-9 After Joseph Wolf, *Cynopithecus niger*, in a Placid Condition, 1871–72. Wood engraving; proof. Reworked by the engraver according to Darwin's notes.

> crept under him. Finally he went and sat on the Turtle. Darwin was much amused, and asked for a drawing of the incident."[37]

Darwin undoubtedly appreciated Wolf's accuracy and attention to detail, but he also valued his flexibility in portraying animals as he himself perceived them. The request to reposition the ears in the renderings of *Cynopithecus niger* was one example of this.

Darwin had initially written to Wolf in March 1871, requesting drawings of a "monkey's face while laughing." He suggested the Barbary ape would have been "incomparably the best" for the purpose; but, as he reported, it "is dead." Instead, he suggested that Wolf attempt to draw specimens of *Cynopithecus niger*, although he was concerned that its "permanent transverse wrinkles on the face" might make the expression of laughter difficult to see. "When Sutton the Keeper allows this monkey to play with his hair," Darwin wrote, "it chuckles or laughs, and keeps moderately still. The face then becomes a good deal wrinkled, and as far as I could see under disadvantageous circumstances, the skin is especially raised and wrinkled around the eyelids."[38] Wolf prepared the drawing as requested, but when interviewed later about the illustration, he admitted, "I never believed that that fellow was laughing, although Darwin said he was. I am not one of those who place absolute belief in 'authority.' "[39]

Wolf's remark about Darwin's "authority" sheds light on the relationship between Darwin and some of his illustrators. Unlike Wolf, Woolner had volunteered his illustration of the "Woolnerian tip" and limited his observations to a specific area of knowledge. Wolf, however, Darwin commissioned, asking him to illustrate a range of subjects about which he may have had no particular expertise. In total, Palmer records eleven sketches of animals that Darwin requested, ten of which are preserved in the Darwin Archive.[40] All are rendered in conté crayon heightened with white, using the same paper stock, and signed with Wolf's stylized "JW" monogram.

Two of these, *Head of a Snarling Dog* and *Cat in an Affectionate Frame of Mind*, correspond to engravings in *Expression* that are credited to another artist, Thomas William Wood (fl. 1855–72). Unlike photographs from the period, which are frequently inscribed with observations and analysis, annotations on the drawings and prints in the archive frequently instruct the artist to make corrections. For example, in Wolf's renditions of *Dog Approaching Another Dog with Hostile Intentions*, and *The Same in a Humble and Affectionate Frame of Mind* (figure 8-10), Darwin wrote in the margin, "Please do not destroy this drawing." "The hair is generally smooth on the loins, & this makes the roughness on the back & neck more apparent." There ought to be no collar and the hair erect on the neck." "Head rather lower & more in line of the back."[41] Evidently, Darwin felt himself a better judge of behavior in animals than his artist colleagues. Ultimately, Darwin decided not to publish even revised versions of Wolf's drawings, instead commissioning substitutes from the artist Briton Riviere (1840–1920).

Wood, who provided eight of the engraved figures in *Expression*, was an illustrator frequently used by Darwin's publisher, John Murray. His versions of the *Dog* and *Cat* figures (figures 14 and 10, respectively) conform to Wolf's overall plan for the composition, but change the posture of the animals slightly. In *Cat in an Affectionate Frame of Mind*, Wood also changed the background. Wolf's drawing depicts a cat arching its back as it rubs up against the leg of a table, while in Wood's version, published in *Expression*, the cat is shown rubbing against the leg of a person[42] (figures 8-11, 8-12). The substitution may have been intended to reinforce the idea of "affection," which would have seemed comparatively unconvincing directed at an inanimate object. Nevertheless, Darwin apparently felt Wolf's drawings had to be altered before

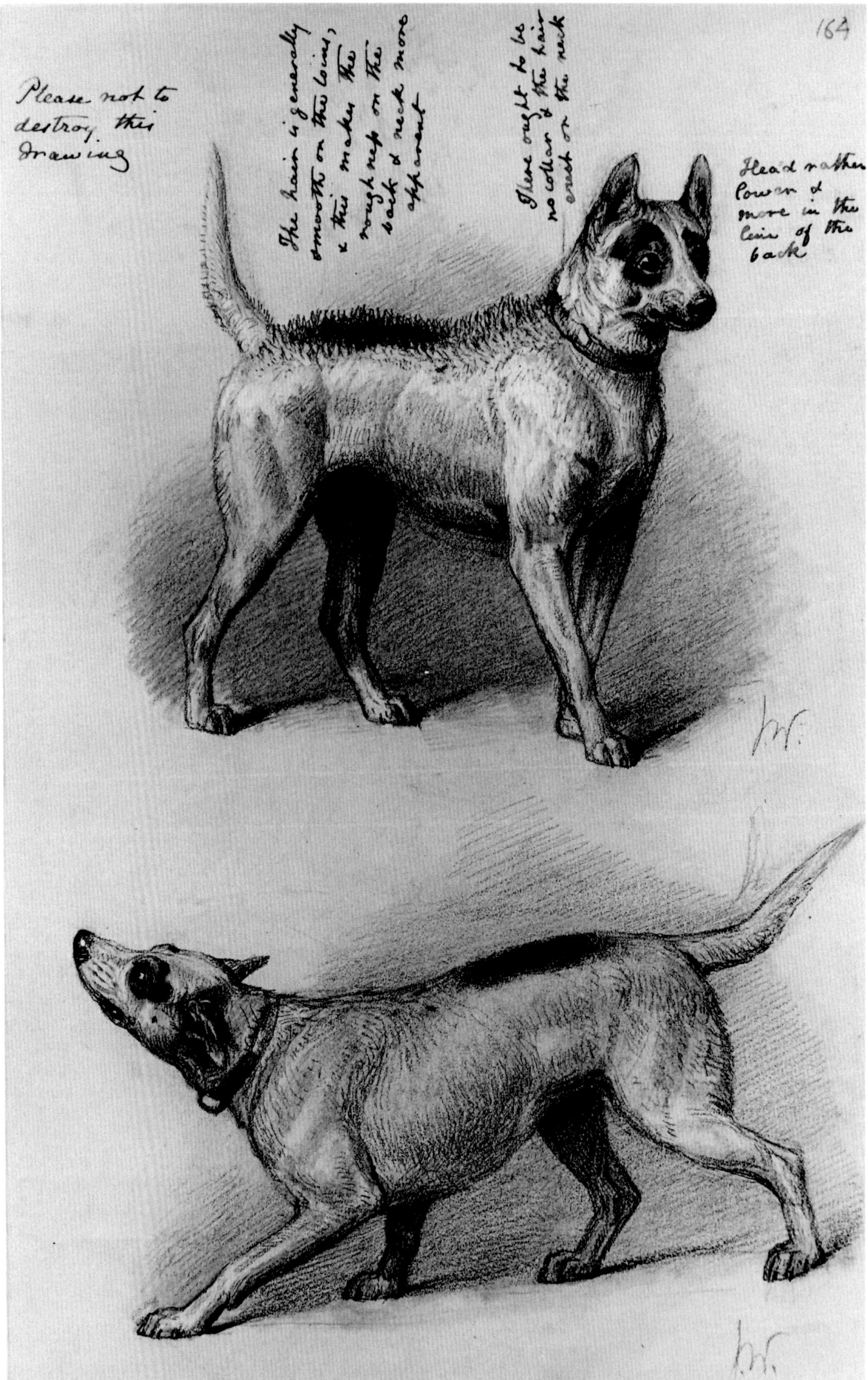
164
Please not to destroy this drawing
The hair is generally smooth on the loins, & this makes the roughness on the back & neck more apparent
There ought to be no collar & the hair erect on the neck
Head rather lower & more in the line of the back

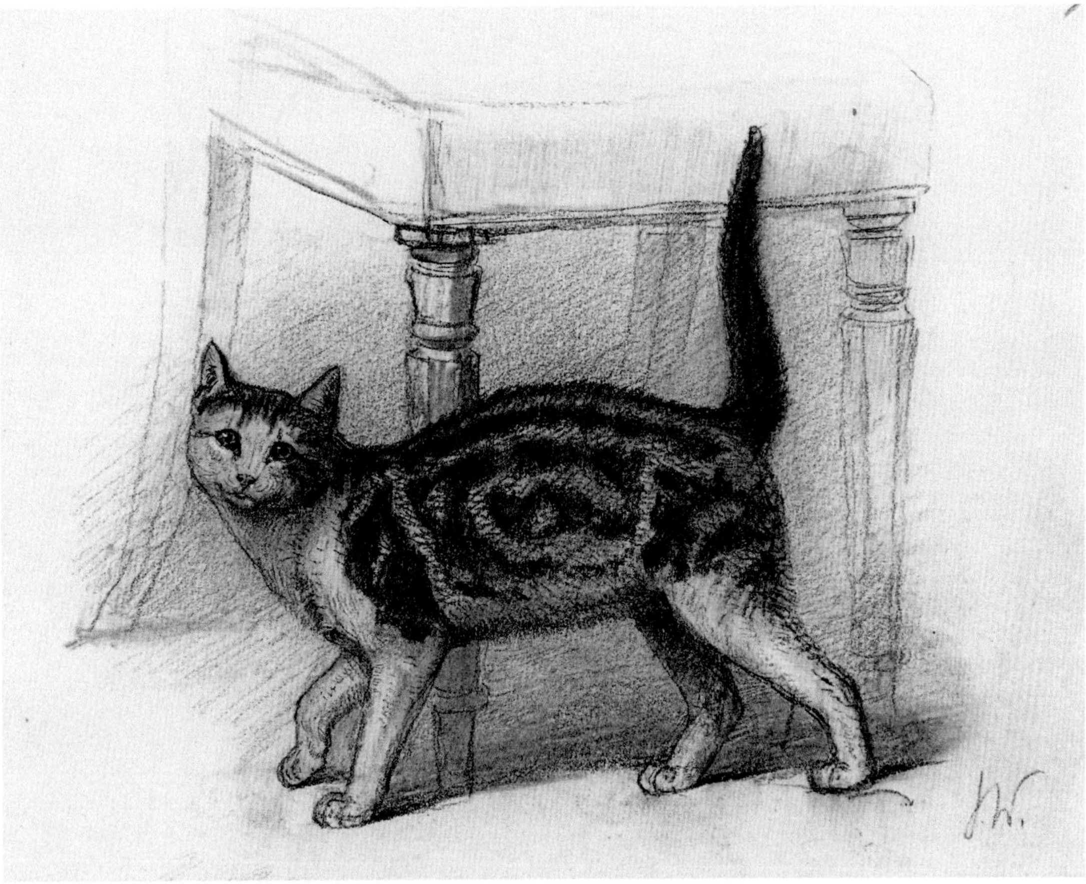

8-11 Joseph Wolf, Cat in an Affectionate Frame on Mind, n.d. Conté crayon, heightened with white. Wolf depicted the cat rubbing against a chair leg.

Opposite:
8-10 Joseph Wolf, Dog Approaching Another Dog with Hostile Intentions, and The Same in a Humble and Affectionate Frame of Mind, 1871–72. Conté crayon, heightened with white. The postures in Wolf's drawings weren't quite what Darwin was looking for. He marked them up and then assigned the project to another artist.

they would be suitable for publication. Although there is no written record of Darwin's discussions with Wolf about the drawings, it may be surmised that Wolf was either unable or unwilling to alter them to match Darwin's understanding of how the expressions of dogs and cats should be represented. Wood's revised drawings differed enough from Wolf's original conception that he may have simply disavowed their authorship.

Abraham Bartlett (1812–97), whom Palmer credited with having introduced Darwin to Wolf, reportedly described Wolf as having "an eye like photographic paper, it will seize on anything!"[43] However, Wolf was often required to draw from his imagination or to subordinate his observations to the wishes of a patron. In Darwin's case, this meant altering the ear tips of *Cynopithecus niger*, but in his biography of Wolf, Palmer described several other instances in which Wolf was forced to compromise in his depictions of animal behavior. In one case, he had been asked by an unnamed author to illustrate an orphaned mouse, said to have been raised by a mother cat. Purportedly based on an event he

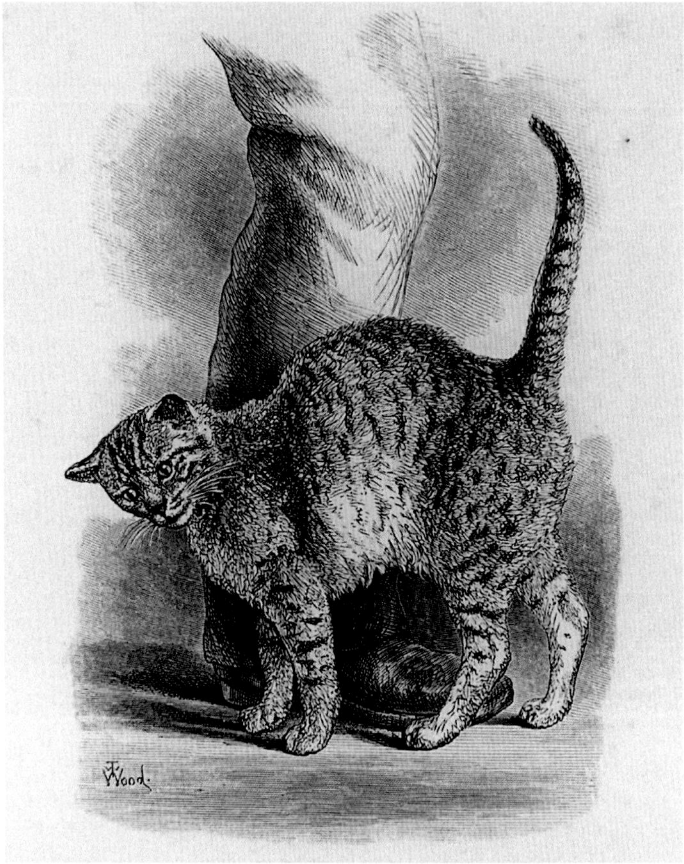

8-12 Thomas Wood after Joseph Wolf, Cat in an Affectionate Frame of Mind, c. 1871–72. Wood engraving. Wood felt a human leg was a more appropriate object of affection.

had observed himself, the author called for an illustration of the mouse suckling from one of the cat's teats.[44] Wolf was convinced that such behavior had never occurred. Nevertheless, he agreed to produce the drawing according to the author's instructions.

Despite tensions of this sort, Darwin and Wolf were genial collaborators. Darwin occasionally visited Wolf at his studio on Berners Street in London, near Oxford Circus.[45] Wolf recounted an anecdote that he said typified the tone of these visits:

> Among the artist's birds at that time was a particularly tame piping Bullfinch, which had learned, among many accomplishments, to distinguish the note of his master's bell from the others. At the first tinkle, he would fly to a chair-back near the door of the studio, where he would sit and bow and pipe to his favourite visitor, but would attack any person he mistrusted. One day Darwin called; and the Bullfinch, not liking the look of his long white beard, flew straight at it, pulling with all his little might, while the old man laughed and chuckled.[46]

Privately, Wolf referred to men of science as "dry sticks," who "work with their noses a few inches from their desks."[47] Darwin, however, was different. He "was not like many great men of science who would put you down with a look or a sentence," Wolf recalled. "A child might have talked to that man. He was wonderful in that respect."[48]

The Tail Wagging the Dog

The most famous of the artists Darwin engaged to produce illustrations for *Expression* was Riviere, whom the magazine *Connoisseur* described in an obituary as "the successor to Sir Edwin Landseer in the role of animal painter to the British public."[49] During the years 1868–71, when Darwin discussed his desire for illustrations of dogs with him, Riviere was employed by the satirical magazine *Punch*. His introduction to Darwin came just as Riviere was beginning to perfect the narrative style for which he remained popular the next twenty years.[50]

Initially, correspondence between Darwin and Riviere was relayed by their mutual friend, Elinor Mary Bonham-Carter (n.d.).[51] In June 1871 Riviere sent Darwin a long letter describing the characteristics of emotional expression in dogs, focusing particularly on the arrangement of the ears during expressions of pleasure.[52] He believed that dogs put their ears back when expressing pleasure and prick their ears up when alerted. This puzzled Darwin, who believed that the drawing back of ears in affection was related to the tendency to bite playfully while uncovering the lips.[53] Ten months later, in April 1872, Darwin asked Riviere if he would be willing to provide illustrations of the expressions he described, for inclusion in *Expression*. Riviere readily agreed and offered to draw directly on woodblocks for convenience.[54]

More than a month later, Riviere wrote apologizing for the delay, offering to coordinate the cutting of the plates directly with the engraver Darwin had retained, James Davis Cooper. "Expression hangs upon so subtle a difference of lines," Riviere wrote, "that what may be right in the drawing can easily be made wrong in the engraving."[55] Darwin approved the first drawing, *Dog Approaching Another Dog with Hostile Intentions*, with minor alterations. He rejected the second, *Dog in a Humble and Affectionate Frame of Mind*, in which the dog was supposed to be shown with a wagging tail. Riviere explained:

> With regard to the second dog I endeavoured to carry out your written remarks & in one important respect did so contrary to my own belief & experience (viz, in putting the head down instead of up) I trust that a dog seldom or never puts its head down when it sees its master unless it is afraid of him. A dog never puts on the expression you want except

in the presence of a man & always looks at the man's eye & so always puts its own head up.

To give the effect of a wagging tail is impossible in a *drawing*, which can only illustrate a climax. In *painting* it can only be hinted at & the observer must know what is meant before he looks at the picture. Setting on one side the "wagging tail" I think that the accompanying rough sketch gives an idea of the expression you want & should you think so I should have much pleasure in drawing it on the wood for you.[56]

A compromise was struck. Darwin accepted Riviere's advice regarding the elevation of an affectionate dog's head, and Riviere provided an updated drawing (figure 8-13). Darwin published the revised drawing, with additional modifications, as engraved figure 6 of *Expression* (figure 8-14).

Earlier, Darwin had attempted to obtain a similar drawing from Thomas Wood, who declined on the grounds that he was "not very familiar with dogs."[57] Darwin also referred in his correspondence with Riviere to a photograph and two related drawings of a dog he had obtained by "a German artist, lately dead, whose name I forget."[58] These materials have since been lost.

As noted, Joseph Wolf was also among those Darwin approached about the humble and affectionate dog, but Darwin was unhappy with the version Wolf submitted; like Riviere, he resisted Darwin's request

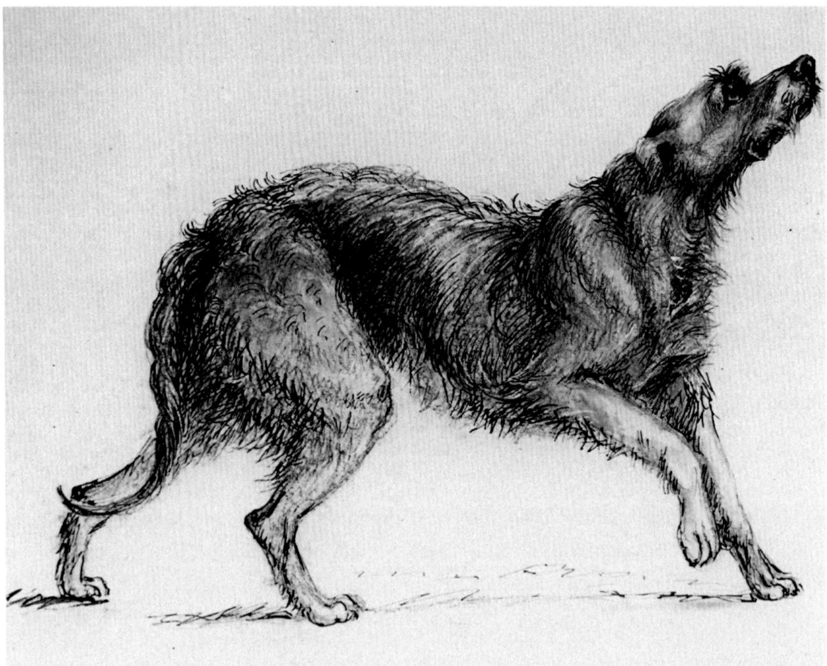

8-13 Briton Riviere, Dog in a Humble and Affectionate Frame of Mind, 1872. Ink and pencil heightened with white on gray card. Riviere refused to budge on the question of whether an affectionate dog lowers its head; Darwin acquiesced.

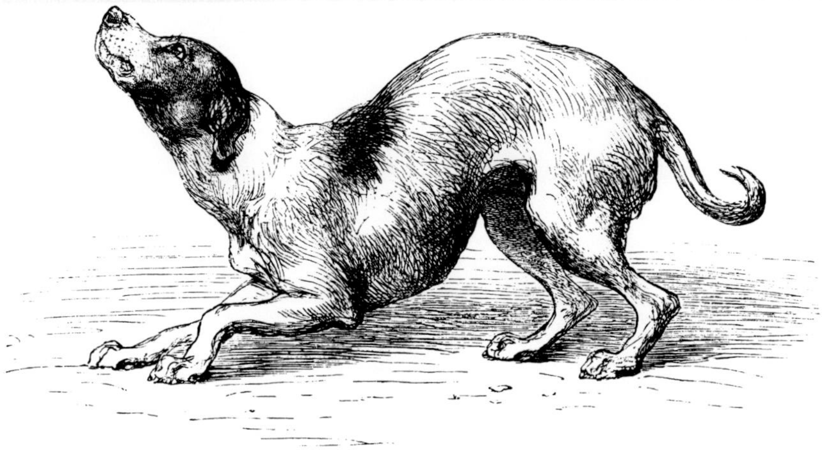

8-14 After Briton Riviere, Dog in a Humble and Affectionate Frame of Mind, 1872. Wood engraving. Figure 6 of *Expression*. Darwin published one of Riviere's drawings, updated and with modifications.

to lower the dog's head. Darwin registered his criticism in the margin of Wolf's rejected drawing: "Head rather lower and more in line of the back."[59] However, with his deadline approaching and a plurality of opinion against him, Darwin relented slightly. Riviere's published figure is a compromise of sorts; the head of the dog dips down at the neck and rises up at the nose, creating the effect of a serpentine curve through the length of its body. Ultimately, Darwin capitulated to the combined experience of his artists, though he initially tried to superimpose his own understanding of canine expression on their work.

The original sketch Riviere sent to Darwin appears related to his oil painting *Fidelity* (1869), which shows a loyal dog nuzzling its dejected master[60] (figure 8-15). Like *Fidelity*, the sketch depicts the same breed of dog, with a relatively straight back and ears pulled back. In each rendition, Riviere attempted to convey the wagging of the dog's tail by blurring its tip slightly. The published version of *Dog in a Humble and Affectionate Frame of Mind* also makes use of this conceit, although the effect is barely visible, as Riviere had warned Darwin it would be.

Nevertheless, Darwin's desire to convey the wag of a tail indicates a curious ambiguity in the representation of time in the *Expression* engravings. This movement would have been difficult to convey photographically, and the blurriness Riviere employed to achieve this effect would have been unsuitable in a scientific photograph. Darwin used photography to isolate phases of expression. His photographs were ostensibly instantaneous, highlighting the fragmentary nature of expression and enabling scrutiny of its muscular mechanics. In an engraving, however, he was able to indicate the passage of time by having Riviere render the tail indistinctly. Ironically, the wagging tail in the engraving

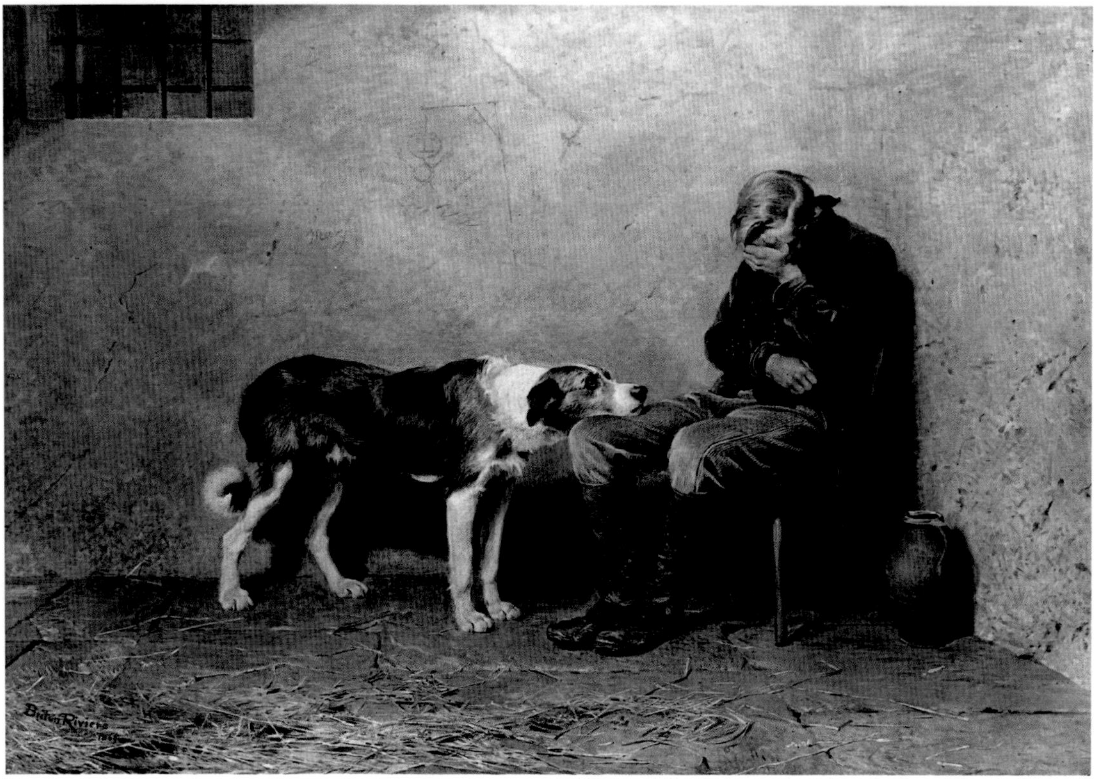

8-15 Briton Riviere, *Fidelity*, 1889. Oil on canvas. Darwin's dog seems to have been cribbed from Riviere's painting.

resembles the indistinct, superimposed forms obtained by a long photographic exposure of a moving object.

Engravings are confined to the first and last parts of *Expression*, the sections focusing on expression in animals, and the human expressions of fear, terror, and insanity. Photographs are reserved for illustrations of humans, beginning with the expressions of infants and concluding with grown men. Organizing the book in this way was in part a practical matter. Limiting the number of photographs helped to reduce production costs, because engravings were much less expensive to print than photographs. Moreover, Darwin evidently found it more difficult to obtain effective photographs of animals than photographs of people. Animals are relatively impatient and unpredictable, and the materials in use at the time required lengthy exposure times. And whether intentional or not, the pictorial divide between engravings and photographs added dramatic emphasis to the sections on "normal" human expression. The engravings provided a visual counterbalance to the photographs and helped to integrate photography into the book in a compelling way.[61]

The Darwin Archive contains sketches for many of the illustrations published in *Expression*, but there are no drawings by Thomas Wood in the archive. Wood drew directly on woodblocks, and as a result, his drawings were destroyed when the blocks were incised. James Davis Cooper, the engraver, invented a system for copying drawings onto printing plates using a photographic transfer process. Although Darwin wrote in his introduction to *Expression* that this process ensured "almost complete fidelity" in reproduction,[62] photographic reproduction was just one of the services Cooper offered his clients. The accounts ledgers in the John Murray Archive show that Cooper was paid a total of sixty-eight pounds and fifteen shillings for the project.[63] This was a large sum, compared to a little more than ninety pounds paid to the engravers who worked on *Descent*.[64] *Descent* contained seventy-six engravings, while *Expression* contained only twenty-one. Cooper charged extra for his photographic transfer process, so the higher price of the *Expression* engravings suggests the process was used for the book.

Even if Cooper used his photographic technique to transfer drawings to woodblocks, the accuracy of the engravings would not have been guaranteed. The "fidelity" Darwin referred to was the accurate translation of drawings to engraved plates. This is a curious tautology, implying that the original drawings are the standard by which the engravings should be judged. "Fidelity" for Darwin was not synonymous with objectivity. It was truth born of consensus, forged in dialogue and give-and-take with colleagues. As demonstrated by the correspondence he exchanged with artists and the numerous alterations he made to the drawings, Darwin viewed truth in illustration subjectively. This attitude would have a significant effect on his attempts to acquire photographs to illustrate *Expression*.

9

Darwin's Art Photographer

Oscar Rejlander, Darwin's favorite photographer

After nearly three years of collecting pictures to illustrate human facial expressions, Darwin had managed to acquire only eleven of the thirty photographs he would ultimately publish in *The Expression of the Emotions in Man and Animals*. Frustrated in his attempts to find a single source of action photographs, he had instead put together an eclectic group of pictures of uneven quality. None of the photographers he approached seemed capable of producing the range of images he needed. The photographs he had acquired, gathered opportunistically from printsellers and photographic studios, failed to illustrate many of the specific behaviors he planned to discuss in his book. It was at this stage, around a year before *Expression* was sent to press, that Darwin met the Swedish expatriate photographer Oscar Rejlander.

Darwin's collaboration with Rejlander fundamentally changed the project. Darwin considered *Ginx's Baby* (chapter 7) one of his most important pictures, but the Swedish photographer contributed much more: in his imagination and creativity, he was different from the others Darwin had contacted. He practiced his craft with the sensibility of a painter or draftsman. Most of the other photographers Darwin approached produced popular imagery, especially carte-de-visite and cabinet card portraits of distinguished persons and celebrities. Rejlander engaged in some of this type of work, but his main interest lay in the aesthetic possibilities of the medium. Producing comparatively few pictures, he aspired to have his work recognized for its artistic merit. He exhibited at public exhibitions and was perhaps the only photographer engaged by Darwin who showed at the Royal Academy.[1] Whereas other photographers made literal representations, Rejlander created compositions with narrative and symbolic content.

Some of the photographs Rejlander contributed to *Expression* were from his existing stocks of prints and negatives; others, Darwin commissioned. Shortly after the two men met, Rejlander opened the vaults of his studio, allowing Darwin to draw from fifteen years of works produced for a wide variety of purposes. Approximately eight such images made their way into *Expression*—it is difficult to know the exact number because the dates of some are impossible to establish. Others, though not published, Darwin used for reference.

At least as early as 1871, Darwin commissioned Rejlander to produce additional photographs for *Expression*. Most of the photographs that appear in the book are of this type. In making custom pictures for Darwin, Rejlander became closely involved with the project and shared what he knew about the nature of emotional expression.

To understand what Rejlander brought to *Expression*, it is necessary to examine his career.[2] While an important figure in the history of photography, Rejlander remains surprisingly obscure, in part because the contents of his studio were dispersed after his death. His only original writings known to exist are in the Darwin Archive, a primary source for studying Rejlander's working methods and the maneuverings involved in Darwin's use of photography.

Darwin and Rejlander Join Forces

In Rejlander, Darwin found a photographer who was capable of manipulating his results to fit specific needs. Rejlander did attempt to produce the images Darwin sought using direct, unstaged photography, but where this proved impossible, he occasionally provided simulations to illustrate the behaviors Darwin targeted. His ability to concoct pictures using a mixture of natural and artificial elements proved to be a deciding factor in the completion of *Expression*.

Although they may have met earlier, Darwin and Rejlander were probably introduced in 1871. The Moderna Museet in Stockholm contains two portraits of Darwin by Rejlander that, it has been argued, were produced in the late 1860s[3] (figure 9-1). However, Darwin's son Francis, in his biography *Life and Letters* (1887), estimated these photographs to have been made around 1870, based in part on the appearance of a derivative steel engraving sent to subscribers of the journal *Nature* in 1874.[4] Francis approximated the date of these pictures from memory. Based on correspondence in the Darwin Archive, it is now possible to

Opposite:

9-1 Oscar Rejlander, *Charles Darwin*, c. 1871–72. Albumen print. The opportunity to make this portrait may have figured in Rejlander's compensation.

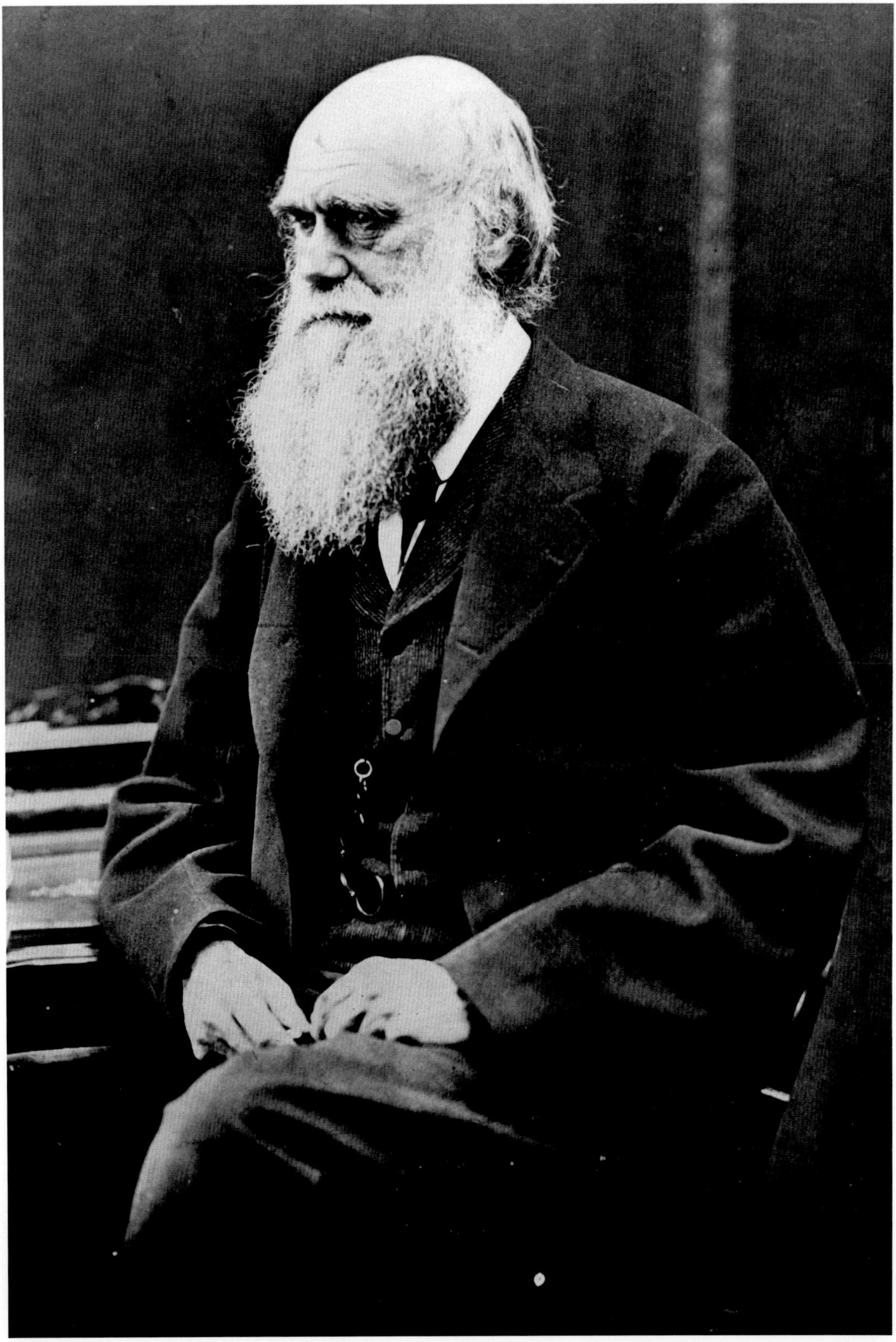

date them more precisely: they were probably made in 1871 or 1872. The first datable reference to their association appears in a letter Darwin sent to James Crichton-Browne in April 1871, in which he wrote, "I am now rich in photographs for I have found in London, Rejlander, who for years has had a passion for photographing all sorts of chance expressions exhibited on various occasions, especially by children, and taken instantaneously."[5] There is no evidence Rejlander photographed Darwin before their collaboration on *Expression*.

The circumstances of their introduction are unclear. The photohistorians Camfield Wills and Deirdre Wills assert that Darwin contacted Rejlander as a result of Darwin's inquiries into the forms of ears,[6] but there is little evidence to support this idea. Rejlander's "Hints Concerning the Photographing of Criminals," in which the subject of ears is addressed, did not appear until 1872, the same year *Expression* was published.[7] That article, which echoes concerns voiced in *The Descent of Man*, was almost certainly inspired by Rejlander's exposure to Darwin. It is more likely the two men were brought together by Julia Margaret Cameron, who received photographic training from Rejlander in the early 1860s and whom she admired.[8] Cameron entertained Darwin at her home on the Isle of Wight during his stay in one of her cottages in the summer of 1868.[9] Knowing Darwin's interest in photographs of expression, Cameron could have recommended Rejlander to him. Or Darwin may have simply wandered into Rejlander's studio on one of his London excursions, as the photographer's Albert Mansions studio stood adjacent to London's Victoria Station.[10]

In either case, the meeting was propitious. In a brief but intense collaboration, Rejlander emerged as the most significant contributor of photographs to *Expression* and Darwin's principal adviser on photographic issues. He presented Darwin with at least sixty-four pictures of expression, more than any other photographer, accounting for roughly one-third of the pictures of expression in the Darwin Archive.[11] Once the book was published, the importance of his contribution was immediately recognized. As one reviewer declared, the partnership of Darwin and Rejlander scattered to the winds all previous attempts at portraying the passions.[12]

In securing Rejlander's assistance, Darwin had harnessed a prominent force in the English artistic photography movement. Frequently described as a "father" of art photography,[13] Rejlander had worked since the 1850s to create photographs emulating popular painting. He was one of the most accomplished photographers in Victorian England, and for a time his influence as an artist, teacher, and author was unparalleled. "Should the history of the art of photography be at any time written," the Berlin publication *Photographische Mittheilungen* proclaimed in 1865,

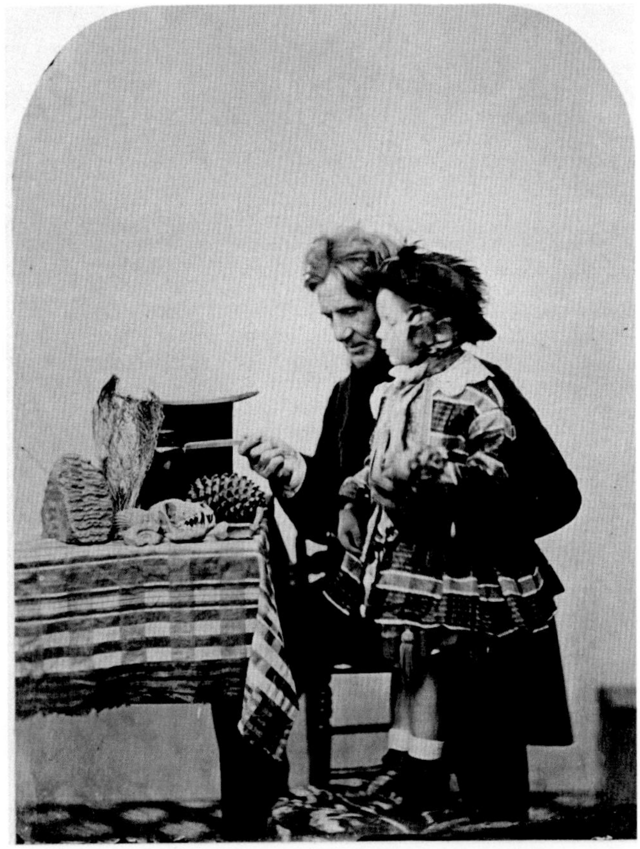

9-2 Oscar Rejlander, *A Young Naturalist*, c. 1860. Albumen print. A prelude to Rejlander's interest in natural history.

"the name of Rejlander will appear as a star of the first magnitude."[14] Although not all writers were as sanguine as this reviewer, even his detractors conceded that Rejlander was one of the most influential photographers of the period. Yet, purists objected to his practice of altering photographs to achieve desired effects, and by the time Darwin approached him, Rejlander's methods had become unfashionable. At first glance, he is an unlikely candidate to have contributed to *Expression*.[15]

Conspicuous as Rejlander was in the photographic community, he was little known in scientific circles. He does not appear to have had any connection with Darwin's publisher, John Murray, nor did he work for other scientist clients. Before making the Moderna Museet portraits, Rejlander is not known to have photographed Darwin, members of his family, or his associates. Perhaps the only augury of his later interest in scientific photography was his genre scene, *A Young Naturalist* (c. 1860) (figure 9-2), commemorating the fashion for collecting natural history specimens at mid-century. Typical of Rejlander's narrative photographs, it portrays a young girl wearing a fancy dress, nestled under the arm of a collector, possibly her father. She stands on a chair (Rejlander often posed small children on chairs) studying an assortment of objects shown to her by the man, including a skull, a pine cone, a bird's wing, and a fossilized sponge.[16]

Art Photography

In adopting the title of "art" photographer, Rejlander was careful to distinguish himself from the position famously rejected by the French critic Charles Baudelaire, "that an industry which could give us a result identical to Nature would be the absolute of Art."[17] On the contrary, Rejlander believed that thoughtful intervention by the operator was required to make an artistic photograph. "Artistic photography" to him implied an affinity with practitioners of the traditional fine arts; the intent of the maker was crucial to successful composition. He devoted himself to photographs containing, as he described them, "so many artistic requisites that the painter has only to translate the whole into colours to make it a real picture."[18] His strict adherence to this view prompted the *Art Journal* to identify him in 1868 as one of the few photographers worthy of the "artistic" label. "For masterly and artistic feeling and treatment [his photographs] are consistently before all in Europe. In a word, they are works of Art, and contain many highly original thoughts."[19]

Rejlander established his reputation with genre scenes, depictions of the female nude, and, initially, ventures into landscape photography.[20] He was highly regarded in portraiture, but he was most often identified with narrative photographs,[21] a category his contemporaries described as "miscellaneous subjects of the 'Wilkie' character, very popular."[22] The influence of the early paintings of Sir David Wilkie (1785–1841) is evident throughout the narrative work, which, like Wilkie's, featured light-hearted genre scenes and tableaux of middle- and working-class subjects. Wilkie, a noted protégé of Edinburgh professor Charles Bell, had been affected by his study of Bell's 1806 *Anatomy and Philosophy of Expression* (see chapter 4). Rejlander frequently submitted these narratives to competitions, afterward producing affordable reductions for sale to the public.

Such photographs were sometimes also described as "group and figure studies," reflecting Rejlander's hope that they might serve as source material for artists working in other media. They are almost always infused with Rejlander's mischievous sense of humor. Among many examples are such photographs as *The Washing Day* (c. 1854–56), depicting a group of women hanging laundry while flirting with a young man; *Play* (c. 1854–56), portraying a young girl playing cat's cradle with an elderly man; or *I Pays* (c. 1854–56), a tavern scene, which the *Pho-*

tographic Journal described as a patron "executing the privilege of every Englishman, 'To grumble, and to pay.'"[23]

Rejlander's narratives were frequently compared to the work of contemporary painters and draftsmen, as with *Drat the East Wind!* (c. 1856),[24] which was said to be reminiscent of the drawings of the illustrator and *Punch* caricaturist John Leech (1817–64), "though it merely represents the querulous distressed face of an old lady rubbing her rheumatic knees."[25] Occasionally Rejlander made direct reinterpretations of popular paintings, as with *Drunken Barnaby Leaving the Tavern*, after William Etty's (1787–1849) work of the same title, or *Tis Dark without, Tis Light Within* (c. 1855–60), depicting a blind woman playing a tune on the violin, inspired by Wilkie's *Blind Fiddler* of 1806.[26] He also produced pastiches of Old Master paintings, such as his widely reproduced *Cherub* after Raphael's *Sistine Madonna*.[27]

As a youth, Rejlander is reputed to have studied painting in his native Sweden and in Rome, where he was employed for a time producing portraits and copying Renaissance paintings.[28] The date of his arrival in England is not recorded, but he lived in Lincoln and Wolverhampton before taking up photography in the early 1850s to complement his work as a portrait and miniature painter. He was almost entirely self-taught, having learned the basics from the noted early photographer Nicolaas Henneman (1798–1867), protégé of William Henry Fox Talbot and former assistant at Talbot's Reading Establishment in Berkshire.[29] This instruction was confined to a single afternoon session in 1852, three hours of training in calotype, and only half an hour in wet-plate collodion, the process that became his stock in trade.[30]

Perhaps as a result of his early experience as a painter, Rejlander claimed that photography should serve, as the photographer Frank Howard (active 1850s-1880s) suggested in a speech before the Liverpool Photographic Society (1853), as a "handmaid" to art by providing studies that artists could copy in their paintings.[31] Rejlander remained equivocal about whether a photograph could be considered a work of art, suggesting but never directly claiming that photography ought to be considered on equal terms with drawing and possibly even painting.[32] In "An Apology for Art Photography" (1863), he wrote that he did not look upon photography "as an ultimate art, or an art depending on itself, or complete in itself, except in the details."[33] Yet elsewhere he wrote, "[I]t is the mind of the artist, and not the nature of his materials, which makes his production a work of art."[34] His evasiveness on this issue may have resulted from perceived opposition from the photographic community, in which there was widespread reluctance to accept photography as an art form. Despite his efforts to soften his stance, however, he was generally perceived as an advocate of photographic art.

Rejlander sold many of his photographs as studies for use by artists, who, like Darwin, were interested in photographic records of various human expressions.[35] Some he used himself as studies for his own drawn and painted compositions. His photographs are likely to have circulated throughout the artistic studios of London, facilitated perhaps by his association with the printseller and publisher Joseph Hogarth,[36] from whom Rejlander rented his first home in London, at No. 5 Haymarket, in 1862.[37]

At the very least, Rejlander would have been acquainted with many prominent artists of Victorian London through his membership in the Artist's Rifles Association, 38th Middlesex Regiment. The "Artist's Company," as it was called, included among its ranks John Everett Millais (1829–96), Frederic Leighton (1830–96), and George Richmond (1809–96).[38] Unfortunately, with the exception of a single album of photographs assembled by the precocious painter Cecil Gordon Lawson (1851–82), Rejlander's studies are not known to have survived in the archives of any of his artist customers.[39]

Rejlander held that photography was better than painting at catching fleeting feelings, emotions, and the passions and considered the relative instantaneity of the medium to be one of its main attractions.[40] His prominence in this area was reinforced by his decision in the late 1860s to reprint details from some of his narratives as separate character studies and to distribute these as individual carte-de-visite prints.[41] These studies, isolating expressive situations and moments of dramatic action, were widely distributed in London in the 1860s and 1870s. Their ubiquity suggests another possible means by which Darwin may have become aware of Rejlander. While searching for pictures at London printsellers, Darwin may have encountered photographs from the Rejlander studio before they ever met in person.

A Plastic Medium

In various writings, Rejlander urged photographers not to use the medium merely to record the appearance of things, but to exploit its plasticity to achieve creative effects. Although he valued the capacity of the medium to delineate scenes in detail and with an economy of labor, he believed photographers should use the resources at their disposal to build compositions of artistic worth. In his own compositions he frequently intervened, both in the studio and in the darkroom, to organize and interpret the information captured by his camera. He described his works as "sun-paintings . . . painted by me with pencils of light about ninety-three millions of miles long."[42] Rejlander applied the same metaphor in his famous allegory *The Infant Photography Handing the Painter Another*

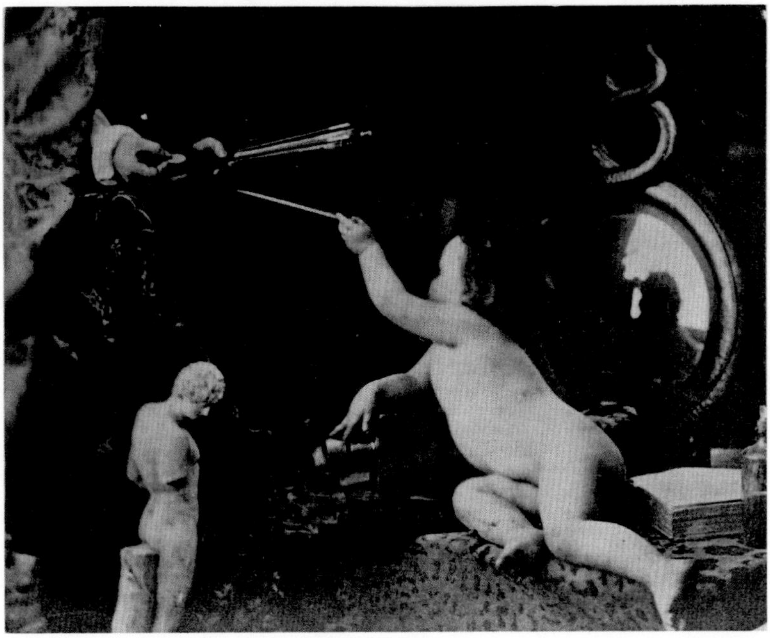

9-3 Oscar Rejlander, *The Infant Photography Handing the Painter Another Brush*, c. 1856. Albumen print. For Rejlander, photography was a tool in the artistic repertoire.

Brush (c. 1856),[43] in which a putto, symbolizing the new medium, passes a paintbrush into the hands of an artist (figure 9-3). The *Photographic News* described the piece: "The hand of the artist holds many brushes; young photography—only a child—gives him one more."[44]

Rejlander believed that photographers should not be content to merely replicate ordinary vision, but should embrace the medium as an artistic tool, imposing an artist's will on a photographic picture. This meant mediating every step of the process: from subject to camera, from camera to negative, and from negative to print. His artistic arsenal included a range of illusionistic techniques, including posing subjects to give the illusion of action, combining information from different negatives to create synthetic compositions, and using pen, ink, and other materials to alter negatives before printing. In his book *Pictorial Effect in Photography* (1869), Rejlander's colleague Henry Peach Robinson summarized this practice: "[A]ny 'dodge, trick, or conjuration' of any kind is open to the photographer's use, so that it belongs to his art, and is not false in nature."[45]

Ironically, Rejlander also recognized photography's potential for mechanical objectivity and occasionally invoked this aspect in his writings. One of his earliest experiments in photographic veracity involved using a camera to test the anatomical correctness of Old Master paintings. He created living replicas of famous pictures using models in the studio and then photographed the arrangement for comparison with the original. The results showed that even the most revered artists were guilty of errors in form and perspective.[46]

The most controversial aspect of Rejlander's activity was the production of composite photographs, which Darwin would certainly have seen on visits to his studio. Rejlander advocated the alteration, and in some case the fabrication, of images in the darkroom. He was renowned for developing a printing system in which separate photographic negatives were superimposed to create pictures assembled from elements of different exposures. Rejlander was not the first photographer to adopt this practice, but he was one of the most proficient. The Scottish photographers Berwick and Annan (firm active mid 1850s) suggested the technique in the *Journal of the Photographic Society* (1855), the same year Rejlander submitted his first composite photograph for exhibition.[47] Similar references appeared in the French journal *La Lumière* as early as 1851, including an article by one Monsieur De Montfort prophetically describing "A Procedure for Grouping a Number of Independently Obtained Portraits in Order to Create a Single Heliographic Composition."[48] But Rejlander was the first photographer to bring composite photography into successful practice, and it was he who is most often credited with popularizing the technique. Whereas others had used composite printing to "improve" photographs, for example, to introduce clouds in a landscape where the camera had failed to register them, or to compensate for underexposed areas in a dim environment such as an architectural interior, Rejlander created completely synthetic pictures. By combining a variety of seemingly unrelated negatives in a single print, he was able to make convincing photomontages of scenes that never actually existed.

Rejlander's ideas about the merits of composite printing and its implications for artistic photography were often criticized. In particular, he was frequently denounced for his famous picture *The Two Ways of Life*, first exhibited at the Manchester Art Treasures Exhibition in 1857 (figure 9-4). A large frescolike work, it was said at the time to have been produced using fifty-seven separate negatives,[49] but is now understood to have been made using approximately thirty-two.[50] Experimental in nature, it was designed to demonstrate that photography could be used to produce compositions as laden with meaning as conventional history paintings. The center of the composition featured Rejlander himself, foreshadowing his later role in Darwin's *Expression* photographs. He is shown clad in a toga, negotiating the temptations of vice and virtue. Due to lack of funds, several of the models depicted in the scene were repeated in various parts of the image.[51] A number were friends and in-laws or, as one writer acerbically described them, "coarse work girls and non-professional male models."[52] A festooned curtain at the top of the picture was actually a gathered cloth, photographed and printed several times slightly off center to approximate the appearance of an ornate proscenium.[53] A tablecloth sufficed to represent the stately drapery of classical Greece, and hand-painted backdrops were used to depict a distant forest, cavernous rocks, and stone lions.

Even its supporters agreed that *The Two Ways of Life* had failed to gain photography equal standing among the plastic arts, as Rejlander had wished. Ironically, the realism of the work may have been its undoing. Many viewers were unsettled by the admixture of reality and fiction in the piece. The *Athenaeum* complained, "some of the details are ill-chosen; the beings of photography are *all clay*, and the sun brings out their imperfections."[54] Similar criticisms persisted well into the twentieth century. Sir Kenneth Clark, for example, dismissed Rejlander as "the most philistine of early photographers," while at the same time praising the effectiveness of his nudes.[55]

The amateur appearance of the subjects in *The Two Ways of Life* troubled some reviewers, while others simply rejected Rejlander's audacity in adapting to photography the compositional methods normally associated with drawing and painting, denouncing such works as trickery. The inclusion of abundant nude female figures further offended some reviewers, although it did attract praise from influential quarters, including the Prince Consort, who invited Rejlander to Buckingham Palace and acquired a copy of the print.[56] The publicity generated from the picture created a financial windfall, such that by 1862 Rejlander had amassed enough money to move his studio from Wolverhampton to London's Haymarket district. His firm became one of approximately 150 active in London at that time.[57] Rejlander promised that his activities in London would be more orthodox than they had been in Wolverhampton. Speaking at a meeting of the South London Photographic Society in 1863, he vowed, "It is very hard, but I must confess it, that I positively dare not now make a composition photograph, even if I thought it might be very perfect."[58]

Darwin probably saw the controversial *Two Ways of Life* on his visits to Rejlander's studio, because Rejlander used the work to advertise his technical abilities. Despite the opposition it received, it still carried the cachet of Royal patronage and was much talked about in the press. Rejlander treated the work as a sort of professional calling card, continuing to submit it to exhibitions throughout the 1860s. As late as 1868, *The Illustrated Photographer* lamented that *The Two Ways of Life* had been trotted out again for exhibition.[59]

An account of the same year confirms that Rejlander maintained a portfolio of approximately eighty of his best pictures, which he used to introduce customers to his work.[60] Although the albums have been lost, a review of the retrospective exhibition of Rejlander's work held at the London Camera Club in 1890 provides some clues as to what they contained.[61] *The Two Ways of Life* was included, as well as separate prints of some of the negatives used to assemble it.[62] The exhibition also contained other examples of Rejlander's composite photography from the 1850s and 1860s, including *An Old Rustic, or The Wayfarer,* and *Drat the East Wind.*[63] Darwin would have seen similar prints, perhaps even the same ones, on his visits to Rejlander's studio.

Darwin possessed several Rejlander composites, including an otherwise unknown photograph called *The Toothache* that Rejlander sent in 1871 (figure 9-5). The composition derives from two separate negatives, printed in combination to give the appearance of a unified scene. To the

9-4 Oscar Rejlander, *The Two Ways of Life,* 1856–57. Albumen print. Famous for the manipulations it contained, Darwin almost certainly saw it.

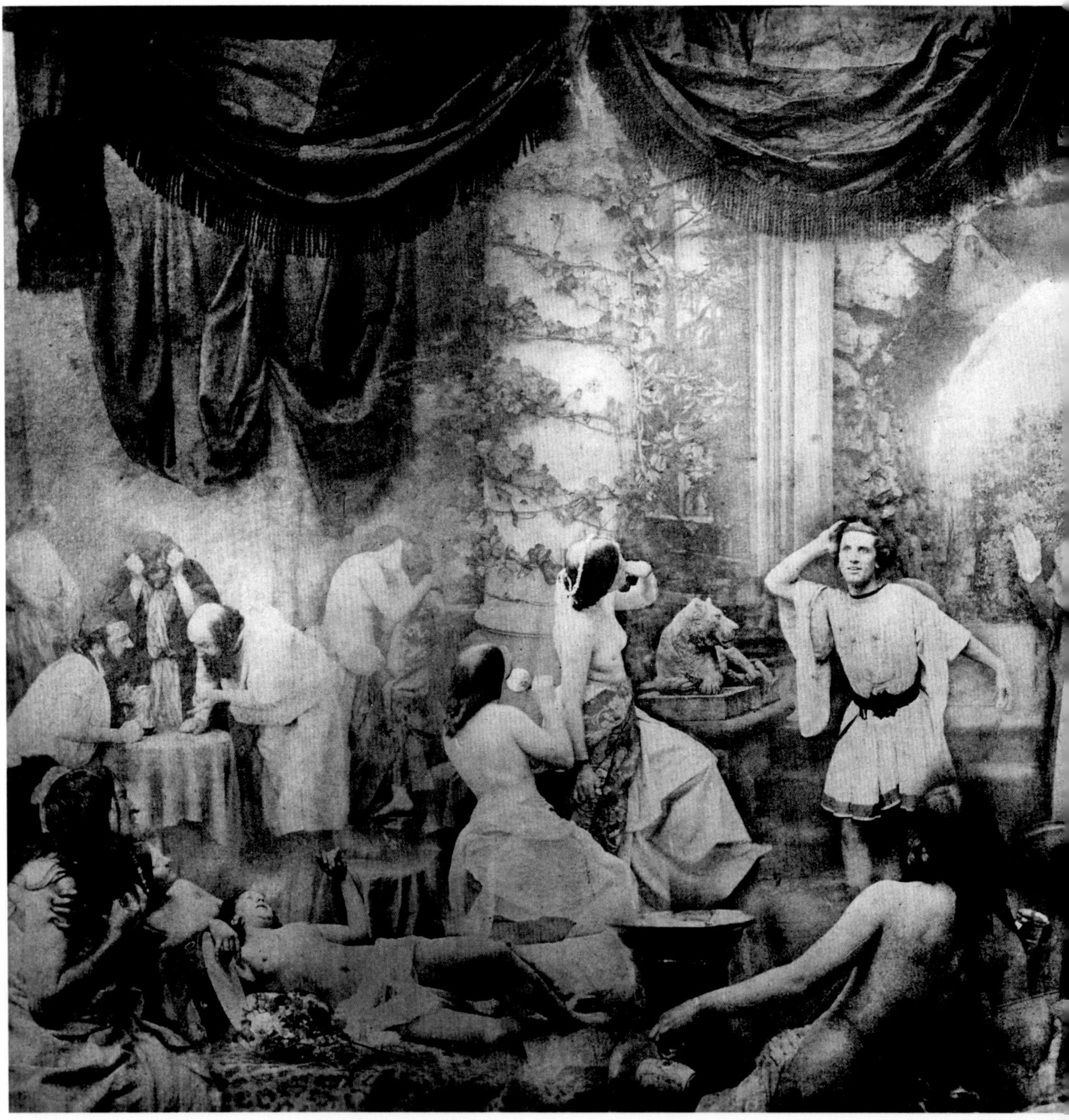

left is a pouting boy, in profile, holding his hand to his face as if soothing a painful tooth. The right depicts a group of children looking on in interest. Originally unrelated, the negatives corresponding to each of these elements were printed so that the children on the right seem to be regarding the suffering child on the left. The version Darwin received appears to have been a proof, because it was not cut down to tidy the edges of the print, and the transition from left to right was left unpolished. The knees of the sufferer blur where the adjoining negative overlaps them, and the lower folds of the girl's dress blend awkwardly

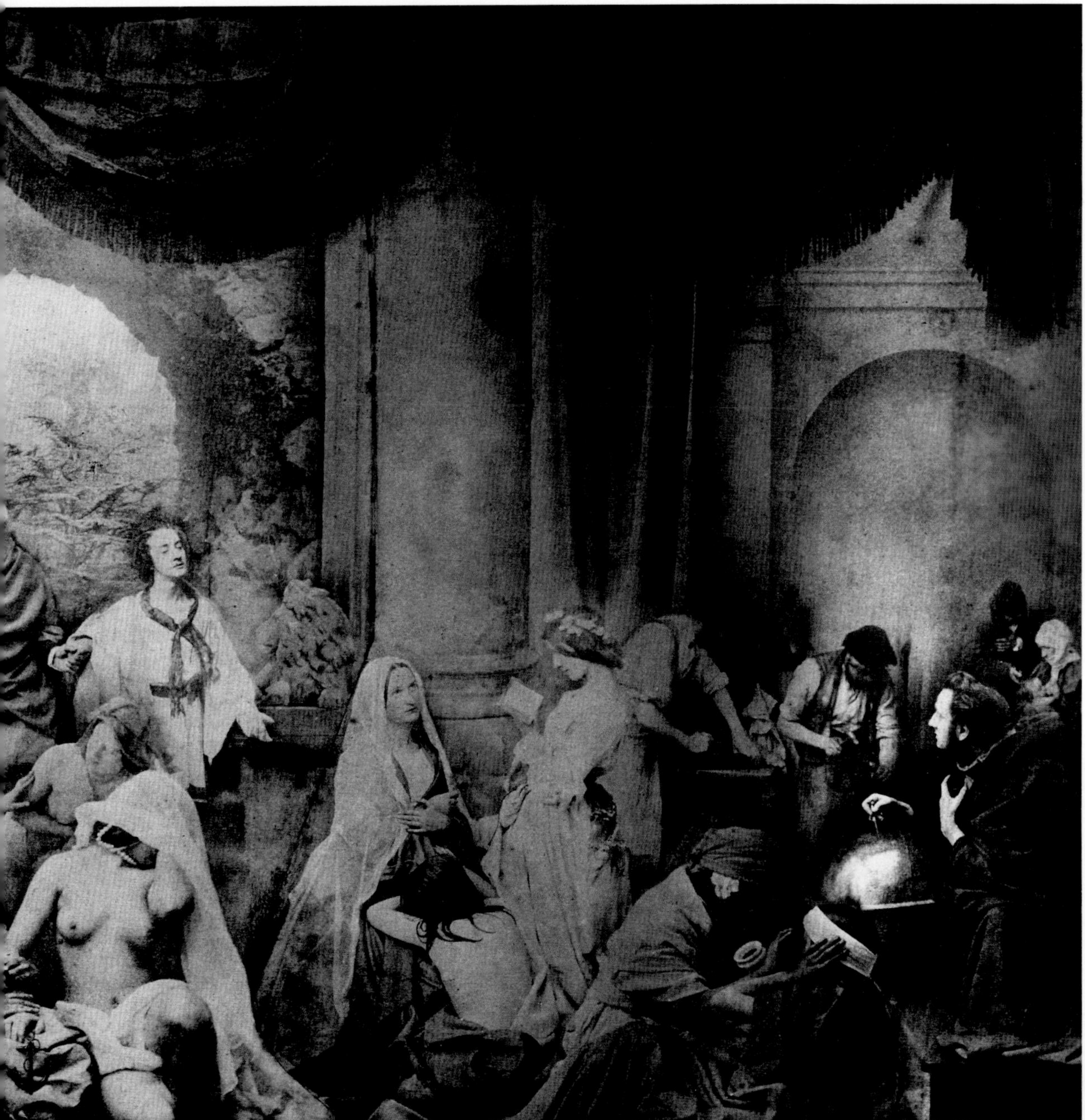

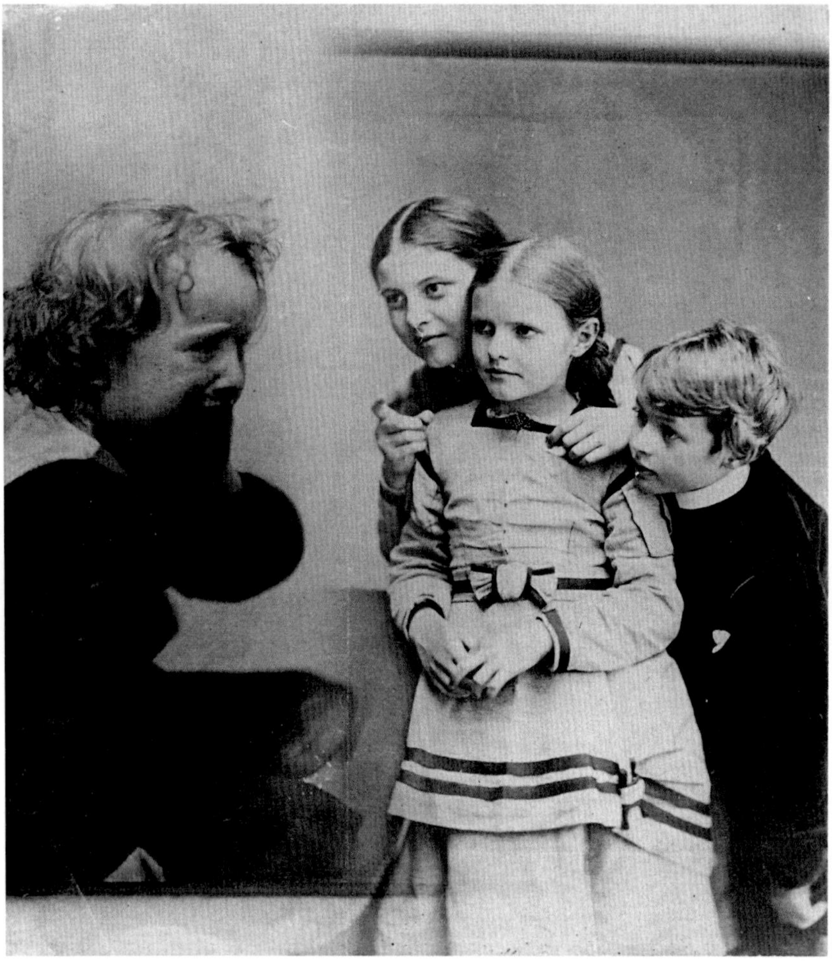

9-5 Oscar Rejlander, *The Toothache*, c. 1871. Albumen print. Rejlander sent Darwin this composite photograph.

into the disjointed planar surface to her right. The two negatives are also out of proportion with one another, making the distressed child appear unnaturally large in relation to the other children.

These incongruities would have made it clear to Darwin that *The Toothache* was a composite print. On this evidence alone, he would have been aware of Rejlander's penchant for altering and manipulating photographs. Darwin decided not to use the image itself, but he did use two variants of its central subject, an older child in fancy dress holding his hand to his face, which he published as plate 1, figure 6, and plate 2, figure 5 of *Expression*. The child appeared in six figures on those two plates.

Reality versus Truth

Rejlander's unconventional approach to photography may seem inconsistent with Darwin's desire for objective representations. However, Rejlander's willingness to compromise between the actual and the illusory is precisely what would have commended him to Darwin. The blend of technical proficiency and ingenuity in surmounting obstacles was just what Darwin needed to illustrate *Expression*. Composite prints such as *The Toothache* may have been artificial, but they were not, as Robinson cautioned, "false." The elements that composed them, such as the pouting boy subject that Darwin decided to publish, were realistic.

Darwin described how he obtained one of the photographs of the pouting boy published in plate 2, figure 5, a "boy about to burst out crying." He wrote that he discovered the photograph in a shop window after he and Rejlander met. The shop may have been Sams' of St. James St., which the *Art Journal* identified in 1868 as the only place in London where Rejlander's photographs were displayed in the front window.[64] Darwin purchased the photograph "and took it to Mr. Rejlander for the sake of finding out by whom it had been made; remarking to him how pathetic the expression was. He answered, 'I made it, and it was likely to be pathetic, for the boy in a few minutes burst out crying.' "[65] Considering Darwin did not recognize the picture as one of Rejlander's when he saw it on display, it may be surmised that Darwin acquired *The Toothache* and related photographs of its central subject after this incident.

Regardless of their pedigree, Rejlander's photographs attracted Darwin because they were based on precise observations of social behavior and were unusually effective in conveying the subtle nuances of human interactions. "Much can be done with even the worst model," Rejlander is reported to have said, while bending the joints of an imaginary doll into an effective pose. "Remember how much expression you may get out of your lay figure if you know how to go about it."[66]

For Rejlander, control was fundamental to artistic achievement. "In all picture compositions the thought should take the first place," he wrote, "and all else be regarded as the language which is to give it expression."[67] Such preconception began in the studio, where the form and content of pictures are comparatively easy to manage. Rejlander often staged episodes for the camera; as a result, some art historians have described him somewhat disparagingly as a theatrical photographer.[68] However, his ability to direct subjects was just one aspect of his effectiveness. Arranging sitters solved some of the problems inherent to wet-plate collodion photography, but photographing in a studio raised others. Although his compositions were often contrived, the gestures and expressions of his subjects were unusually well rendered. As the art historian Elizabeth Heyert has observed, notwithstanding the various means by which he produced his pictures, his figures were remarkably natural and candid and were among the most sophisticated likenesses of the period.[69]

The view that Rejlander was principally a dramatist, common in recent assessments of his art, conflates two separate but related issues. The similarities between his methods and Victorian *tableaux vivants*, as well as his direct connection with prominent figures in English theater, such as John Coleman, the actor-manager of the Wolverhampton Theatre Royal, have been convincingly established.[70] He was certainly influenced by popular entertainment, because the sentimentality of his compositions of the 1850s and 1860s is evident in his attempts to co-opt literary themes. It is often noted that he seldom photographed outside the carefully arranged confines of his studio, where he posed his sitters. However, given the slowness of wet-plate collodion materials, his stagecraft resulted as much from technical necessity as from stylistic inclinations. At the time, the genre subjects he favored could not be photographed candidly. He was forced to fabricate studio scenes that would be made with less contrivance in subsequent years.

Although they were artful, the appeal of Rejlander's genre photographs hinged on how effectively they depicted action and their authenticity of gesture and expression. For this reason, the photohistorian Helmut Gernsheim has described Rejlander as an early "reportage" photographer, who "aimed at slices of real life, but owing to the technical difficulties at this period had to pose his subjects to give the impression of natural scenes."[71] This analysis, apparently at odds with the "greenroom" classification, acknowledges Rejlander's abiding concern with the production of visually credible photographs. However, there are important conceptual differences between Rejlander and the documentary photographers who succeeded him. While his pictures were thoughtfully assembled, most would not have been confused with actual events; they were frequently whimsical in nature and transparent in their attempt to portray fictional scenes.

Rejlander's ability to get the most out of the studio was critical to obtaining satisfactory results. With exposures lasting as much as a minute or more, artificial conditions were the only practicable means of simulating action. Posing was essential. The longer sitters could be held stationary, the more effectively their features would be captured on the negative. By coaxing subjects to hold their poses, Rejlander was able to produce sharp portraits with few blurry details. This helps to explain the abundance of self-portraits in the material he provided for Darwin.

Rejlander also considered careful illumination fundamental in making effective photographs. He designed his studio to control and intensify available light. Although the plans for his Albert Mansions studio, where he produced the photographs Darwin commissioned, have not been preserved, the design of his previous studio at Malden Road is well documented. Known as the "tunnel studio" for its unusually long and narrow structure, it was glazed with transparent and translucent glass panes and baffled with blackened screens. These elements were arranged to eliminate all but essential light from striking his subjects. Because electric lights were not yet available, this design was critical in concentrating daylight to shorten the length of exposure. Rejlander claimed that effective photography could only be achieved "if all the light which had no part in the formation of the photographic image could be either cut off or shut out of the room."[72] The studio enabled him to selectively increase illumination to reduce exposure times and control the density of the negative.

In 1863, details of his tunnel studio were published in the *British Journal of Photography*. "Those who have enjoyed the pleasure of examining Mr. Rejlander's productions," the editors remarked, "cannot have missed observing the very effective manner in which his subjects are usually lighted."[73] Illumination was designed to enable Rejlander to achieve painterly effects such as chiaroscuro (bold contrast of light and dark). Wharton Simpson, editor of the *Photographic News*, explained:

> [I]t will be seen that the sitter is lighted from the side and top side; front light, and direct vertical light being entirely avoided. The camera is in comparative darkness, enabling the operator to focus without the aid of a dark cloth over his head. The eye of the model also looks into this darkness, by which is gained the double advantage of comfort to the sitter and expansion of the pupil of the eye, giving it more depth and expression.[74]

Although he considered such effects unnatural, Robinson agreed: "The camera was in comparative darkness, enabling the operator to focus without the need of a cloth. The eye of the sitter also looked into darkness, a great advantage when we consider the dazzling effect of some of the studios of the time."[75]

The columnist Alfred Wall considered Rejlander's ability to make his sitters comfortable a key to his success. Writing in 1886, Wall pointed out that unlike contemporaries who asked their sitters to assume a falsely pleasant demeanor, Rejlander strove to develop a genuine rapport with his subjects:

> He was genial, friendly, talkative, sympathetic, full of life, sprightly, and amusing—made you feel at home with him, made you as anxious to please him as he was to please you, robbed you of all feeling of restraint and strangeness, occupied so much time with each sitter . . . in the end, [he] achieved photographs that were pictures—portraits which were not mere front elevations of men and women, but faithfully represented sentient beings full of life and feeling; not petri-

fied specimens of humanity, but faces and figures in action, thinking, speaking, or about to speak, with smiles caught in the act of coming and going—not fixed, meaningless grins—real portraits of real people, with their individuality stamped upon them in action, pose and expression.[76]

This evaluation is typical of the praise Rejlander received for his portraits. The editors of the *British Journal of Photography* described them as displaying "that vast variety of character shown in action, pose, or expression which we observe in men and women performing the ordinary functions of their everyday lives, but so seldom see in the glass room or the album."[77] The *Art Journal* agreed, calling his photographs "singularly forcible and effective."[78]

The Lens Turns to Darwin

Ironically, the promise of a portrait commission may have figured in the compensation for Rejlander's work on *Expression*. Although he was not especially active in celebrity portraiture, the opportunity to photograph a prominent figure like Darwin must have been irresistible to him. He may have agreed to undertake the project in part in exchange for rights to a portrait. Rejlander was paid a total of seven guineas for the photographs he gave to Darwin,[79] a respectable amount, but appreciably less than that paid to the other artists who worked on the project. Thomas Wood, for example, who supplied eight drawings used in the book, was paid twenty pounds, nine shillings, and eight pence, more than two pounds, ten shillings per picture. Other artists received similar compensation: Arthur May, who supplied two drawings, received two guineas, and Joseph Wolf, who also supplied two drawings, received four.[80]

Calculating the amount due to Rejlander was complicated. He provided a reservoir of useful photographs, but their production involved less labor than equivalent drawings. Some were printed from negatives that had been made years earlier and, as multiples, and without an exclusivity agreement, could easily be reprinted and sold to other customers for further profit. The payment issue was raised in the first letter Rejlander sent Darwin in April 1871: "You ask for the bill—you cannot pay for goodwill. For all the photographs you have had I will lump it at [3 pounds, 3 shillings]. And for each plate that you may want to make use of I will charge only [10 shillings]."[81] Darwin accepted this suggestion, adding to the sum in ensuing months.

Rejlander knew a portrait of Darwin would sell readily and could lead to further commissions. The income would supplement the sum he received for his *Expression* photographs and may partially account for the modest amount he accepted for the work. For his part, Darwin, who was notoriously dissatisfied with pictures made of him, was generous in his praise of Rejlander's photographs of him. In May 1871, he was already recommending his portrait by Rejlander to the editor of an unidentified journal: "I can have no objection to the portrait & memoir of myself in your Review as proposed. The best photographs of me have been taken by Mr. Rejlander, & as it will save you trouble I send you one."[82] In 1874, Darwin arranged for a version to be sent to subscribers of the prestigious journal *Nature*, as Francis Darwin noted.[83]

Rejlander made at least three photographs of Darwin in the same sitting. Two of these, now at the Moderna Museet in Stockholm, are conventional three-quarter-length portraits taken in partial profile. A third, in the Darwin archive, is unusual[84] (figure 9-6). Cropped tightly to eliminate all but the bust of the figure, it depicts Darwin viewed from slightly behind his right shoulder, in gentle contrapposto with his face in full profile. This version, seldom seen, suggests another possible dimension to Rejlander's portraits of Darwin. Unlike the other two portraits, in which Darwin wore his hair over his ear as he is usually shown, in the archive version his hair is combed away slightly, exposing most of the outer structure of his ear.

It is tempting to imagine that the picture reflects Darwin's interest in the form of human ears. It was not unlike Darwin to use himself as a specimen. Yet he did not include photographs of himself or his family in *Expression*; that would have been unbecoming a scientist of his stature. The closest he came was a wood engraved illustration of a *Small Dog Watching a Cat on a Table*, which appeared as figure 4 in *Expression*. The historian of science Janet Browne has identified the animal as Darwin's dog Polly.[85] In the text, Darwin wrote, "When a piece of brown biscuit is offered to a terrier of mine and she is not hungry (and I have heard of similar instances), she first tosses it about and worries it, as if it were a rat or other prey; she then repeatedly rolls on it precisely as if it were a piece of carion."[86] Darwin ascribed the image in the text to Oscar Rejlander. If Browne is correct, then presumably Polly accompanied Darwin on his London jaunts.

Conversations with Darwin about the properties of ears may have prompted the article Rejlander submitted to the *British Journal of Photography* in 1872, suggesting the best procedure for photographing criminals. He described a standard method for making comparative photographs of ears, suggesting sitters be arranged at a 70° angle to window light emanating from behind. This appears to have been the technique Rejlander used for the profile portrait of Darwin. Remarking on the unique characteristics of the lobe, aural canal, and the shape of ear carti-

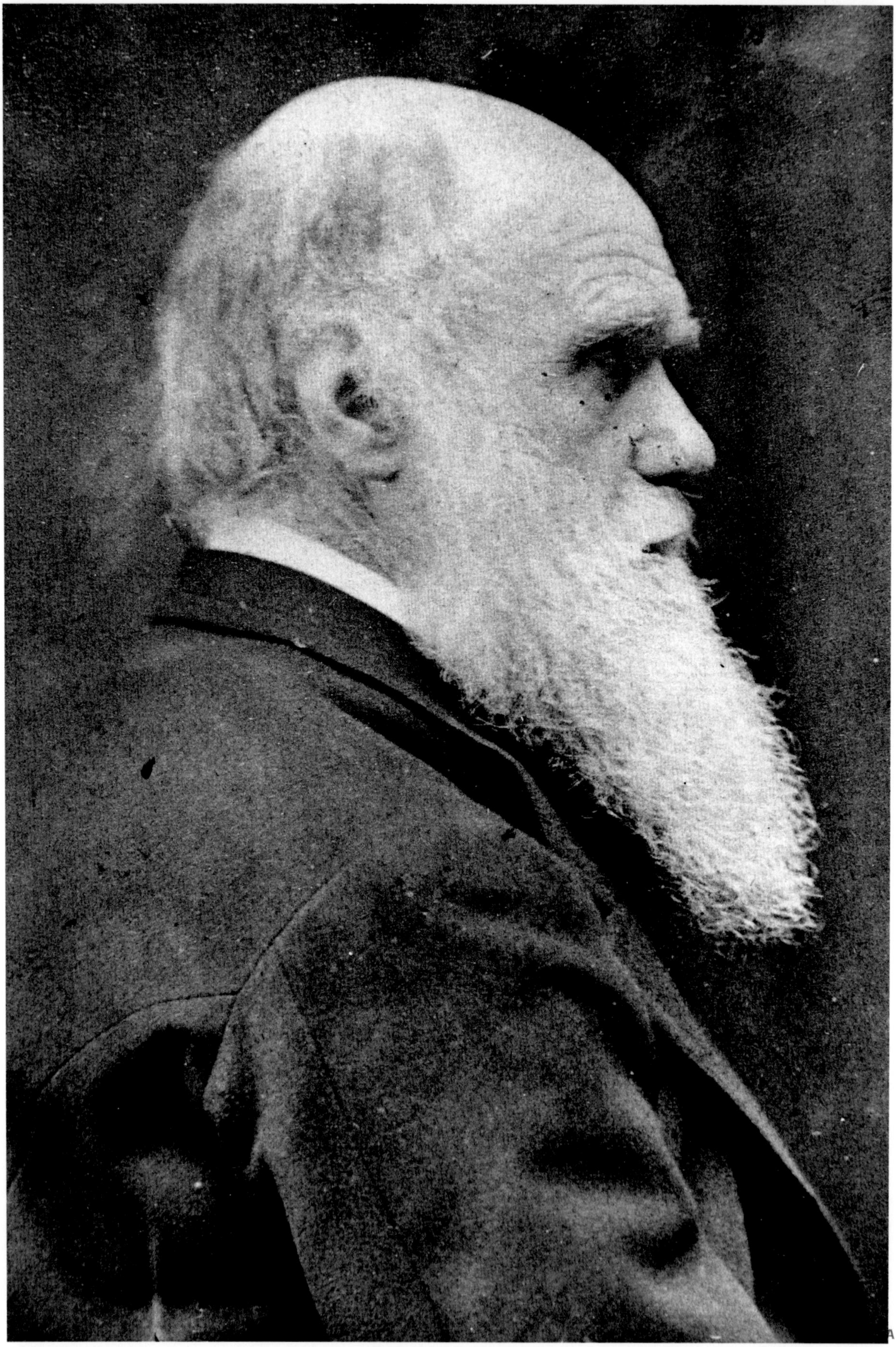

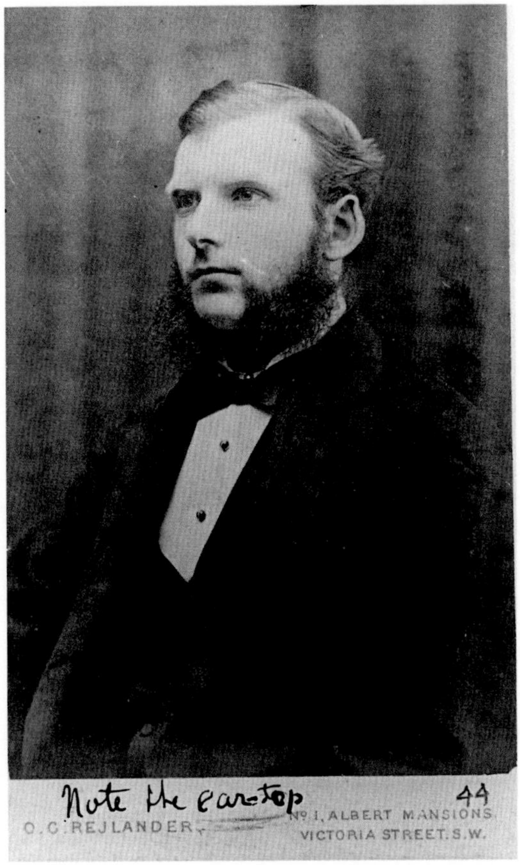

9-7 Oscar Rejlander, *Portrait of a Man, to Show the Unusual Shape of His Ears*, 1860s. Albumen print. Rejlander scanned his inventory to find sitters with ears like primates.

lage, Rejlander jokingly recalled his experience photographing various subjects, including the Duke of Wellington, who, according to him, had no discernible ear lobe. "I remember that the first time I saw his Grace and noticed his thin, pointed ear running into his cheek I said to myself—'There is a much better chance of your having a ring put into your nose than into your ear.'"[87] He went on to echo Darwin's arguments in *Descent*, concluding that ears have no obvious sexual appeal. "I have studied ears for nearly forty years, and although I know several persons with well-shaped ears, I know of only four persons who have *beautiful* ears."[88] As if to prove the point, he sent Darwin a picture of a man with an unusually shaped ear top; Darwin's response is not recorded (figure 9-7).

Opposite:
9-6 Oscar Rejlander, *Profile Portrait of Charles Darwin*, c. 1871–72. Albumen print. Darwin appears to have allowed his own ears to be studied for evolutionary clues.

Exploring Rejlander's Inventory

Of the sixty-four Rejlander photographs contained in the Darwin Archive, roughly a dozen appear to have been made before their collaboration. For example, an uncropped version of the image *Young Girl*

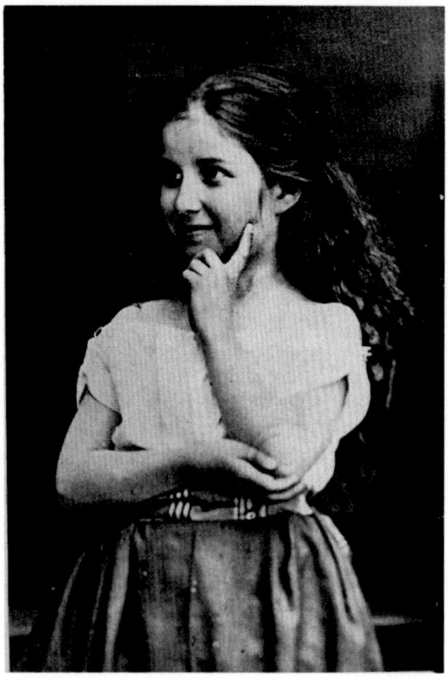

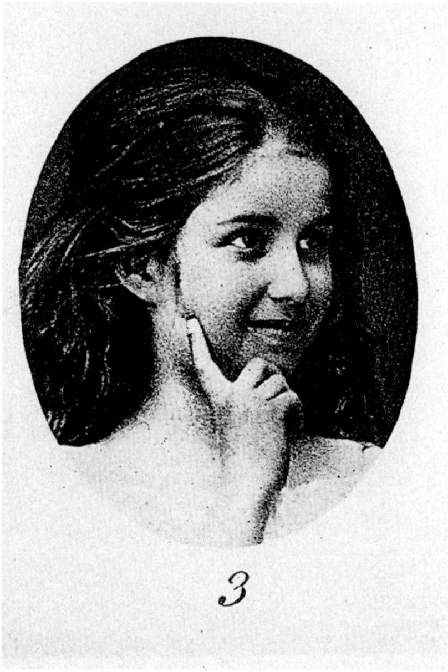

9-8 Oscar Rejlander, *Young Girl Standing with Hand to Mouth*, 1862. Albumen print. This photograph was already ten years old when Darwin received it.

9-9 Oscar Rejlander, Young Girl Standing with Hand to Mouth, 1862. Heliotype. Plate 3, figure 3 of *Expression*. The image as it appeared in the book.

Standing with Hand to Mouth (plate 3, figure 3 of *Expression*) is contained in an album of photographs by Rejlander at George Eastman House that is considered to date from 1862, nearly ten years before the two met[89] (figures 9-8, 9-9).

The Archive also contains a series of three pictures of the photographer's wife, Mary, comforting a crying baby.[90] This group is closely related to an image in the J. Paul Getty Museum called *Babylonian Song*, which, like the Eastman House image, is also considered to date from 1862[91] (figure 9-10). Although Mary Rejlander's apparent age in these photographs suggests that the actual date of this print may be slightly later, it is likely the series was made long before the *Expression* project began. The reverse of the mount is inscribed in pencil with a playful lullaby, "Nursery and baby, bus, bus, Baby and nursery, hush, hush." The Getty also holds an untitled portrait of the boy depicted in *Expression*, the same child seen suffering in the *Toothache* (plate 1, figures 4 and 6, and plate 2, figures 4–7). Like *Babylonian Song*, these were probably produced in the 1860s.[92]

The Darwin Archive contains numerous images gathered from Rejlander's stock. The quantity is impossible to determine, because the source of Rejlander's photographs is not always clear. Among these was a well-known photograph of the 1860s, called *She's Looking at Me, the Dear Creature*, a photograph that Wall described in his 1886 biography (figure 9-11). The image depicts an old man brimming

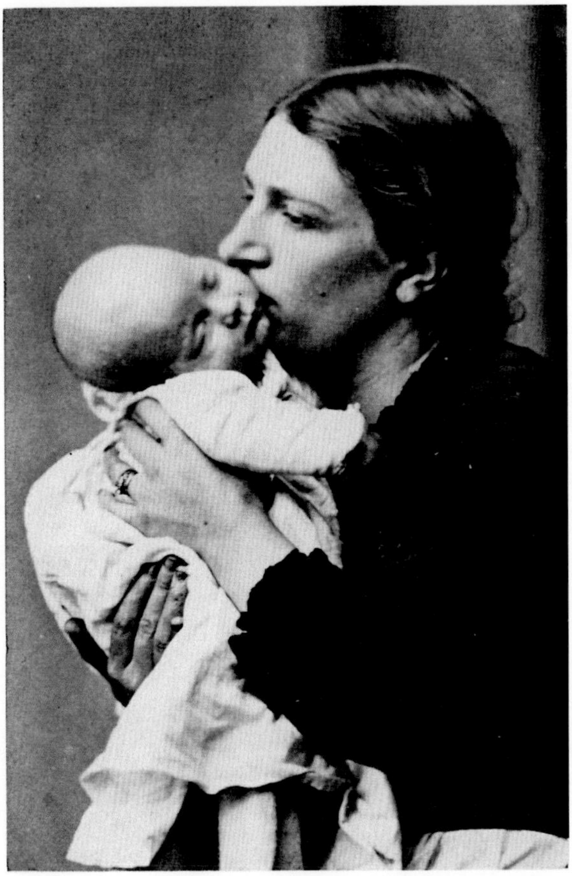

9-10 Oscar Rejlander, *Babylonian Song*, c. 1862. Albumen print. Rejlander sent Darwin three versions of this picture, which he sold with a nursery rhyme on the back. The woman is Mary Rejlander.

with pleasure after a pretty young woman has winked at him. As Wall described it:

> It represented a face of supreme ugliness at a window, resting upon and framed by huge hands, the elbows resting angularly upon its sill. A grin of unmistakable self-gratification and conceit, irradiating the phiz, added to its comicality, and made the story began by its title complete. The artist, in lighting his face, strongly illuminated it, and in the development had seemingly the same end in view, that of giving the greatest force to the facial features and markings. A dark background, which is not, however, a flat one, projects the head well forward, and adds power to the idea of the odd creature's seeking the admiration he regards is due his personal charms.[93]

Among other pictures that Darwin possessed from earlier in Rejlander's career are a photograph of a *Woman Wresting Herself from a Man's Lustful Advances* (figure 9-12), apparently a pastiche of William

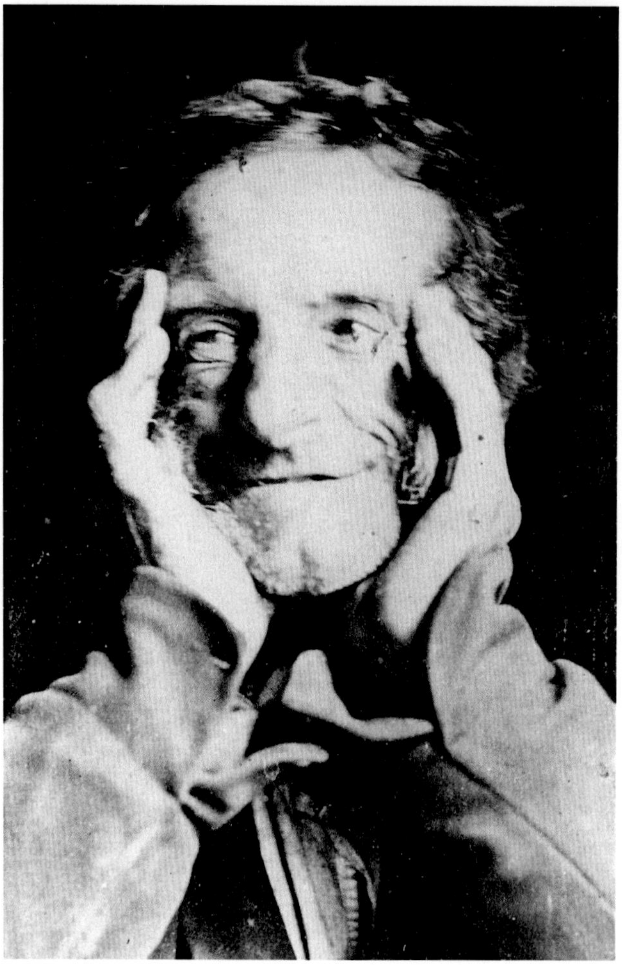

9-11 Oscar Rejlander, *She's Looking at Me, the Dear Creature*, c. 1862–68. Albumen print.

Hogarth's paintings *Before* and *After*;[94] the photograph *Who a' You?* (figure 9-13) portraying a young tough boy with fists raised ready to fight; and another of a child chewing absently on her pinafore.[95] Wall described the latter as a girl, "Half shy, half sly, and archly smiling, while biting the end of her pinafore, which she also unconsciously tugs at with both hands . . . a bit of real girl-life and feeling such as everybody has witnessed and can at once appreciate"[96] (figure 9-14). It is not clear to what extent Rejlander directed his sitters to pose in these pictures, but Darwin annotated Rejlander's *Lustful Advances*, "This was acting & so better not be engraved." Evidently he had considered using it.

In one of the photographs taken from Rejlander's inventory and reproduced in *Expression*, Darwin informed his readers of the narrative pretense on which it had been made. Describing plate 5, figure 1, he explained, "The accompanying photograph by Mr. Rejlander,

Opposite:
9-12 Oscar Rejlander, *Woman Wresting Herself from a Man's Lustful Advances*, c. 1860–65. Albumen print. Darwin considered using this photograph, which was probably inspired by William Hogarth.

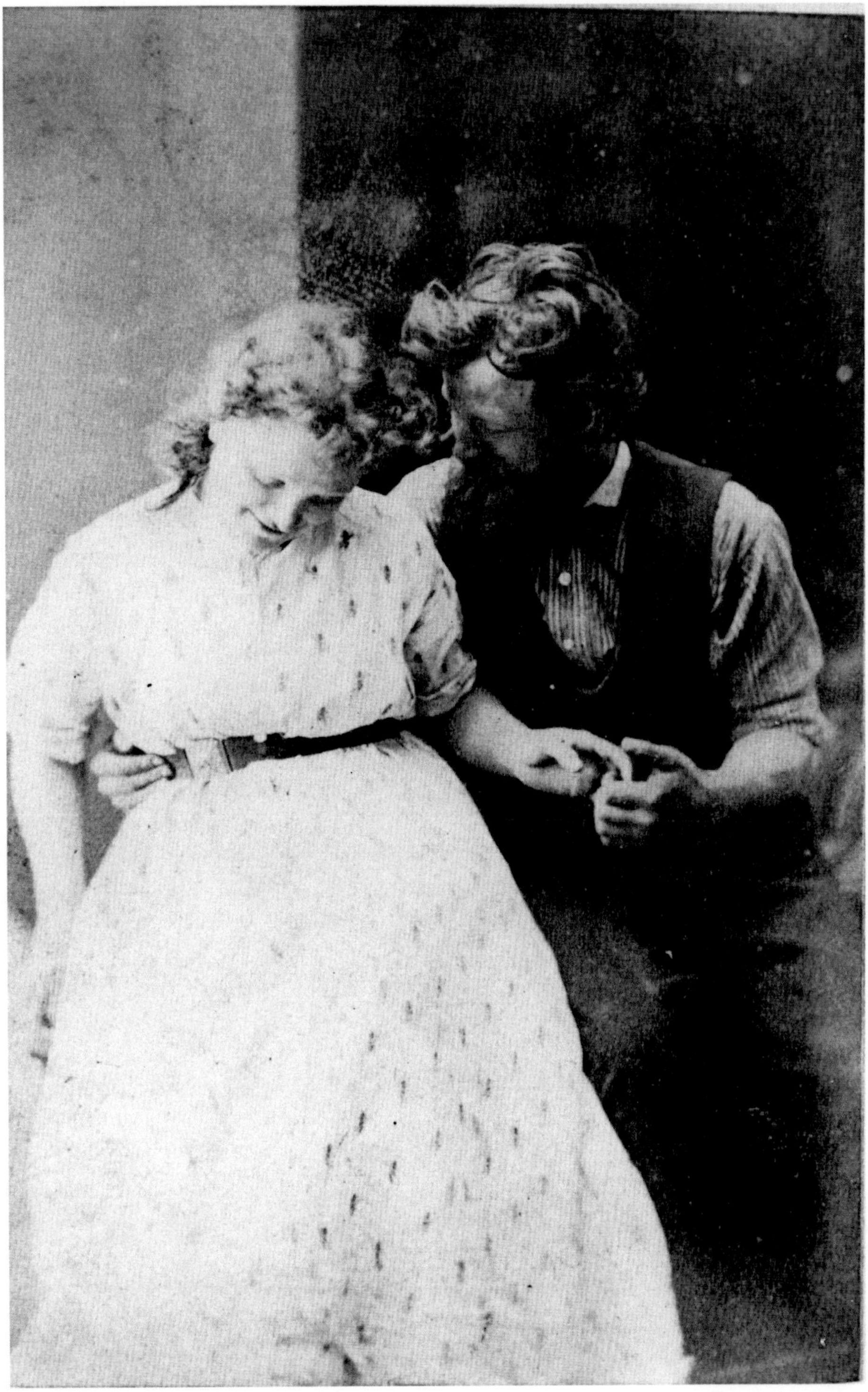

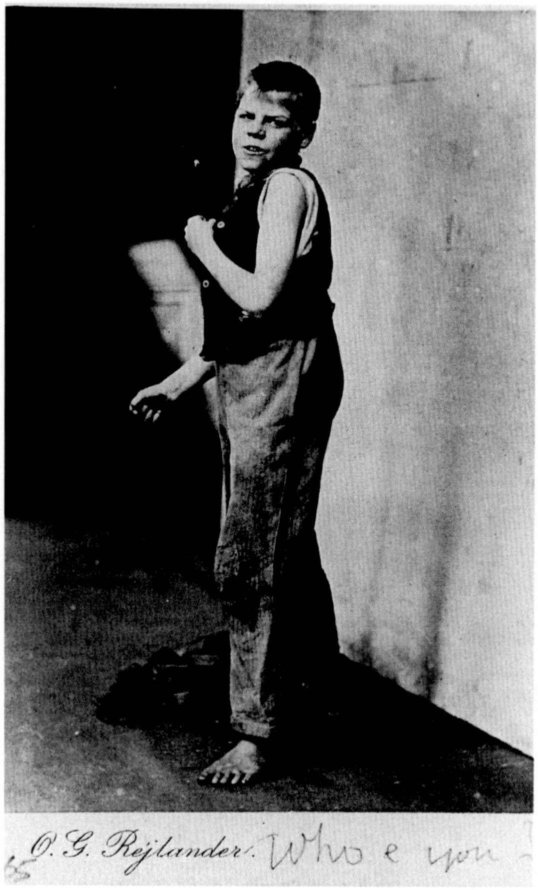

9-13 Oscar Rejlander, *Who a' You?* c. 1871 or earlier. Albumen print. A boy showing his survival instincts was especially interesting to Darwin.

shows . . . disdain. It represents a lady who is supposed to be destroying the photograph of a despised lover."[97] That the woman was reportedly tearing up a photograph, a curiously reflexive expression of scornfulness, would have had little bearing on the scientific value of the image. Nevertheless, the insertion of a simple storyline did provide context for the illustration that was not otherwise apparent. Intentionally or not, the explanation served to legitimize the image for his readers, while providing a poignant if subtle reinforcement of the prevalence of photographic imagery in modern life.

The presence of early prints by Rejlander in the Darwin archive confirms that, at least initially, Darwin approached him as a collector, much as he did with other photographers from whom he obtained photographs. He began by sifting through Rejlander's stocks, looking for pictures that revealed certain aspects of expression. There is evidence that Rejlander helped him to find appropriate prints: in a letter to Darwin in November 1871 he refers to "going through his instants" looking for pictures he might find useful, adding that he "had similar

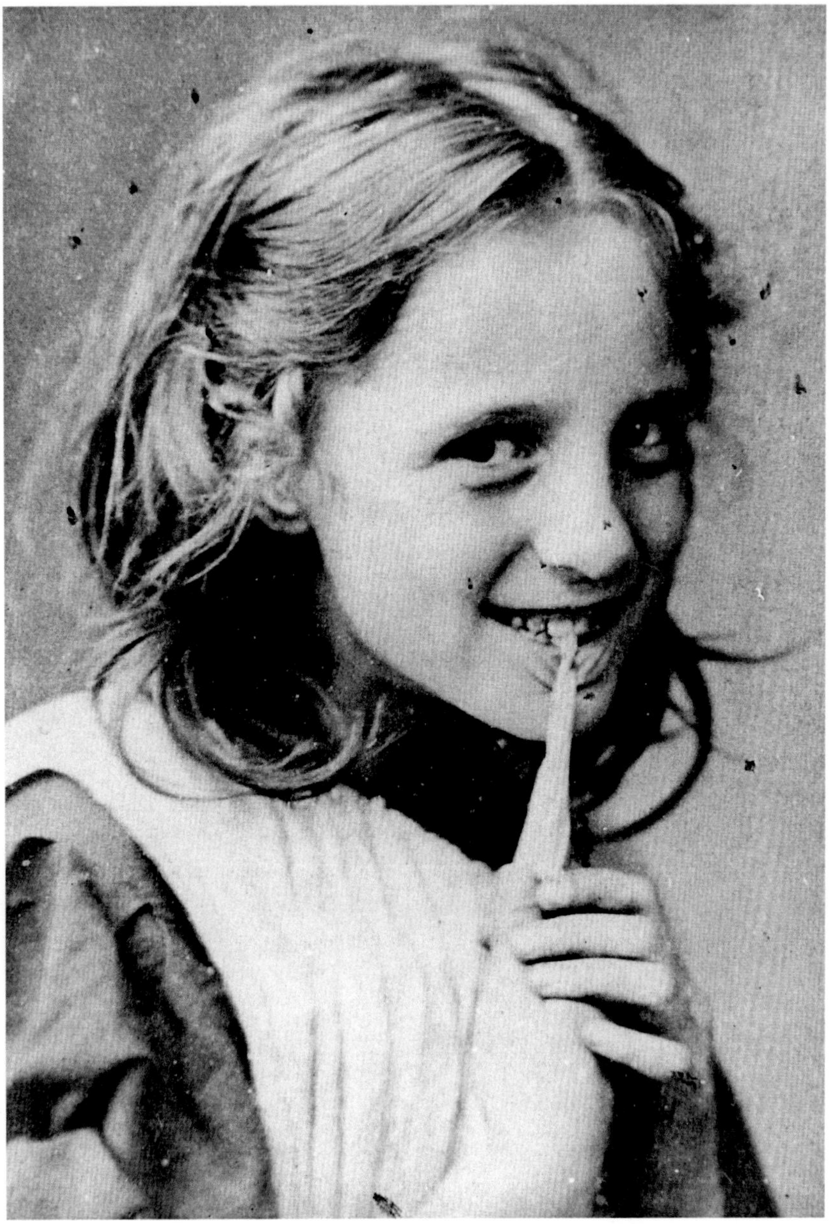

9-14 Oscar Rejlander, *Girl Chewing Her Pinafore*, c. 1871 or earlier. Albumen print. One commentator described this as "a bit of real girl-life and feeling."

inclinations with your own but have not had time to carry them out."[98] Though not specifically intended to illustrate particular expressions, the photographs Rejlander prepared from preexisting negatives were those he considered his most relevant. Such images, made according to Rejlander's understanding of facial expressions, reflected the photographer's artistic interpretations of expressive behavior.

Documents in the archive and elsewhere suggest that Darwin visited Rejlander's studio on at least three separate occasions and that the two men exchanged no fewer than four cycles of correspondence. Darwin's letter to James Crichton-Browne in April 1871, in which he describes himself as having "found" Rejlander, confirms one meeting, as does the anecdote recounted in *Expression* of the *Boy about to Burst Out Crying*, in which he describes himself bringing a photograph to Rejlander to find out by whom it had been made.[99] A further, unconfirmed encounter is described in a letter addressed to John Murray's assistant Robert Cooke in May 1872, in which Darwin wrote that he would arrange to visit Rejlander in six or seven weeks and would send negatives to the printer at that time.[100] They may have met on more than just these occasions, as their collaboration continued for more than a year, and when they did meet it is likely that it was at the Albert Mansions address. Further meetings are implied in their correspondence. In one letter, for example, Darwin refers to conversations he had with Rejlander regarding the suitability of the heliotype process for reproduction,[101] an issue that Darwin did not pursue until April 1872.[102]

The tone of the letters between Rejlander and Darwin suggests a congenial relationship. On biological issues the two saw broadly eye to eye. Thanking Darwin for the loan of two books early in their association, Rejlander wrote: "[I] cannot believe in the creation of an adult at any one time any more than of a plant jumping up full grown into existence." In the same letter he copied a section of one of these unidentified works that he considered particularly unbelievable: "'And the eyes were opened and the light did not affect them—and could at once discern all of the distances—and the gastric juice was in the stomach but did not corrode the lining—and the liver and kidneys were active on nothing—and the blood was there injected'—it's getting absurd!"[103]

Though he did not specifically endorse Darwin's theory of evolution by natural selection in these remarks, he did make clear his disdain for the creationist position. Unfortunately, only Rejlander's correspondence to Darwin has been preserved, as Rejlander's papers were dispersed after his death, and the letters Darwin sent to Rejlander have not been found.

Rejlander was a good-natured man, whose warmth and enthusiasm undoubtedly endeared him to Darwin. As a professional photographer he was helpful for his knowledge of human nature, honed by years of careful observation. Darwin probably viewed him much as he viewed the sculptor Thomas Woolner. Rejlander welcomed individuals of all walks of life and all temperaments to his studio. He was a keen student of human behavior, who prided himself on using photography to create mood and convey the personalities of his subjects. His considerable knowledge of the idiosyncrasies of expression, gathered in the field and tempered with experience, provided an invaluable resource to Darwin. As A.H. Wall described him, "Rejlander, with his fine *os frontis*, keen bright blue-grey eyes, and bearded chin: full of pantomimic action and genial cheerfulness, gives us information of value whenever he speaks."[104]

10

Rejlander's Performances

Posing for Darwin's pictures

Rejlander's inventory of photographs was an invaluable resource for Darwin, but it was not enough. Since 1869 or so, he had been struggling to find critical images to round out his book. His survey of museums, print shops, booksellers, and the accumulated wisdom of friends and associates failed to uncover the compelling images he needed to complete the project. A new strategy was called for, and Rejlander seemed just the man for the job. He was to provide new pictures, made to order, illustrating the things they discussed: crying, disgust, indignation, astonishment, and other emotions. For this to work, Rejlander would have to fully comprehend what Darwin was trying to achieve. And he would have to apply all his skills, pushing the limits of what could be obtained with slow, wet-plate collodion materials. Spontaneous, natural photographs would be ideal, but where that failed, expressions could be acted out. The creation of these pictures was an organic process, full of give and take. The history of the most famous photograph Rejlander created for Darwin, *Ginx's Baby,*is described in chapter 7. But many more images emerged from their association.

In 1871, Rejlander sent Darwin seven pages of notes on expression, titled "Odd Odds and Ends by O.G.R." The notes provide a rare glimpse into the workings of the Rejlander studio—because of its posthumous dispersal, few unpublished documents survive to describe its day-to-day operation. "Odd Odds and Ends" is the most extensive Rejlander manuscript to have been preserved (figure 10-1; see also appendix).[1]

Odd odds and ends by O. G. R. 1871

No. 1

I was teaching my Wife how to "develope" a collodion photogr. — saying "now pour this evenly over the plate — there — now move the plate gently up & down" — She moved her head to & fro but held the plate quite level — I saw her curls swaying + but the hand was still — so intent she was in listening that the connection betw. the hand and brain was suspended — in fact the brain rocked without communicating the action to the hand — She often talks in her sleep —

+ She stood so for nearly half a minute before I awoke her I was so interested &c

"He looked very sheepish" = some one baffled &c When to express that the exulting party puts his thumb to his nose with extended fingers to imitate a long face. ? (The origin of gestures)

Some rub the palms of their hands briskly together at any success in joke or otherwise.

Complacently stroking the beard raising the chin & lips

A searching look — half-shutting the eye lids fixing the eyes steadily on the object.

Indignation — challenging — Head erect, thrown back nostrils expanded — superciliary muscles projecting lips compressed — or if accomp. with contempt, the corners of the mouth drawn down — the attitude ready for combat —

"He looked as dropped from the Skye" — ~~"Crest fallen"~~. a vacant look, mouth open ~~Arm~~ hanging as lifeless at the sides — &c

— "Crest fallen" — A man sitting both hands on the knees, head drooping — jaw drooping — the eyelids extended they eyes rather diverging being fixed on no object —

"Odd Odds and Ends" reveals the depth of Rejlander's involvement in theoretical aspects of the *Expression* project and confirms his considerable visual acuity. It contains short observations on characteristics of expression. Some of these relate directly to behaviors Darwin addressed in his book, such as the hand and facial movements used to express particular emotional states. Others are sociolinguistic in nature, describing the ways in which certain states of mind can be consciously communicated. A final category of observations is more anecdotal, detailing incidents that Rejlander considered interesting from an evolutionary point of view. Darwin probably requested this information as part of his effort to gather facts about expression. He routinely solicited biological information of this sort from friends and correspondents, as he did with Thomas Woolner.

The notes are related in stream-of-consciousness fashion and are not organized thematically. Given the size and style of the document, it may be supposed that at least part originated from Rejlander's artistic daybooks (professional diary).[2] This would account for the loose structure of the notes, because daybooks would have been written chronologically. Keeping a professional diary was common practice at the time and would have been an ideal source for the type of information Darwin sought. If the notes were not copied from daybooks or other written notes, they would constitute a remarkable feat of memory, because they include nearly 100 separate observations on a wide range of subjects, including but not limited to expression. It is also possible that they were composed primarily from daybooks but supplemented with spontaneous recollections. Darwin may have asked Rejlander for his opinion on certain key problems, prompting Rejlander to consult both his records and his memory to formulate an answer.

There is evidence, for example, that Rejlander took notes on his conversations with Darwin. The back of one picture in the Darwin Archive (a Rejlander photograph titled *At the End of the Brighton Review*) contains pencil notations which were later erased and crossed out. Most of these are nevertheless still legible, reading "Uneasy / Hands over the head alarming / Hands in all sorts of / positions & under / armpits—sitting coyly / Laughing" and "so so / Sudden change of / behaviour—disgust / Sleep."[3] The clipped meter of the notes is similar to those in "Odd Odds and Ends," although the information there is more complete. The passage on alarm, for example, includes a more detailed description of the various actions associated with that emotional state:

> Astonishment—simple—head & neck extended mouth open as if the eyes & ears were not sufficient to receive the impress.
> Astonishment with wonder—hands showing the palms—raised the

Opposite:
10-1 "Odd Odds and Ends" by O.G.R., 1871. Manuscript page. Rejlander sent Darwin seven pages of handwritten notes on expression (see appendix for transcription).

eyes, wide open eyebrows arched, the mouth open but rounded for the interjection Oh. head thrown back——

Astonishment—agreeable—hands raised as before, eyes dilated eye-brows raised head rather on one side slightly bent and a suppressed smile.[4]

Earlier in the document he advised: "The hands!!! Raised over the head imploring. Palms out or down at the sides expostulating—extended horizontally—open to receive."[5]

Other remarks in "Odd Odds and Ends" echo the notes on the back of the archive photograph. For example, there are numerous observations on sleep, such as the case of a certain "Mrs. W.," who is said to have seen her husband writing upside-down while he was sleepwalking.[6]

Many of the notes parallel observations reported in *Expression*, especially the muscle movements associated with common expressions. Several correspond to Rejlander photographs published in the book. Rejlander's description of "indignation," for example, relates to plate 1, figures 1 and 2: "Indignation—challenging—head erect, thrown back nostrils expanded—superciliary [*sic*] muscles projecting—lips compressed—or if accomp. with contempt the corners of the mouth drawn down—the attitude ready for combat."[7] It is not known if the notes and corresponding photographs were made simultaneously, or if they were made before, or as a result of, discussions with Darwin. However, they do confirm that their interaction was, qualitatively, similar to Darwin's interactions with the draftsmen who contributed to the project (see chapter 8). At the same time that he directed Rejlander to produce photographs of certain subjects and in particular ways, he also seemed to have viewed him as a collaborator.

Darwin reviewed "Odd Odds and Ends" and highlighted five sections relevant to his research in colored pencil. The marked passages describe particularities of various expressive behaviors:

"The pouting child both lips out if angry—as well—the underlip only if sad, called 'the dripping-pan'" (page 3)

"speaking with the teeth clenched in a growl-like voice—(male)" (page 3)

"Forefinger—upright—perpendicular = warning" (page 3)

"An Uneasy cough—or cough to show unwanted / indifference" (page 4)

"A Sneer—a disguised contempt accompanied with an awkward forced smile on one side of the face" (page 7)

These passages reflect issues of particular concern to Darwin, such as coughing, which he described in *Expression* as a "familiar example" of a reflex action in which several groups of seemingly unrelated muscles act nearly simultaneously,[8] or clenching the teeth in anger, about which he explained, "the appearance is as if the teeth were uncovered, ready for seizing or tearing an enemy, though there may be no intention of acting in this way."[9] It was in this context that Darwin cited an unidentified plate by William Hogarth in his text, almost certainly plate 1 of the series *Rake's Progress* (1735) (see figure 1-4).

There are numerous parallels between the sections Darwin highlighted in "Odd Odds and Ends" and the photographs Rejlander supplied. Darwin considered pouting of the lips when sad to be characteristic of young children, and included an illustration of the "dripping-pan" (protrusion of the lower lip when pouting) described in Rejlander's notes in plate 2, figures 5 and 7 of *Expression* (see figures 9-5, 10-8). An inscription on the back of a variant of these images, in Rejlander's hand, testifies to its veracity: "'Buttons' had slapped his face. I was just about taking a portrait but put this boy up instead. This . . . was a genuine pout with a sob. . . . Tho' moving it has a more genuine expression from within—or involuntary—the pout with both lips are more of a sulk or anger. In a genuine pout there is an additional muscle visible under the lips."[10] This observation is particularly interesting because it reflects Darwin's concern to differentiate muscle movements associated with spontaneous expressions from those that occur when an expression is artificially posed. As the psychologist Paul Ekman has noted, clinical studies have shown that real smiles, for example, involve the simultaneous contraction of two facial muscles,[11] while simulated ones may be distinguished by the disproportionate contraction of one, with little or no contraction of the other.[12]

Similarly, Rejlander's observation on the asymmetrical nature of sneering, an awkward smile on one side of the face, is represented as plate 4, figure 1 of *Expression*. Darwin described the subject as "a lady, who sometimes unintentionally displays the canine on one side, and who can do so voluntarily with unusual distinctness"[13] (figure 10-2). Modern clinical evidence corroborates this observation. Ekman and others have found that voluntary expressions do tend to be less symmetrical than involuntary ones.[14] Darwin's commentary makes this distinction clear, implying but not specifying that the photograph represents a simulated expression.

From Rejlander's perspective, however, the photograph was a private joke; the subject was his wife, Mary. According to A.H. Wall,

> Rejlander frequently made her [Mary] his model, and he could not well have had a more graceful one, or a face more mobile in expressing various emotions. He used to watch it very attentively. I have often

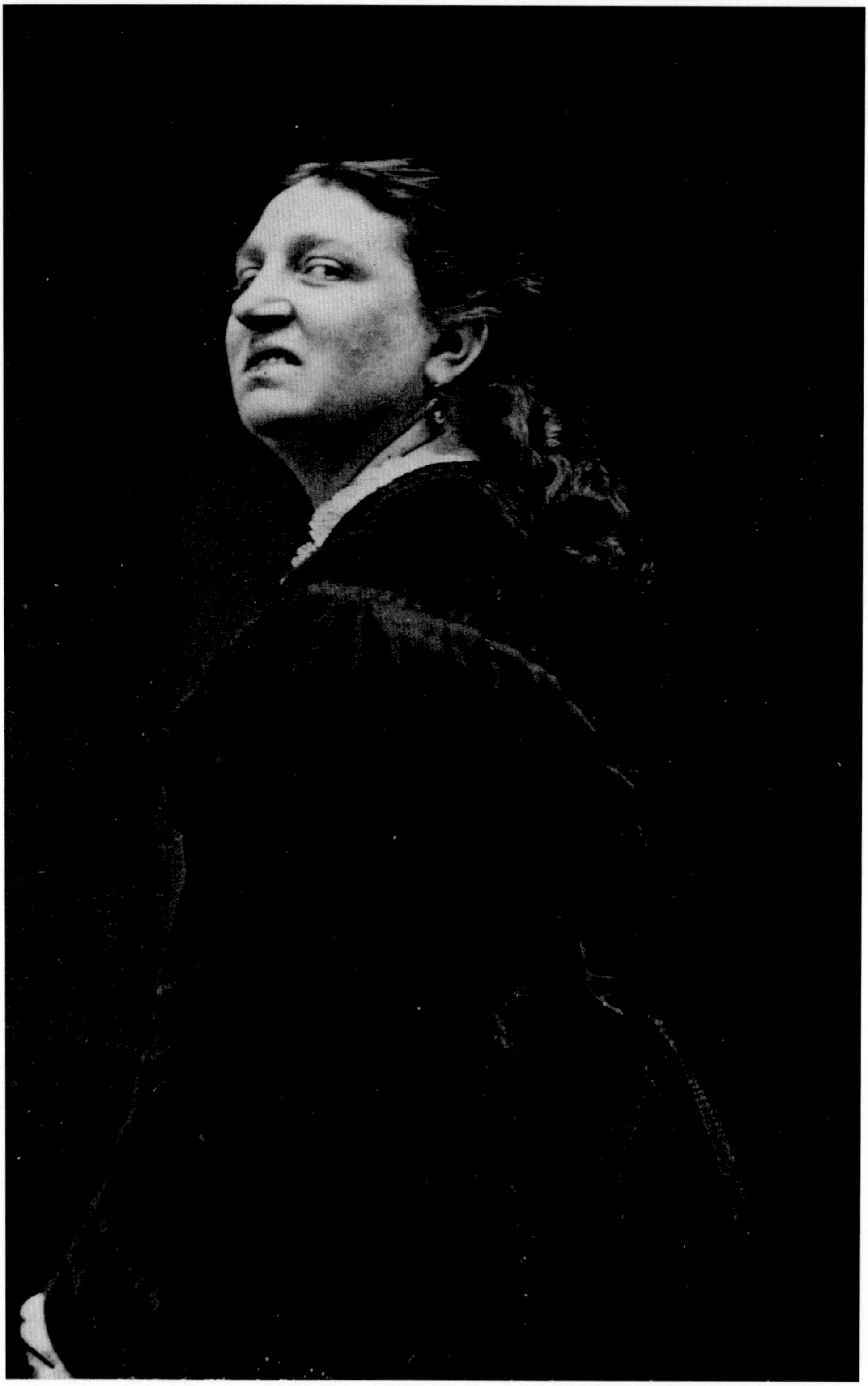

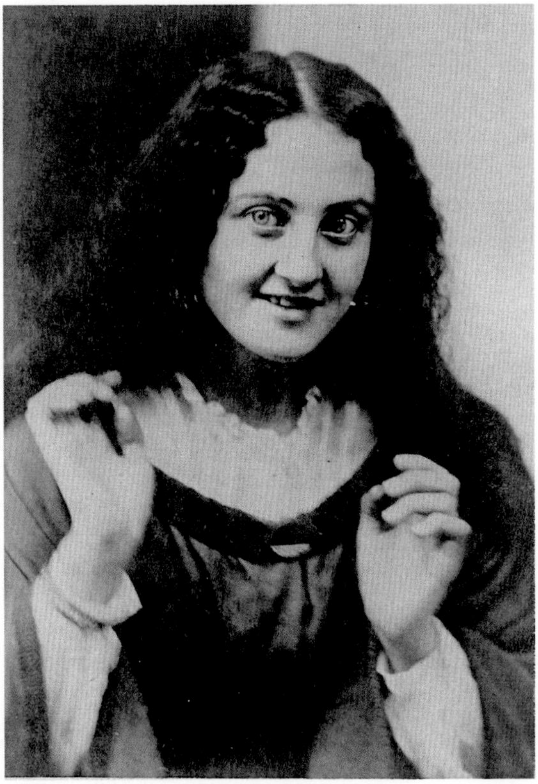

10-3 Oscar Rejlander, *Surprise* (Mary Rejlander), 1864–69. Albumen print. Mary joined Oscar in dramatizing expressions.

> seen him doing so, and when the conversation called up some expression more than usually forcible or eloquent, would make a mental note of the subject of conversation and its treatment, in order that he might hark back to it on some occasion when the camera was *in situ* and a new subject wanted. . . . This particular face was not, however, usually changeful.[15]

Rejlander sent Darwin at least one other photograph of his wife from his catalogue, showing her acting surprised (figure 10-3). It was apparently taken in the same sitting as a variant portrait of Mary in the Harry Ransom Center of the University of Texas at Austin, which has been dated from 1864–69.[16]

Much of the information in "Odd Odds and Ends" was anecdotal. Rejlander was, after all, not a scientist, and he freely mixed useful information with hearsay about strange facts and curiosities. He included dubious stories, such as that relayed to him by one Mr. Dawson of King's College, who was said to have seen a child born with a human head and a dog's body, complete with wagging tail.[17] The notes also contain descriptions about a dog that could speak its master's name, a barn-door cock with two horns, and a hunch-backed, club-footed child prone to squinting.[18] Rejlander also made cultural observations,

Opposite:
10-2 Oscar Rejlander, *Sneering* (Mary Rejlander), 1871–72. Albumen print. A private joke: Mary expressing disdain for her husband. The lower half of the image was cropped for publication.

for example, that while the Scotch are likely to say "Hm! Hm!" in conversation, Danes are more likely to say "So So," with a corresponding sucking noise. Furthermore, while the English normally say "aah!" Swedes favor, "jah! Jah-ah?"[19]

Like Darwin, Rejlander drew on his experience with friends and family in considering the elements of expression. For example, the first observation in "Odd Odds and Ends" describes teaching his wife how to develop collodion photographs:

> I was . . . saying "now pour this evenly over the plate — there — now move the plate gently to & fro but hold the plate quite level" — I saw her curls swaying but her hand was still — so intent was she in listening that the connection between the hand and brain was suspended — in fact, the brain rocked without communicating the action to the hand. (She stood so for nearly half a minute or so before I awoke her, I was so interested.) She often talks in her sleep.[20]

Although it has long been assumed that Mary Rejlander was active in the production of photographs attributed to her husband, this passage provides the first firm evidence that she was in fact trained in photographic techniques. This raises the intriguing possibility that the pictures of expression for which Oscar Rejlander posed could have been photographed by Mary Rejlander. In the absence of a cable release or a mechanically timed shutter, he would have relied on an assistant to aim the camera and cap the lens. Rejlander had no other known assistants; thus, it may be argued that the photographs depicting Oscar Rejlander that appear in *Expression* should be attributed not to the artist himself but at least in part to his wife.

Rejlander as Specimen

Oscar Rejlander posed for twenty-two of the photographs of expression preserved in the Darwin archive.[21] A number of these were made using a stereo camera, so in those cases there are two related images of the same expression. Nearly all correspond closely to published plates in the book, although some are rejected variants of images actually used. For example, the archive contains four self-portraits illustrating astonishment, one of which was published as plate 7, figure 1 of *Expression*. The other three were bundled in an envelope together with a photograph of the same expression posed by a different sitter, and marked in Darwin's hand, "Astonishment, *not* for engraving."[22]

Rejlander is the subject of six of the photographs published in *Expression*, in plates 5–7. These are three-quarter length portraits showing him in partial profile simulating fear, disgust, indignation, indifference, and surprise. He is dressed in a velvet smoking jacket and, in some cases, a thick felt hat. Darwin acknowledged in the text that these were simulations, describing the photographs on plate 6, for example, as "successful acting."[23] In a letter to Darwin in April 1871, Rejlander explained that practical reasons had prompted him to pose for the pictures himself: "It is very difficult to get, at will—those expressions you wish—Few have the command or imagination to appear real. In time I might catch some. So I have tried *in propria persona*—even cut my moustache shorter to try to please you in the last—ha hah."[24]

This was not the first time he had used himself as a model. In 1863, he confessed to using photographs of himself to produce his popular portrait of the Italian revolutionary Giuseppe Garibaldi: "I was rudely taken to task. It was intimated that I had paid my model 1s. 6d. per hour, when I myself was the model, and never realised 6d. per hour!"[25]

Unfortunately, there is no way to know exactly how Rejlander's commission was structured and to what extent Darwin directed him to make photographs in particular ways. Although it can be surmised that Darwin did provide some guidance, it is unclear whether Rejlander developed his own interpretations. Rejlander's report to Darwin that he had decided to photograph "in propria persona" reveals that it was he, at least, who chose the subjects. Rejlander was encouraged to express his opinions about the elements of expression, as the notes in "Odd Odds and Ends" confirm.

Rejlander was, in any case, an excellent subject. His balding pate afforded an unusually good view of the actions of *occipito frontalis*, the large, bilateral muscle stretching from the eyebrow region to the top of the head (figure 10-4). This was particularly useful for illustrating expressions in which the forehead is strongly implicated, such as disgust, resignation, and indignation, all of which Rejlander acted for the camera. This simple fact, curiously ignored by scholars who have written about Rejlander's expressive self-portraits, may have been a deciding factor in the decision to use himself as the subject of numerous illustrations. Depictions of baldness had only become common in European art earlier in the nineteenth century, propelled in part by the phrenological theories of Johann Caspar Lavater and his school. The Romantic painter Benjamin Robert Haydon (1786–1846) was among those who celebrated baldness as a window to cranial shape and would almost certainly have appreciated Rejlander's prominent, unencumbered brow. "A large, broad, and powerful frontal bone," he wrote, "is essentially an intellectual characteristic for a standard human head, and will be found in all the most elevated and celebrated human beings."[26]

In addition to the self-portraits Rejlander sent, two other adult subjects appeared. One was his wife Mary. The other, who has not been

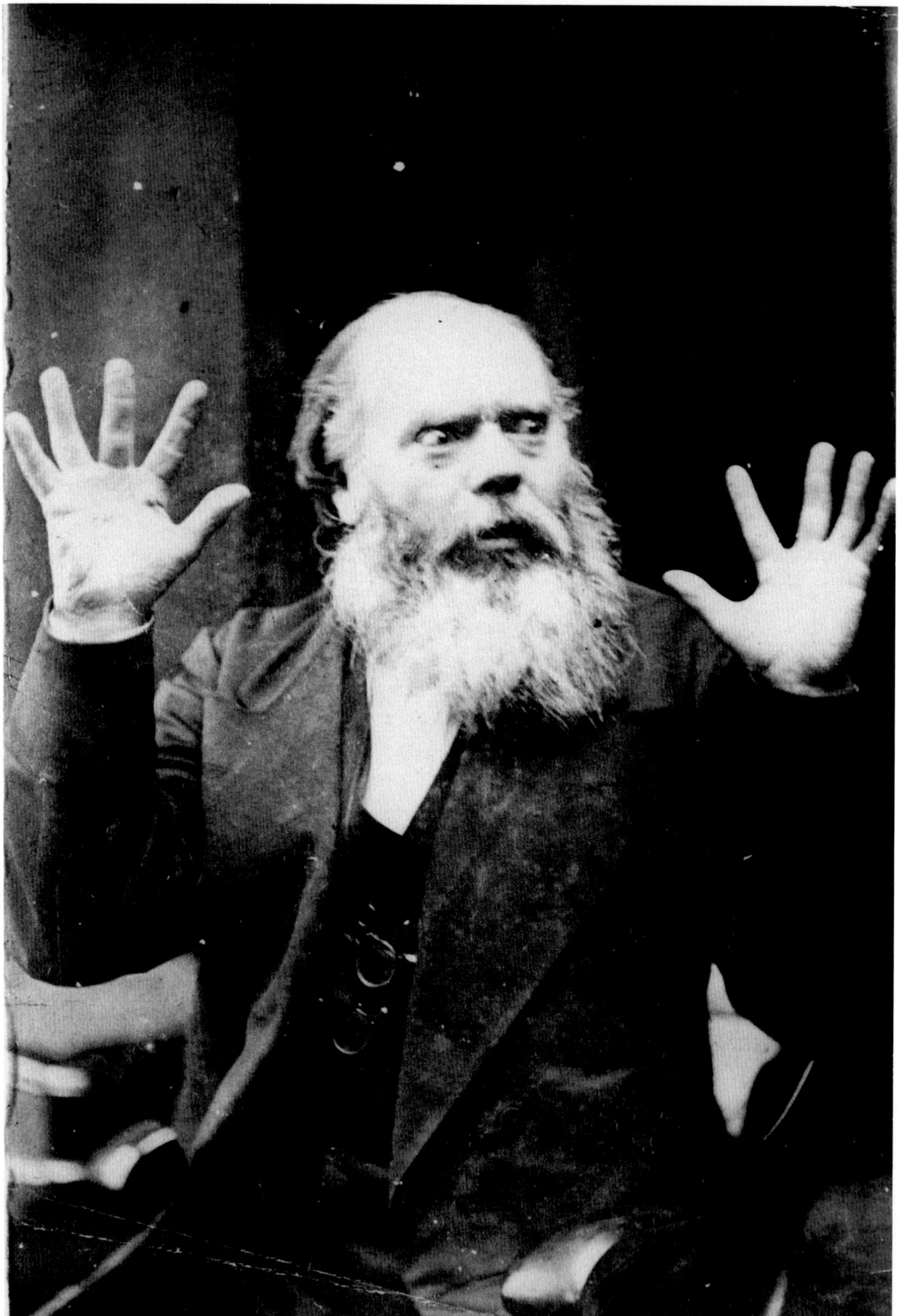

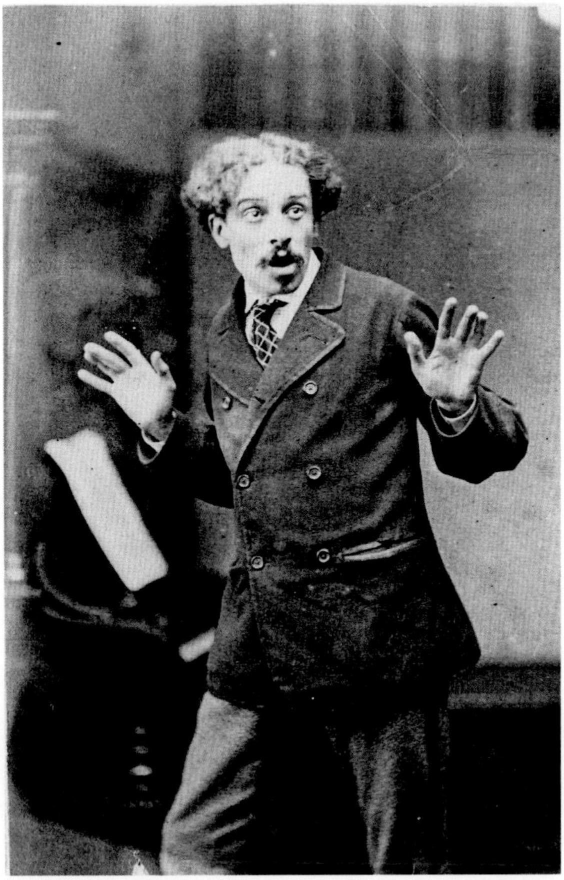

10-5 Oscar Rejlander, *Astonishment*, 1871–72. Albumen print. A younger man posed for some of the shots.

identified, is a younger man who appears only once in the book[27] (plate 6, figure 1). He was not one of Rejlander's usual models. Darwin seems to have been unimpressed with his thespian talents. On the back of one image depicting the man, he wrote, "Acting—seems rather more like start than fear"[28] (figure 10-5).

One of the rejected images is marked "Excellent,"[29] but the published variant is annotated, "This is the best on account of hands" and "For astonished hands."[30] Such notations reveal the conflicting standards Darwin used to evaluate his photographs. He was concerned that they should be genuine and that they should clearly illustrate certain behaviors. Mindful of the small scale of the photographs and the limited resolution of the heliotype process he would use to reproduce them, Darwin selected images in which expressive features were extremely clear and at times even exaggerated. So, while he rejected one of the photographs of astonishment as "acting," he accepted the other for emphasizing the placement of the hands.

Opposite:
10-4 Oscar Rejlander, *Astonishment*, 1871–72. Albumen print.

The tension between these competing criteria is evident throughout Rejlander's self-portraits. He was required to strike a balance between accurate portrayal and illustrative emphasis. For example, on the back of one of his self-portraits representing disgust, he inscribed, "This one I sent before. I think the expression is *telling*—the one on the other side will help—by its exaggeration—to qualify *this* one."[31] Darwin accepted Rejlander's recommendation, publishing the image as plate 5, figure 3. Together with plate 5, figure 2, it is one of two Rejlander self-portraits of similar expressions included on that plate.

The reason Darwin included two photographs of the same subject adopting a single expression is not explained in the text. He may have thought that the exaggerated nature of plate 5, figure 2 rendered the image suspect, and that a second, corroborative photograph was necessary to reinforce it. Oddly, in figure 3 Rejlander posed wearing his felt hat, whereas in figure 2 he posed without it. The hat effectively deemphasized the muscles of the forehead region and may help to explain its inclusion. But it is evident the two photographs were not meant as strict equivalents, because the positions of the hands and the degree of contraction of the facial muscles differ in the two pictures. They may represent different degrees of the same expression, a possibility not elaborated in the text.

The photographs for which Rejlander posed were taken in at least two separate sittings. In some he appears relatively unkempt, with hair that looks flat and oily (figures 10-6, 10-7). In others, presumably taken later, his hair is presented more stylishly; these are the versions Darwin decided to publish. The arrangement of the hair was not a trivial matter; given the long exposure times required for these photographs, it is unlikely the energized, tussled appearance of the hair in *Indignation* (plate 6, figure 2), for example, resulted from shaking or tossing of the head. Instead, the hair would have been arranged to look dynamic, possibly even stiffened with pomade to provide an illusion of movement.

One of the striking innovations Rejlander employed to produce Darwin's photographs of expression is evident in the archive self-portraits: the use of separate, sequential exposures of a stereo camera to create two images without having to reload his camera. Rejlander did not embrace the vogue for stereo views that blossomed in the 1860s and 1870s and is not known to have used a stereo camera outside of this project. In part he may have been deterred by the slim profit these views obtained, because it was a high-volume, low-margin business. Moreover, those images originally made with a monocular camera (composite prints, e.g.) could be turned into stereos only by artificially adjusting the angle of view to create the offset, three-dimensional effect stereos are known for. The process was troublesome, and the results unsatisfactory.

Opposite:
10-6 Oscar Rejlander, *Indignation*, 1871–72. Albumen print. Darwin rejected this photograph, perhaps because of the appearance of the hair.

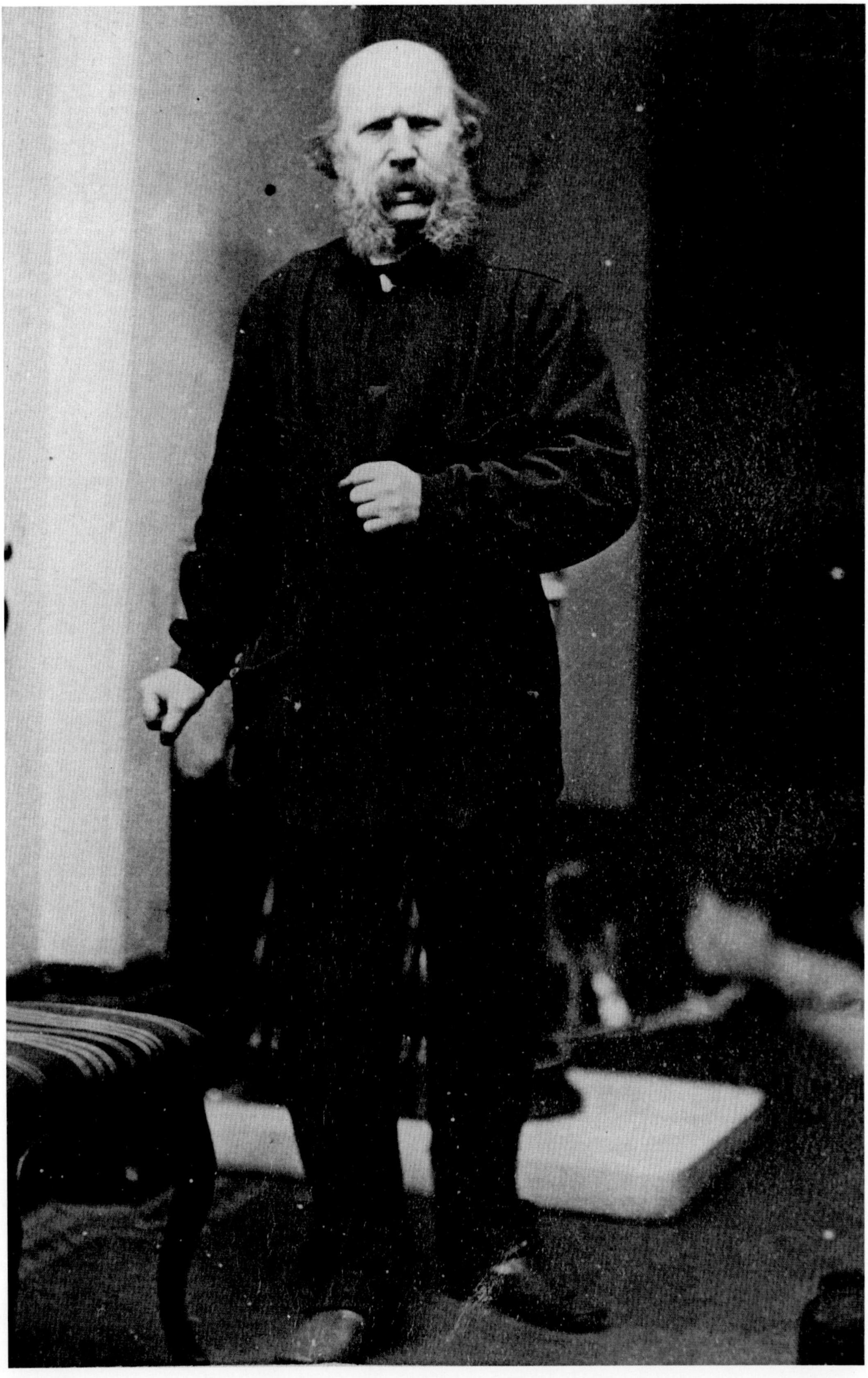

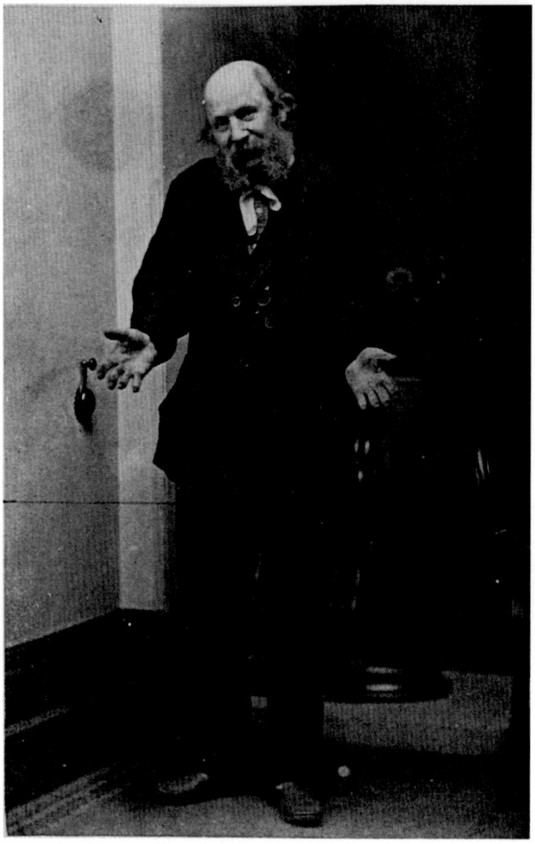

10-7 Oscar Rejlander, *Shrugging the Shoulders*, 1871–72. Albumen print.

Even when he did use the stereo camera for Darwin, Rejlander did not use it the way it was designed. In the photographs he sent Darwin, the camera was adapted, covering one lens and then the other to create separate exposures on a single plate. In a number of the images, including four depicting *Indignation* and *Disgust*,[32] stereo plate marks are clearly visible. The published versions of these images in *Expression* (plate 2, figures 2 and 3) are variants of these known stereo views, as are several related, trimmed versions. Because these were almost certainly taken in the same sitting as the known stereos, it is likely that most if not all of the Rejlander self-portraits were made using this unusual method. Rejlander also used the adapted stereo procedure to produce a sequence of six photographs of at least one other subject, the pouting boy in profile in plate 2, figures 5 and 7 of *Expression*[33] (figure 10-8).

Sequential imagery was not unknown in photography at the time. For example, the inventor François Willème's system produced "photosculptures" in the round, by translating a series of staggered profile photographs into realistic plaster busts.[34] This was the type of sculpture Francis Turner Palgrave had condemned to Thomas Woolner (see chap-

Opposite:
10-8 Oscar Rejlander, *Pouting Boy* (profile), 1871–72. Albumen prints. By covering one lens of his stereo camera and then the other, Rejlander improvised a primitive movie camera.

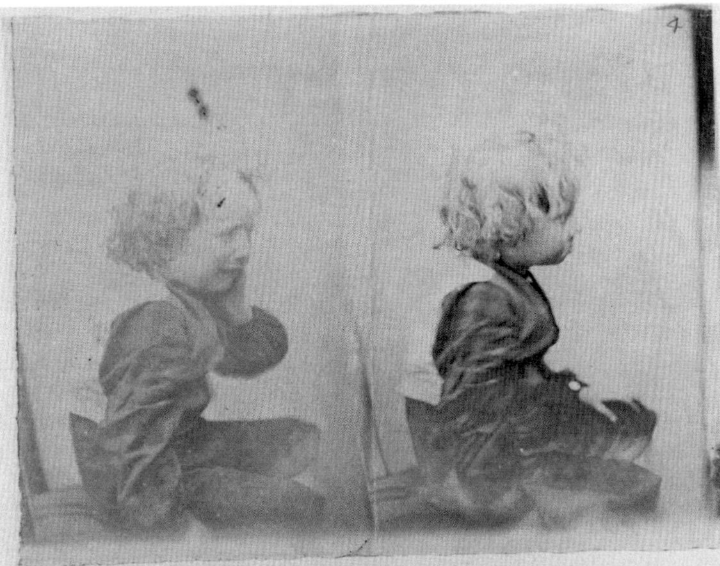
4

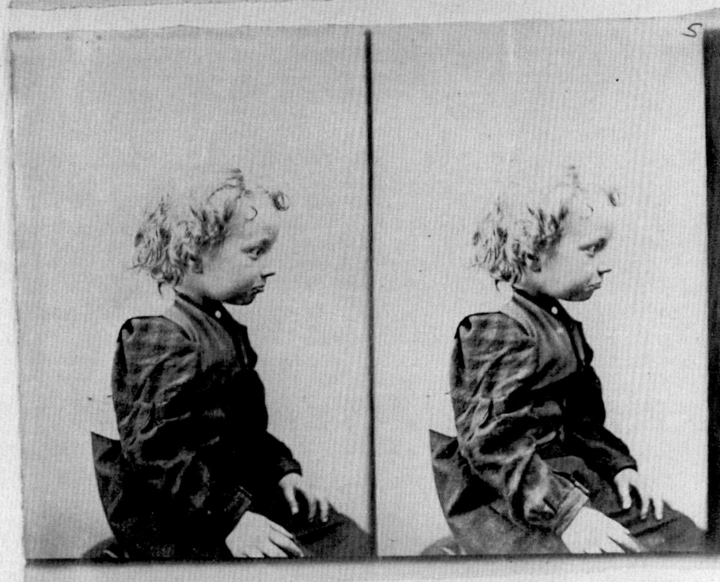
5

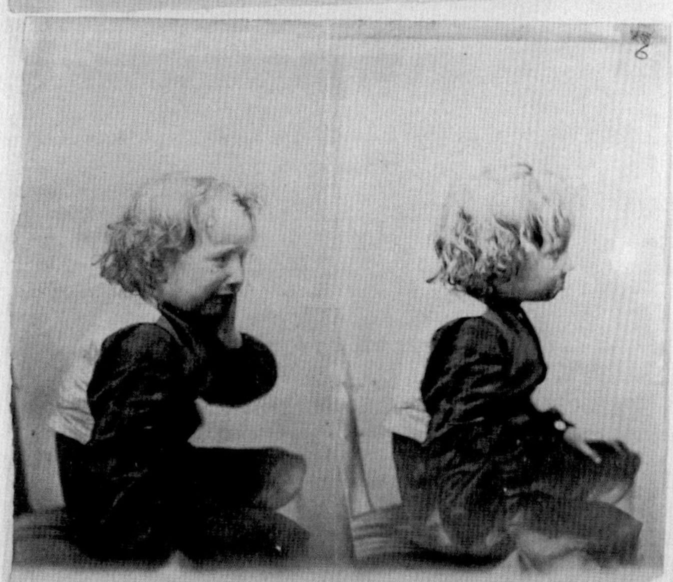
6

ter 8). However, while Willème's technique involved sequential exposure through space, Rejlander's is the first known example of sequential exposure through time. Using his adapted stereo camera, he was able to isolate fragments of activity unobtainable using a conventional wet-plate camera.

There is no evidence that the innovation of Rejlander's adapted stereo images, which were not published in their sequential form, was recognized by other photographers. However, Rejlander did publish an article "On Photographing Horses" in the *British Journal of Photography* (1873), the year after *Expression* was published, in which he described a related scheme. He proposed to have "a battery of cameras and 'quick-acting' lenses ready charged and loaded. The signal given, the rider starts some distance off the focussed point, and at the moment of passing it—bang! With a strong wrist and sleight of hand, the exposure and covering is done—not 'in a trice' but at once."[35]

This recipe for producing sequential images of horses bears striking resemblance to that adopted by noted expatriate British photographer Eadweard Muybridge, who had begun his experiments with horses in motion in Sacramento, California, the previous year, but switched to using a battery of cameras at approximately the same time as Rejlander's article.[36] As a result, one might reasonably contend that Muybridge, frequently identified as a key contributor to the development of cinematography, shared a direct lineage with the *Expression* project.

While Rejlander's photographs for *Expression* set new standards in scientific photography, they were produced without strictly adhering to mechanical objectivity. Rejlander believed manipulation was needed to produce convincing illustrations for the book. By posing for some of the illustrations himself, he was able to control the behaviors he sought. In this his approach was not unlike that of Guillaume-Benjamin Duchenne de Boulogne with his *Mécanisme* photographs, in which electrical stimulation was used to induce expression. Like Duchenne's, Rejlander's photographs contain a blend of contrived and genuine expressions.

The pictures Rejlander produced for Darwin were among his last works. Enfeebled and anticipating his death, he penned a final message to Darwin in December 1874. He explained that he had arranged to have a taxidermist prepare a tableau of stuffed birds for him. "I dare say I shall not see you again," he wrote, "but I have a good thing to give you: A group of sparrows, i.e. a hen with a little one on either side—with fluttering wings—feeding—one at a time. I saw it 15 years ago. I sketched it then, but have this year got it represented like nature."[37] A month after writing this message he died of Bright's disease, a kidney disorder that plagued him throughout the last years of his life.

In the rest of the letter Rejlander seems delirious, and it is not clear if the birds he described actually existed. The promise of a gift still served as an affectionate farewell, marking the end of an extraordinary collaboration.

A newcomer to scientific inquiry, Rejlander had managed to produce Darwin's coveted photographs of emotional expression. The methods he and Darwin employed were made necessary by the otherwise insurmountable obstacles presented by wet-plate collodion photography. The empirical ambivalence with which the photographs were presented resulted not from deceitful intentions, but from the conceptual vacuum in which they were produced. The *Art Journal* eulogized,

> Late years have shown that more can be done than we at one time thought possible, and that results are obtainable from lens and camera, which are not merely imitations and copies from still nature, but productions of mind and thoughtful study, and which, when gazed on, raise emotions and feelings similar to those awakened at the sight of some noble sepia sketch, the handiwork of a good draughtsman. Of Mr. Rejlander's pictures (for such we may justly call them) we have no hesitation in saying that they are full of beauty and full of mind.[38]

11

Alice, Eugenics, and the Spirit World

The aftermath of Darwin's experiments

After *Expression* went to press, Darwin retreated to more comfortable ground. While his final scientific books are recognized as important in their respective fields, they lacked the metaphysical punch of *Origin*, *Descent*, and *Expression*. In his remaining years he turned his attention to the garden, which he made into a private botanical laboratory. In 1875 he published *Insectivorous Plants*; in 1876 *Effects of Cross and Self Fertilisation in the Vegetable Kingdom*; in 1877 *Different Forms of Flowers on Plants of the Same Species*; in 1880 *On the Power of Movement in Plants*; and finally, in 1881, *The Formation of Vegetable Mould through the Action of Worms*. *Expression* was to remain his most avant-garde work.

Scientifically, *Expression* pushed evolution to its limits, suggesting that human thoughts and feelings were, like fingers and toes, the result of evolutionary pressures. Artistically, it also represented an extreme. Darwin would never again publish a book or paper with original photographic illustrations. Nor would he produce a book in which original illustrations played such a vital role. *Insectivorous Plants* and *Forms of Flowers* contain black-and-white wood engravings of flowers, some based on beautiful, colored drawings in Darwin's possession (now preserved in the Darwin Archive). *Cross and Self Fertilization* was illustrated with a single diagram. *Movement* includes 196 figures, nearly all plots of plant movements calling to mind the motion studies of the French physiologist Etienne-Jules Marey. It was left to others to extend the lessons of Darwin's revolutionary photographs.

Expression had an enduring influence on countless authors, pro- and anti-Darwinian in their outlook. Oscar Rejlander's photographs became a benchmark of nineteenth-century scientific photography, as scientists of various disciplines, artists, photographers, and book designers absorbed them. Art historians, too, fell under their spell. The influential British art historian Aby Warburg (1866–1929), for example, read *Expression* as a student in Florence in 1888 and cited it as an influence throughout his career.[1] Warburg's copy of the book, now at the Warburg Institute in London, is inscribed, "Finally a book that helps me!" [*Endlich ein Buch das mir hilft!*].[2] Warburg saw *Expression* as enlarging art history to include the transmission of emotion in pictures. The quality of "pathos," as he described it, the evocativeness of art, became a linchpin to his theories of art. Warburg passed his love for the book to the German art philosopher Ernst Cassirer (1874–1945), who incorporated Darwin's pictorial ideas into his *Philosophy of Symbolic Forms* (1923–29).[3]

During Darwin's lifetime, *Expression* enjoyed mixed success. John Murray had warned Darwin that photographic illustrations could hurt profits. The cost did, in fact, prove damaging to sales. *Expression* sold well in its first few months but crested quickly. In 1878, six years after publication, records at the Murray firm indicate that only 172 copies were sold in England. This dropped to 96 in 1879 and rose to 133 and 167 in the following two years.[4] By 1882, Murray had 1,180 copies on hand; at that rate, enough to satisfy demand for the next eight to ten years. Cost was certainly a factor in this precipitous decline, but there may have been a more compelling reason. The illustrations that in 1872 gave the book a cutting-edge look now seemed dated; they were not as sharp and clear as photographs had grown to be. The color was also wrong: Darwin's heliotypes were brown in tint, while newer photographs were purple. The figures were dressed in clothes that were out of fashion; Rejlander, with his velvet smoking jacket, looked outmoded. Older pictures, which Darwin had obtained from studio inventories, seemed even more dated. Girls in Darwin's book wore dresses that women now recognized from their childhoods, baby clothes did not match modern sensibilities, and the actor with the handlebar moustache looked unfashionable to contemporary eyes. The pictures that had served to make *Expression* so compelling when it first came out became an anchor from which the text could not escape. Photography had enabled Darwin to illustrate his book with expressions performed by real individuals of his day. However, as time passed, the particular appearance of those individuals began to assert itself, and the message of Darwin's theories on expression became obscured by syntactical qualities.

With his last book, *Action of Worms*, Darwin did return briefly to pho-

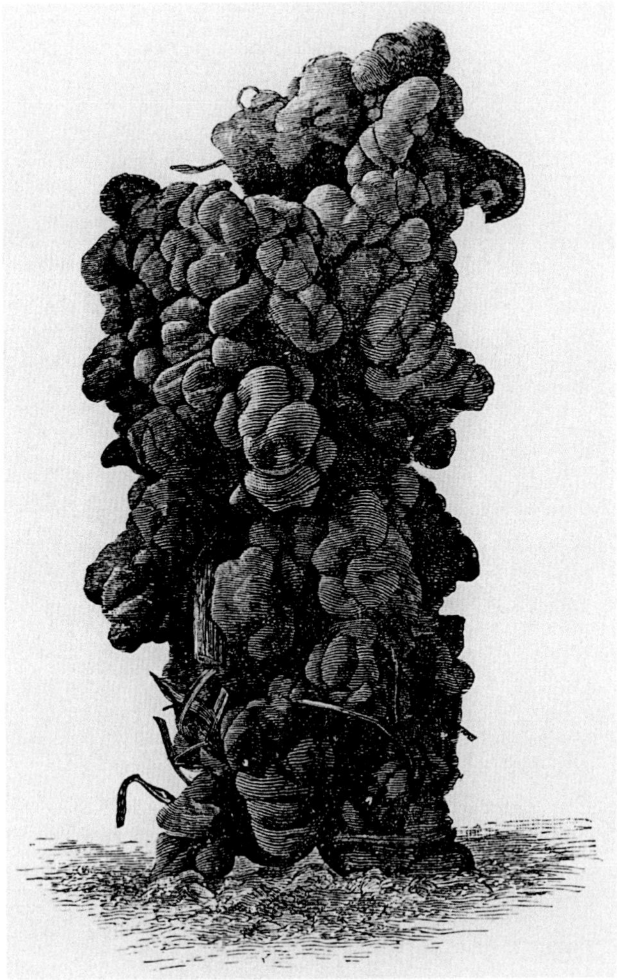

11-1 Circle of George King. A Tower-like Casting, Probably Ejected by a Species of Perichaeta, from the Botanic Garden, Calcutta, c. 1881. Wood engraving; copy of a photograph. Figure 3, *Formation of Vegetable Mould through the Action of Worms* (1881). The pictures of excreta published in *Worms* marked Darwin's last venture into scientific photography.

tography. Like *Expression*, the book marked the end of an inquiry he had pursued since the beginning of his career. He had presented a paper on worm activity before the Geological Society of London in 1837. *Worms* contains fifteen wood engravings, including three made after photographs of earthworm castings. These bizarrely convoluted forms—one weighing more than a quarter pound, Darwin assured—are among his most unusual illustrations. Isolated, enlarged, and with powerful graphic strength, they resemble twisting organic cities (figure 11-1). The engraver's technique of building forms through hatched horizontal lines of varying densities amplifies the effect of gracefulness. The reason Darwin chose photography as the basis for these engravings is unknown. But with *Expression*, and increasingly after that book was published, photography carried the cachet of authenticity.

Darwin explained in the text that George King (1840–1909), superintendent of the Botanic Gardens in Calcutta, was his most useful source of information for *Worms*. King sent samples of real castings, which Darwin was able to measure and weigh for himself. However, the engravings show the castings in situ, entwined with leaves and other detritus, and presumably resting on the ground where they were made. The photographs on which these engravings are based have not been found, so there is no way to know how accurate the engravings are. King most likely sent them to Darwin; whether he made them himself is a matter of conjecture.

Photography would have been a good way to record the exact appearance of castings in the field, without employing a skilled draftsperson. On the other hand, to make such photographs did require special equipment, because castings are generally small, and a lens capable of extreme close-ups was necessary. Although technology had improved enormously in the nine years since *Expression*, photography was still a demanding craft. By the time *Worms* was published in 1881, easy-to-use gelatin dry plates were available and enlargement was more practical than it had been. But a skilled individual was still necessary to take and process pictures.

Expression through a Looking Glass

After *Expression* was published, the author and photographer Lewis Carroll (Charles Lutwidge Dodgson) wrote to Darwin offering to help with future photographic investigations. Initial correspondence between the two has been lost, but a letter preserved in the Harvard University Library contains one of Darwin's replies. "I am much obliged for your kind note & offer," he wrote. "If you are quite sure that you can spare the unmounted photo., I shd. very much like to possess it, although I am doubtful whether I shall ever make any actual use of it. I half scruple accepting your very kind offer & I remain yours very faithfully Charles Darwin."[5] Carroll replied on 26 December 1872, noting in his diary that he had composed several letters that day, including one to "Mr. C. Darwin whose book on *The Expression of the Emotions* I am reading, and to whom I have given a print of *No Lessons Today*"[6] [figure 11-2]. The image, preserved in the Darwin Archive, may have been one of Carroll's favorites because he made numerous copies, including a nearly exact duplicate now at Princeton University.[7]

The subject was Flora Rankin, whom he photographed at Elm Lodge, London, in July 1863. Darwin's copy of the picture is faded and yellowed with age. On the back it is inscribed in pencil, possibly by Francis Darwin: "by the Rev. C. L. Dodgson / Girl about 10 years old, made to smile by being asked whether she would never like more

Opposite:
11-2 Lewis Carroll, *No Lessons Today*, 1863. Carroll sent Darwin one of his most expressive photographs.

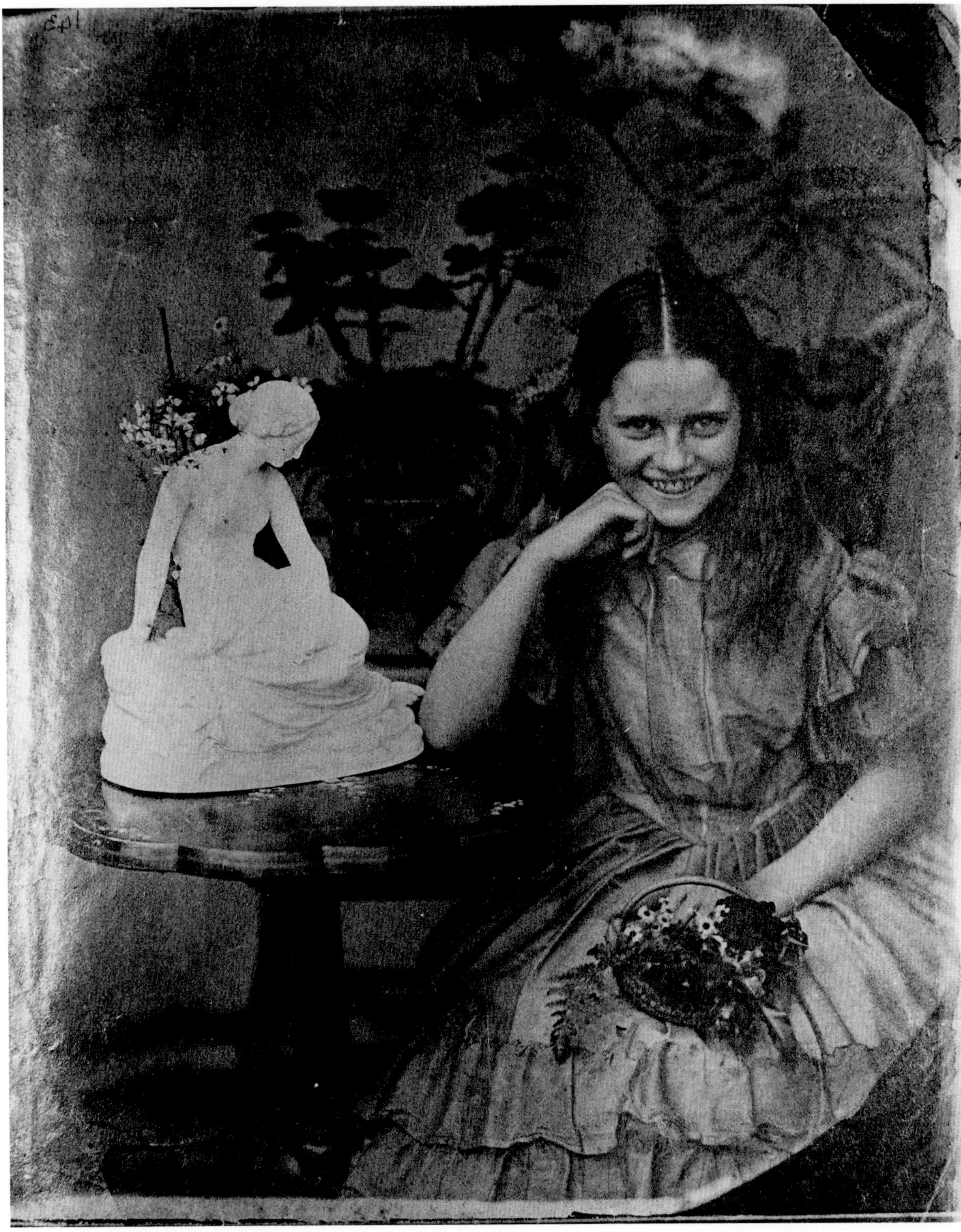

uniforms." The girl is shown three-quarters length, sitting, wearing a pleated frock, with a small basket of flowers in her lap. Her right elbow rests on a small round table with her hand propping up her chin. The small allegorical statue of female virtue on the table to her right (possibly by Carroll's friend, the sculptor Alexander Munro [1825–71]) is not identified. In the background are what appear to be potted plants, probably drawn on a painted scrim. The image Darwin received is untrimmed, possibly because Carroll did not go to much trouble to produce it. Otherwise, it may have been left untouched intentionally to give Darwin the fullest possible view of the scene.

The most conspicuous feature of the photograph is Flora's expression. She faces the camera head on, her face lowered slightly, exposing her upper teeth in a broad, even smile. Her appearance resembles the pictures of "moderate laughter" on plate 3 of *Expression*, especially George Wallich's picture of his daughter (figure 2).

The circumstances described on the back, indicating that Flora was made to smile by being told an amusing anecdote, served to legitimize the image. As seen in photographs Darwin collected in the run up to *Expression*'s publication, the circumstances of production of Victorian photographs were often noted. This served to validate them and explain behaviors captured by the camera, such as Flora's beaming smile. It would also have been useful for Darwin to know the degree of the girl's pleasure; a contented expression, he understood, may manifest itself rather differently than a joyful one. Having been made nearly ten years previously in 1863, *No Lessons Today* shared the circumstances of its production with other photographs Darwin owned. Like many of them, it was made before he started the *Expression* project.

Rejlander and Carroll knew each other well, and it may have been Rejlander who told Carroll of the scientist's interest in expressive photographs. Carroll first met Rejlander in 1863 and visited his studio on a number of occasions. *No Lessons Today* was made at this time, when Carroll was heavily under Rejlander's influence. Rejlander probably would have known the picture and may have commended it to Darwin. Alternatively, Carroll may have learned of Darwin's projects through associates at the University of Oxford. Carroll knew scholars in Darwin's circle; for example, an album of photographs by Carroll in Christ Church College Library contains portraits of Darwin's associates the biologist Thomas Henry Huxley and the geologist Adam Sedgwick (1785–1873).[8] Alexander Munro, who was close to several of the Pre-Raphaelites, may also have introduced them through Thomas Woolner. Carroll photographed Woolner in the summer of 1860; it had been Woolner's suggestion that John Tenniel (1820–1914) be retained to illustrate Carroll's *Alice's Adventures in Wonderland* (1865).

Darwin did not travel much late in his career, but he was fortunate to have a close friend who did. Thomas Henry Huxley was often dispatched to report on the meetings of learned societies, probe associates for information, and represent the Darwinian view in debates and lectures. Still, it is something of a surprise that in 1875, Darwin enlisted his help to investigate suspicious photographs being touted by a medium as evidence of the spirit world. He knew, as much as anyone did, how photography could be bent to the will of the practitioner. He was sure the medium's photographs were fakes, and he asked Huxley to find out how they were made.

On 18 January 1874, Darwin wrote to Joseph Hooker at Kew Gardens to tell him about a séance he attended:

> We had grand fun one afternoon; for George [Darwin] hired a medium, who made the chairs, a flute, bell & candlestick & fiery points jump about in my Brother's Dining Room in a manner that astounded everyone & took away all their breaths. It was in the dark, but George & Hensleigh Wedgwood held the medium's hands & feet on both sides all the time. I found it so hot & tiring that I went away before all these astounding miracles or jugglery took place. How the man could possibly do what was done, passes my understanding. I came downstairs & saw all the chairs &c. &c. on the Table which had been lifted over the heads of those sitting round it. The Lord have mercy on us all, if we have to believe in such rubbish. F. Galton was there & says it was a good séance.[9]

There is little about this light-hearted letter to indicate Darwin thought anything more of it. But he was clearly perturbed. His brother-in-law Hensleigh was a good man, full of interesting ideas, but he was easily led. He was beginning to believe in the spirit world, and it was more than Darwin could bear. Huxley was pressed into action. He attended a séance by a man named Williams and quickly discovered that the medium had tricked the attendees to hold each other's hands when they thought they were holding the medium's. Darwin wrote back on January 29:

> My Dear Huxley,
>
> It was very good of you to write so long an account. Though the séance did tire you so much it was, I think, really worth the exertion, as the same sorts of things are done in all the séances, even at Crookes' [Sir William Crookes, physicist and spiritualist (1832–1919)]; & now to my mind an enormous weight of evidence would be required to make one believe in anything beyond much trickery. It is a very sig-

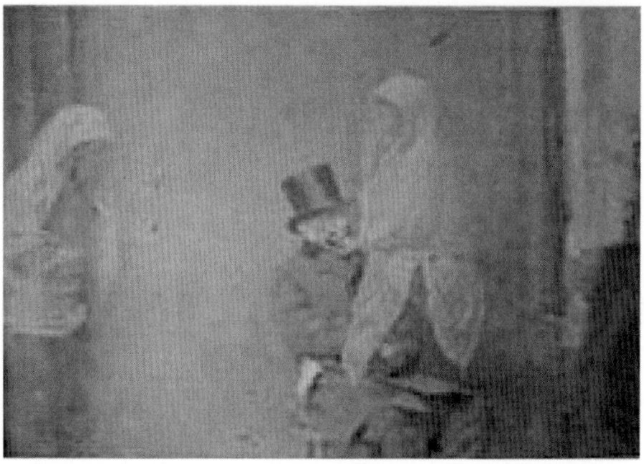

11-3 F. Hudson, *Spirit Apparition during a Seance,* 1874. Albumen print. Darwin asked Huxley to disprove Hensleigh Wedgwood's spirit photographs.

> nificant fact that William now regularly goes to Crookes. I am pleased to think that I declared to all my family the day before yesterday, that the more I thought of all that I had heard happened at Queen Anne St. [his brother Erasmus's house], the more convinced I was that it was all impostors. I would not have thought that H. Wedgwood would so easily be humbugged; viz. that is this Williams managed to get the two men on each side of him to hold each others hand instead of his, & was thus free to perform his antics. I am very glad that I imposed my will on you to attend.[10]

The Huxley Archive contains three carte-de-visite photographs of spirit apparitions apparently made during this séance[11] (figure 11-3). They were made by simply exposing the same plate twice in different circumstances. The first exposure, presumably made earlier, registered the ghost; the second recorded the attendees. Double exposure was a technique Rejlander knew well and occasionally used in his photographs.

The usual procedure for making spirit pictures was to plant a preexposed plate in the camera before guests arrived. The medium or his photographic assistant made a great show of loading the camera with a fresh, unexposed plate. By sleight of hand, a previously prepared plate, usually preexposed with an image of a shrouded apparition, took its place in the camera. Alternatively, a secret compartment in the camera could be used. When the medium photographed the party participating in the séance, the two exposures were superimposed, and the apparition seemed to be floating above them. Enlisting photography in the cause of spiritualism was particularly galling to Darwin, because he had used it to bolster his positions in *Expression.* As charlatans with cameras began plying their trade, they deflated the notion that there was something uniquely believable about photographs.

William Crookes, the most famous spiritualist of his day, was noted for his use of photography. He worked with a teenage medium named Florence Cook (1856–1904), claiming that through her he was able to materialize the spirit of a person named Katie King repeatedly from 1872 to 1874, the years immediately after the publication of *Expression*. He created forty-four photographs, supposedly of Cook and King, shortly before he announced that the spirit had ceased to visit her, in May 1874. Huxley attended his séance just as Crookes was claiming to have photographed the ectoplasm of Katie King. These photographic "experiments" received enormous publicity, attracting admirers among the faithful and scorn among the scientific set.

Pain and Pleasure in Italy

Some researchers seized on Darwin's study in *Expression* as inspiration for further investigations. The anthropologist Paolo Mantegazza (1831–1910), appointed the first chair of an anthropology department in Italy at the Istituto di Studi Superiori in Florence, sought to emulate it. Mantegazza's technique was actually a hybrid of those used by Darwin and Guillaume-Benjamin Duchenne de Boulogne. Subjects were artificially stimulated to create expressions. However, unlike Duchenne, who used electrical probes to cause facial muscles to contract (see chapter 6), Mantegazza, working with the Giacomo Brogi photographic studio, created real discomfort in his subjects. They were made to produce different expressions of pain by being exposed to unpleasant situations, each keyed to a different sense. He flashed bright lights at them, made them drink bitter herbs, exposed them to exceedingly loud noises, made them sniff ammonium sulfate, and crushed their fingers to varying degrees. The results were then photographed. In fairness, it seems, or perhaps because he found few volunteers for his stimuli, Mantegazza endured many of the experiments himself (figure 11-4).

In 1872 Darwin received fifteen cabinet-card photographs made according to Mantegazza's protocols.[12] They probably accompanied a letter of 23 December 1872 in which he described his research. "Your latest work charmed me," Mantegazza pronounced. "You have made out of physiology a new science. What is Lavater by comparison?"[13] He volunteered to act as the translator for *Expression* in Italian and concluded with some of his own observations on phenomena Darwin described. He noted the clenching of the teeth under circumstances of extreme duress, and the tendency to salivate when stimulated in lust.[14]

Mantegazza's photographs were published in *Atlas of Pain* [*Atlante della Espressione del Dolore*] (1876),[15] a book that featured comparisons

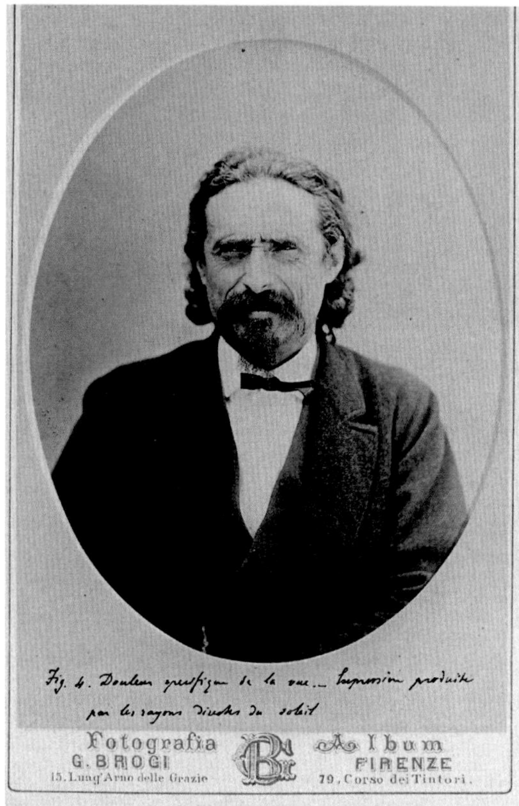

11-4 Giacomo Brogi Studio, *Pain Specific to Sight—Expression Produced by Direct Sunshine*, c. 1872. Albumen print. By applying stresses to people of various races, Paolo Mantegazza hoped to unlock the secrets of the expression of pain. He posed for this image himself.

between expressions of living humans and historical depictions. Mantegazza sent Darwin a copy of this, too, when it was published.

Mantegazza pushed Darwin's ideas in a direction Darwin had not intended. Darwin was committed to the idea that all humans, regardless of ethnic origin, express emotion in the same way. Mantegazza accepted this view, but believed there were subtle differences between expressions in different races. A smile may cause the same muscles to contract in a black man as in a white man, he reasoned, but it may be that some muscles are more prominent in one race or the other. To prove his hypothesis, he performed his pain-inducing experiments on a man of African origin, taking him as characteristic of "negro" physiology; these, too, he sent to Darwin (figure 11-5).

Francis Galton's Composite Photographs

Like Mantegazza, Darwin's cousin Francis Galton (1822–1911) invoked Darwinian theory to try to explain differences between the races, and also like Darwin, he was interested in what photography could reveal about human nature. However, his slant represented a retrenchment of

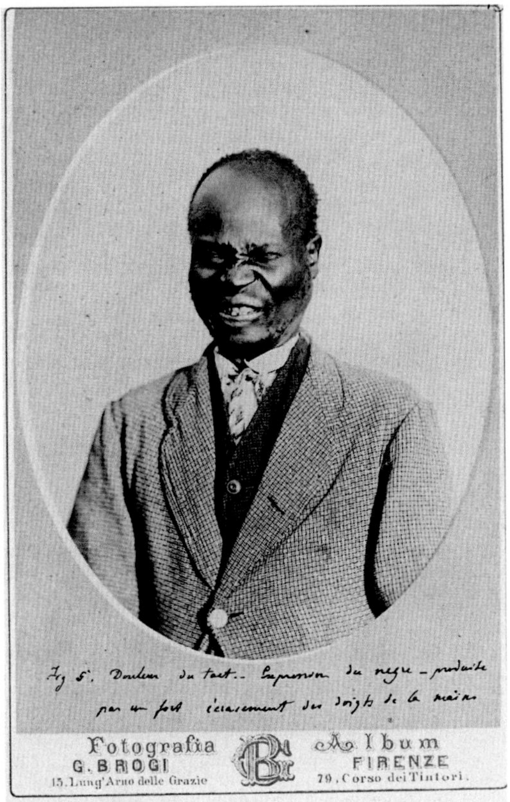

11-5 Giacomo Brogi Studio, *Pain of Touch — Expression of a Negro Produced by a Strong Crush of the Fingers*, c. 1872. Albumen print.

sorts, because he played off ideas popularized by physiognomists some hundred years earlier. Galton believed character is written in personal appearance and that people with similar inclinations share common facial features. This, he maintained, could be demonstrated through photography. Whereas Darwin attempted to accumulate photographs of spontaneous expressions, Galton tried to create artificial meta-portraits conveying the commonalities between individuals with similar characteristics.

In his book *Inquiries into the Human Faculty* (1883), Galton examined members of the same family, officers and privates in the Royal Engineers, criminals, and diseased people with and without tuberculosis (figure 11-6). He gathered portraits of individuals in each category and then combined them in composite prints in an attempt to discover ideal types. Galton photographed prints to get negatives, which he then sandwiched one on top of the other and printed on a single sheet of photographic paper in perfect registration. The cumulative effect of superimposing so many exposures was the deemphasis of outlier characteristics and the enhancement of commonalities. If ten of the people photographed had large noses and two had small ones, for ex-

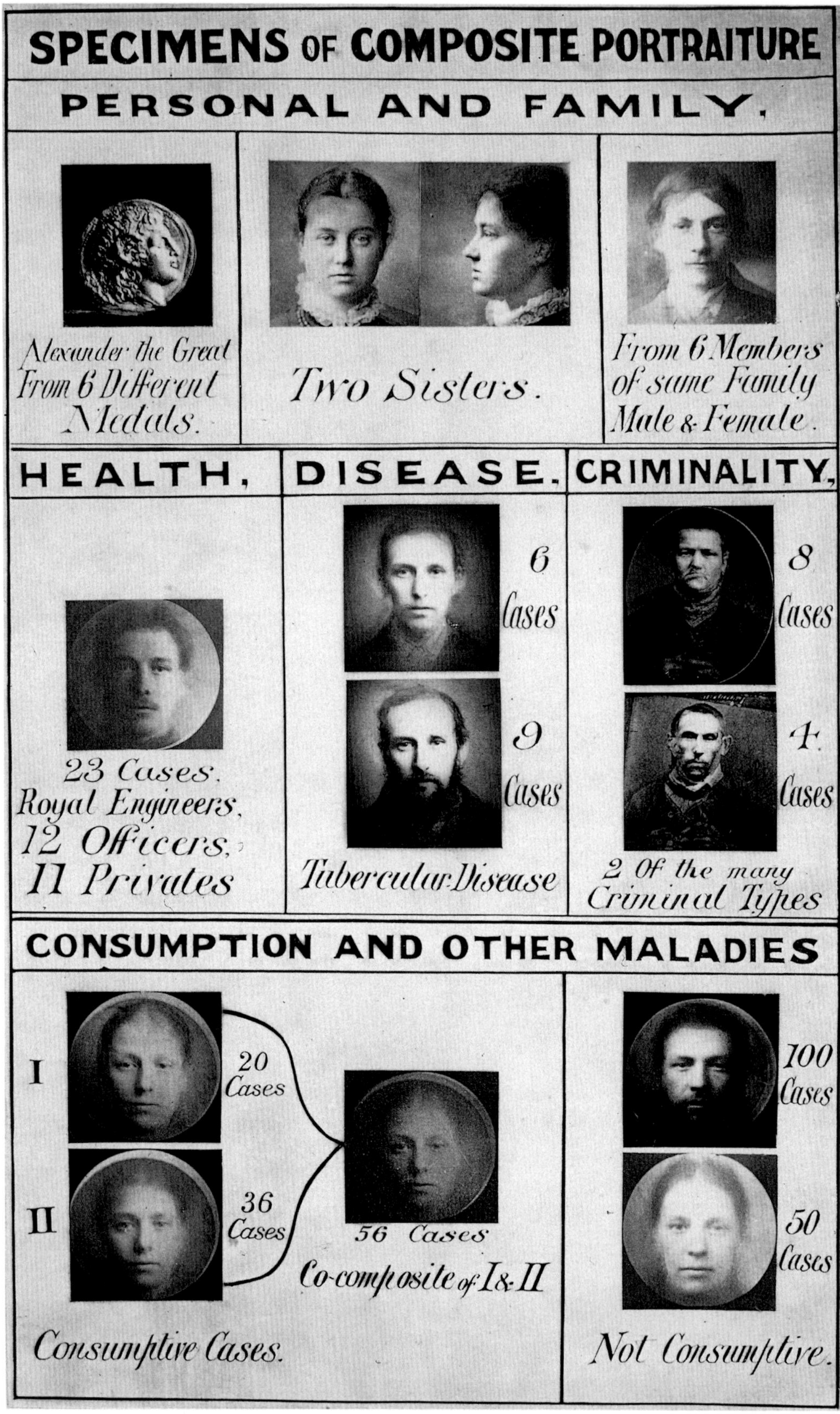
SPECIMENS OF COMPOSITE PORTRAITURE
PERSONAL AND FAMILY.
Alexander the Great From 6 Different Medals.
Two Sisters.
From 6 Members of same Family Male & Female.
HEALTH.
DISEASE.
CRIMINALITY.
23 Cases. Royal Engineers. 12 Officers, 11 Privates
6 Cases
9 Cases
Tubercular Disease
8 Cases
4 Cases
2 Of the many Criminal Types
CONSUMPTION AND OTHER MALADIES
I
20 Cases
II
36 Cases
56 Cases
Co-composite of I & II
Consumptive Cases.
100 Cases
50 Cases
Not Consumptive.

ample, the composite photograph would have a clear large nose, with only a ghostly hint of anything smaller. The shape of the head, size of eyes, nose, mouth, and chin, and even the location of hair could all be analyzed according to this system. When men and women were combined, the resulting portrait was androgynous. When criminal faces were combined, the result was a thug displaying all the features feared by the general population.

Galton made no meaningful distinction between behavioral and physical differences. He also did not differentiate between people who had been born with a trait and those who had developed it, such as those who had caught communicable diseases. To Galton, everything about human nature is inherited, and one need look no farther than his composites to see confirmation of this fact. Never mind that his pictures were self-fulfilling prophecies. If social prejudice inclined to convict criminals (or innocents, for that matter) of a certain appearance, then this is what would be conveyed in his composites, not the innate personality of people who looked that way. Moreover, Galton relied on others for his images, even in one instance combining numismatic portraits of Alexander the Great. That one sculpted portrait might have been based on its predecessor or that social predilections for certain characteristics might influence artistic renderings does not seem to have mattered to him.

Galton's composite technique enjoyed a brief vogue in the 1880s. One of the hotbeds of Galtonian photography was Harvard University, erstwhile home of Louis and Alexander Agassiz, Asa Gray, and others Darwin knew. H.P. Howditch of Harvard Medical School sent Galton three of possibly the most ambitious composite portraits ever made. The Boston firm of Allen & Rowell participated in combining 449 portraits of "American male college students," and a complementary 287 portraits of "female American college students." Doctors were either not as giving or not as abundant: only twelve "Boston doctors" were game to be composited. In New York, Dr. William Noyes sent Galton a portrait from the Bloomingdale Asylum, made from eight superimposed cases of melancholia. The Boston photographer William Notman (1826–91) provided the raw imagery. These pictures and others are now preserved with Galton's papers at University College London.[16] They constituted a dead end in empirical photography. Composite portraits were fun to make, but they had little to recommend them scientifically. After the 1890s, the practice largely disappeared.

Most of Galton's sitters are anonymous, as is fitting for a line of research based on commonalities. However, at least two of the sitters shown in *Inquiries* were associated with a familiar figure.

Henry Peach Robinson, Rejlander's friendly rival and the author of

Opposite:
11-6 Francis Galton, Specimens of composite portraiture, c. 1882. Albumen prints, from *Inquiries into the Human Faculty*, 1883. Galton sandwiched negatives together to create idealized portraits.

highly regarded photographic manuals, contributed them. In a letter dated 3 May 1882, Robinson wrote enclosing family portraits. "I am afraid they will not be much use to you as there are only three," he remarked, "but they are all the daughters I possess and my son is away. They were all taken at a sitting against the same rest. They very much object to having the worst aspects of their faces taken!"[17] Galton duly melded Robinson's photos into one of his composite photographs. Robinson replied on 19 May:

> My Dear Sir,
>
> I am exceedingly obliged for the very curious and interesting composite pictures of my children you have sent me. Knowing the faces so well it caused me quite a surprise when I opened your letter. I put one of the full faces on the table for the mother to pick up casually. She said, "When did you do this portrait of Edith? How *like* she is to Maud! Or *is* it Maud, I never thought they were so alike before." It has puzzled several people to say whether one of the profiles was intended for Edith or May. Then I tried one of them on a friend who has not seen the girls for years. He said, "Well, it is a Robinson for certain but I don't know which." I should very much like to have another set of prints, or the negatives, whichever would give you least trouble, and if ever I can be of use to you photographically please do not hesitate to let me know."[18]

Galton's exchange with Robinson marked the end of an era. Rejlander had died in 1875; Darwin followed him in April 1882, the month before this letter was mailed. With this correspondence, Robinson, one of the last surviving members of Rejlander's circle, briefly joined the scientific enterprise. He was an ambassador of mid-Victorian photography, when everything was possible and nothing was easy. The ghost of Rejlander briefly possessed him. Robinson, like Rejlander when he began to work with Darwin, had never contributed before to scientific experiments. But in his spirit, perhaps even against his own better judgment, he was willing to give it a try. Darwin and Rejlander bequeathed their ideas for others, like Galton and Robinson, to tinker with. Their most important legacy lay in the integrity, novelty, and sheer incandescence of their ideas.

Conclusion

Expression was the high-water mark in Darwin's evolutionary theories; it also contained the most sophisticated program of illustrations in any of his books. While colleagues wrestled with what the book meant for photography and scientific illustration, Darwin quickly moved on to other subjects, and his determination to use photography eased. Prominent thinkers such as Galton, Carroll, and Mantegazza continued to explore photographic illustration in their research, and Darwin carried on acquiring photographs on subjects unrelated to emotional expression.

In some quarters, *Expression*'s influence lasted uninterrupted for generations. In Duchenne's laboratory at Salpêtrière hospital, for example, Darwin's book became a source of inspiration for much work that followed. Duchenne's protégé, the neurologist Jean-Martin Charcot (1825–93), began extensive photographic experiments of his own, and one of his students, Albert Londe (1858–1917), became a major figure in the chronophotography movement of the 1880s. Another star student, Sigmund Freud (1856–1939), would later claim Darwin's *Expression* to have been a significant influence on him. But such pockets of influence were surprisingly rare. The book caused considerable excitement when it was published, but unlike Darwin's other works, it quickly faded from view. In the late twentieth century it was rediscovered and is now recognized as a landmark in the various fields it engages. Yet it languished for most of the first half of the twentieth century.

The positions Darwin took in *Expression* became unfashionable for fifty years or more. Darwin's universalist stance assumed there were more similarities than differences between peoples of the world. This was not only uncompromisingly egalitarian, but also based on a belief in biological determinism that was widely questioned. When debates over nature versus nurture took hold in the early twentieth century, scientists turned their attention to how parents may affect their children's development. This was anathema to Darwin, who felt that parents were of marginal importance in the formation of expressions, which he considered innate. When academic opinion swung back to consideration of biological factors in the 1960s and 1970s, *Expression*'s reputation blossomed once more.

Darwin's photographs also had a substantial effect on the reception of the book. What looked fresh and exciting in 1872 looked dated by the 1880s. The grainy, low-resolution look of the heliotypes and their sepia color appeared antiquated; when Francis Darwin reprinted the book in 1890 he used the same pictures but printed them using the more modern photogravure technique in a neutral black and white tone. However, there was no disguising the high Victorian camp of Rejlander's chapeau and velvet smoking jacket. Darwin had surmounted numerous technical obstacles to obtain his illustrations, but as photographic technology improved, the pictures began to seem strained. One could obtain clearer photographs using the new gelatin dry plates; photographs made with the old wet-plate materials looked primitive. But more important, as rules about scientific photography tightened, scholars became increasingly uncomfortable about compromises made in their production. Darwin's work helped elevate photography to the status of medium of record in scientific experimentation. And yet, because his photographs were gathered in the eclectic, opportunistic way described in this book, they were susceptible to criticism from readers unacquainted with their historical context. Such criticisms have persisted to the present.

Darwin and Rejlander have often been criticized for compromises they made in preparing their illustrations. The art historian MaryJo Marks's complaint is typical of criticisms leveled:

> The book was an ingenious but scientifically flawed attempt to reconcile natural and cultural explanations for bodily expression. . . . Darwin's book implies authorial control is undone by cultural conventions but also by the repetitively automatic character of instinctual, chemical, and mechanical processes. Theatrically simulated, or mechanically simulated, the book's images are of entirely fabricated gestures, repeatable on demand.[19]

Marks challenges the validity of the photographs Darwin used and questions his use of photography as a legitimizing tool, arguing that he used photography as camouflage to disguise the subjectivity of his methods. There is no question that Darwin described photography as superior to drawing, whereas the photographs he actually used, especially those made by Rejlander, reflect arbitrary decisions made by Darwin and his collaborators. In this sense, they are clearly no better than equivalent prints or drawings, though Darwin suggested they were.

The core of Marks's argument lies in the interpretation of what Darwin meant by the term "superior." Marks's claim that Darwin's use of photography was dishonest is anachronistic, because she applies current views of photographic objectivity, rather than views that were prevalent at the time. Darwin did want his readers to be convinced by his photographs. Yet, the distinction between "evidence" and "illustration" is blurred in the book because there was no precedent for the acceptance of photography as scientific data. There was no protocol for the use of empirical photography, precisely because photographers often found it necessary to manipulate their work to enhance not only the visual appeal but also the clarity of their images.

In many ways the publication of *Expression* marked the birth of scientific photography. It could not conform to rules about its use, because it was part of the creation of those rules in the first place. Darwin considered photography superior to drawing because it provided a window on events difficult to observe with the naked eye. However, he did not claim it provided automatic representations of expression, nor did he suggest that photographic images should be accepted without scrutiny.

Marks's argument assumes a dichotomy between documentary truth and artificiality in scientific photography that Darwin and his contemporaries did not comprehend.

Most of the photographs Rejlander supplied to Darwin, for example, were not records but simulations, based in part on photographic investigation but mediated by experience, observation, and consensus. The combination of these factors, not photography alone, was the basis on which Darwin judged their suitability. His approach was consistent with the photographer Henry Peach Robinson's maxim, explained in Chapter 9, that "any 'dodge, trick, or conjuration' of any kind is open to the photographer's use, so that it belongs to his art, and is not false in nature."[20]

What is most remarkable about Robinson's prescription is not that it advocates fakery; rather, it lies in the belief that determinations of truthfulness should be made after a photograph is complete. In determining the veracity of a photograph, the issue was not whether it had been made subject to strict procedural standards, but whether the results withstood careful inspection. By this measure, a photograph could be considered truthful if it depicted events that the viewer could mentally verify, or that seemed reasonable according to experience. These are ideas borrowed from the world of printed illustrations—it is the same logic viewers would have applied to lithographs or wood engravings. According to Robinson, every photograph must be evaluated on its own terms; there is nothing inherently right about a picture made using photography. This idea, widespread in photographic thought at the time, is crucial to understanding Rejlander's approach to the *Expression* project.

To Rejlander, the means by which a photograph was produced was less important than its appearance. Just as for centuries individuals had evaluated works of art based on the plausibility of their appearance, Victorian photographers relied on their critical skills to judge the correctness of their images. In an artistic context, particularly in which traditional values of painterly composition were prevalent, such an approach was natural. It was only when the same criteria began to be applied in the scientific realm that they became suspect.

To a twenty-first-century scientist, Darwin's attitude toward illustration may be puzzling. Science is positivist: theoretical conclusions are meant to be based on observable phenomena.[21] The scientific method as we now understand it requires that conclusions fit data, and not the other way around. The retroactive evaluation of photographic truthfulness would seem to subvert this principle.

The type of subjective, *ex post facto* evaluations Rejlander and Darwin applied to their work appear to stand positivism on its head. Instead of merely aiming his camera, taking the picture, and accepting the results as truthful because the variables that governed their production were fixed and controlled, Rejlander edited, altered, and manipulated his results to create pictures he considered correct. Such contrivances cast doubt on the impartiality of the information in his photographs, but from Rejlander's point of view, they were necessary to obtain the best possible results.

In the case of emotional expression, however, concerns about the objectivity of the artist are worsened by the nature of the activity depicted. Expressions are fleeting; as Darwin said himself, the muscle contractions that form expressions occur very quickly and are difficult to observe. His desire for photographs of these behaviors pushed the project beyond the realm of easy verification. The old standard of mental confirmation could no longer be easily applied, because the activities targeted occurred at the edge of visibility.

Whether one looks on Rejlander's photographs as counterfeits or clever facsimiles depends on the degree to which one accepts the rigor of the observations used to create them. For Rejlander, as for Darwin, photography did not supersede human judgment. Marks argues that Darwin "uses advanced technology to supply a thin veneer of naturalism and authenticity: he sees photography conferring an aura of documentary truth on the pictures' artificially constructed expressions."[22] Yet there is scant evidence for this proposition. Later viewers would interpret the photographs in this light, but Darwin's own understanding of photography, like that of his audience, was more utilitarian. From his point of view, Rejlander's photographs were part of a dynamic of fact-finding that included photography, combining it with other vital sources of information. When Darwin began his collaboration with Rejlander in 1871, both men well understood that photographs of expression could be relied on to be precise, but not necessarily accurate.

Darwin's use of photography undoubtedly helped him to capture the attention of his audience, as a photographically illustrated book, particularly one by an author as famous as Darwin, was distinctive. However, scientific photography was not so much an "advanced technology," as Marks describes it, as a nascent one. Other scholars have made arguments similar to the one advanced by Marks. The literary critic Jennifer Green-Lewis notes:

> The work itself is a curious methodological mix. . . . Darwin's use of photography defers to a positivist model in which total knowledge is defined as scopic and determined, while his chosen illustrations (impressionistic, interpretive) undermine their own testimony in their obvious hermeneutic instability. In his efforts to quantify the emo-

> tions, to standardize expression and thus to identify it as a universal "language," Darwin has Lavater as much as Comte for his inspiration. Rejlander's photographs merely substitute for the physiognomist's line drawings, while they appeal to their privileged scientific objectivity as justification for their author's classifications.[23]

The theoretician Carol Armstrong sees things similarly:

> That then, is the overarching, thoroughly circular point of the experiment in photographic illustration conducted by *The Expression of the Emotions in Man and Animals.* Like the conduct of positivism, that experiment is an exercise in tautology, conducted in thoroughly self-reflexive terms according to self-authenticating logic. But again like positivism, in naturalizing its self-reflexivity Darwin's experiment in photographic illustration ultimately hides the circularity of its movement of causation, explanation, and confirmation, the very circularity that causes the semiotic instability of the photograph.[24]

Each describes the seemingly contradictory elements governing the production of Rejlander's photographs: their simultaneous assimilation of internal rules of photographic process and external rules of authorial acceptability. What Green-Lewis describes as "hermeneutic" and Armstrong describes as "semiotic" instability is the hybrid nature of the images themselves.

Both scholars interpret the works as taking advantage of photography's unique ability to stand in place of real events, while at the same time subverting their own authority by representing suspect situations. In other words, although Rejlander's *Expression* pictures played on notions of photographic authority, they disqualified themselves because they were actually photographs of contrived behaviors. Obliquely invoking the deconstructionist concept of a photographic referent as espoused by Roland Barthes and others,[25] this argument is based on perceived qualitative differences between photographs and other forms of illustration.

The deconstructionist view voiced by Green-Lewis and Armstrong provides valuable insights into Rejlander's *Expression* photographs but sidesteps their historical complexity. In weighing these arguments, it is also important to consider whether the photographs should be viewed in isolation or as integrated elements of the book. Green-Lewis and Armstrong describe the *Expression* photographs as if they are closed systems, pictorial islands in a sea of text. Darwin certainly did not view them this way. For him, they were reinforcing strands in a net of information that tied the text to observable phenomena.

The deconstructionist view also grants special status to photographic representation that would not have been acknowledged by either Darwin or his audience. Rejlander, for one, refused to equate photography with truthfulness. This is spelled out in remarks he made before the Royal Photographic Society in 1858, explaining the reasoning behind his early interest in composite photography:

> The want of aerial perspective, particularly in the larger landscape photographs, has been one of the causes that has induced me to try to take a picture separately in different planes, and then print them together, which mode I have seen termed "tricks." An extensive landscape, especially when taken with a large lens, is not true, nor will it ever be so: the focus cannot be everywhere. Here is a fair specimen of Loch Katrine, taken at once: if it had been drawn by a skilful hand directed by a correct eye, there would not have been so much detail, but the proportions would have been different and more true; and the larger the lens, the greater the errors.[26]

Recognizing the limitations of purely mechanical photography, Rejlander defended the manipulation of photographs such as his notorious *Two Ways of Life* by arguing that even perfectly rendered photographs sometimes required correction to best represent what the eye actually sees. This raises an intriguing point about Rejlander's composite photography. Certainly, composite printing of the sort seen in *Two Ways of Life* is artificial. But is it necessarily false? Rejlander did not think so.

This same line of reasoning can be used to explain the interventions Rejlander made in the photographs he presented to Darwin. After all, people do not see the way cameras do. Rejlander understood that photographic vision—squared-off, flattened, and with limitless depth-of-field—is only an approximation of human vision. In addressing this question, he touched on concerns that would not achieve prominence until the end of the nineteenth century. Pictorial photographers, sparked in part by British photographer Peter Henry Emerson's (1856–1936) appeal for "naturalistic" photography in 1889, made compositions with differential focus, reflecting their understanding of natural vision.[27] Like Impressionist painters, they challenged the veracity of photographic realism and attempted to portray what the eye sees, rather than what the uncontrolled lens transmits.[28] Cameras are artificial eyes, they reasoned, but they are imperfect. They are forever doomed to artificiality. No photograph, no matter how rigorously made, can ever replicate what people perceive when they see something for themselves.

The dichotomy between what people see and what cameras capture was the subtext for everything Darwin did to obtain his *Expression* photographs. There is an extraordinary passage in the book in which Darwin tells the story of watching an elderly woman sitting alone on a train.[29] It may well have been an incident that occurred on one of his London picture hunting visits. Sitting opposite her, he notices a slight twinge of her *depressores anguli oris*—the muscles that control the lower lip. Darwin had come to identify these with sadness, so he was surprised to see action in this particular muscle group. When he spotted her, she did not look sad at all, but as he continued to watch, tears began to well up in her eyes and her countenance fell. Darwin's analysis of this episode fills two pages of text.

Most of the discussion is purely physiological, but for one sentence Darwin allows himself to speculate on what might have caused the woman's sadness. A painful recollection, he decides—perhaps the death of a long-lost child. It was just the sort of pain he had experienced himself. It is a moving passage and a subtle reminder that Darwin was attempting through *Expression* to understand no less than the essence of who we are as human beings. Ultimately, it was a mirror he was also holding up to himself. The way we feel, the way we act—these basic questions drove Darwin as he wrote, designed, and illustrated his book. And what do we see?

APPENDIX

"Odd Odds and Ends" by Oscar Rejlander

The following is a transcription of the notes Oscar Rejlander sent Darwin in 1871. Italics have been added to indicate items that Darwin checked in red pencil as particularly noteworthy.

Odd Odds and Ends by O.G.R. 1871

I was teaching my wife how to 'develop' a collodion photgr.—— saying 'now pour this evenly over the plate—there—now move the plate gently to & fro but hold the plate quite level—I saw her curls swaying but the hand was still—so intent she was in listening that the connection betw. the hand and brain was suspended—in fact, the brain rocked without communicating the action to the hand. She often talks in her sleep———

[Marginalia: + She stood so for nearly half a minute or so before I *awoke* her, I was so interested.]

'He looked very sheepish' = some one baffled etc.

When to express that the *exulting* party puts his thumb to his nose with extended fingers to imitate a long face? (The Origin of gestures)

Some rub the palms of their hands briskly together at any success in joke or otherwise.

Complacently stroking the beard, raising the chin & lips

A searching look—half-shutting the eyelids fixing the eyes steadily on the object.

Indignation—challenging—head erect, thrown back nostrils expanded—superciliary(?) muscles projecting—lips compressed—or if accomp. with contempt the corners of the mouth drawn down—the attitude ready for combat————

'He looked as dropped from the Skye' [crossed out: 'Crestfallen'] A vacant look, mouth open and hanging as lifeless at the sides etc.

'Crestfallen' a man sitting both hands on the knees head drooping—jaw drooping—the eyelids extended the eyes rather diverging being fixed on no object.

[page 2]

Aversion disgust—— Much like the expression of suddenly experience a bitter taste———

Mrs. W. told me of having several nights watched her husband go out of bed in the middle of the night (He was subject to epileptic fits) go downstairs and write a letter, in a good hand—*upside down*!

– A certain officer on Parade, quite sober, (within my knowledge)

sheathed his sabre and walked towards home—unconscious of breech of discipline—(He got one days arrest only)————

In writing—omitting words, habitually or letters———

In reading—for the eye to catch letters lower down———

[page 3]

The pouting child both lips out if angry—as well—the underlip only if sad—called the 'dripping-pan'.

[Marginalia: Some people cannot hold the fingers still—Dying people have begun scratching the clothes just before going—Babies toes move with the fingers.]

The sigh deep in thought & feeling paralysing the action of the breathing until a forcible vent is found in a sigh——

A violent sigh from suppressed anger in man

Passion offensive or spite—derision—by wrinkling the nose to some one——

Stamping the foot in anger or command (fem.)

Snapping the fingers in contempt (male)

Striking the hand on a table—

Striking a closed fist in the palm of the other (female)

Speaking with the teeth clenched in a growl-like voice—— (male)

Grief—the inner point of the eyebrows raised

Anger—the inner point of the eyebrows depressed

The hands!!!

Raised over the head imploring. Palms out or down at the sides expostulating—extended horizontally—Open to receive.

Fingers

pointing with the thumb 'over the left' also really in earnest.

Forefinger more direct

Middle finger—common men

Ring finger the same

Little finger Children—Ladies and many well-bred persons— — — — —

Forefinger—upright - perpendicular = warning

closed fist extended—threat—

Forefinger to the side of the nose—knowing or to the forehead—thinking with difficulty—

Forefinger to the mouth = silence

Hand behind the ear = hearing

Forefingers in the ear = not hearing

One hand drawn smartly across the other palm means '*nothing*'.

[page 4]

An uneasy cough—or cough to show wanted / indifference

A whistle to gain assurance.

Sympathy

Yawning (I could make my dog yawn)

Sick at sight

The mouth watering at acids

The teeth 'set on edge' at the slip of a knife

on a plate or filing a saw.

Smacking the lips

Tricks

Biting the nails

Sucking the thumb or the tongue

Blinking—.

While reading etc. keep moving the fingers in search of inequalities—

Draw the air suddenly thro' the nose

The Scotch—Hm! Hm! = response

Danish—So - So

Also a sucking noise = indeed or 'dear oh me'

Country Engl. aah!

Swedish jah! jah-ah!?

[page 5]

The occupier of the shop of this mansion has a bull-terrier which I had not seen before the other night (nor he me—) when (late) I passed the shop and the terrier ran at me head down—I was apprehensive of attack but he wagged his tail and sneered with pleasure to recognise me—How—by scent knowing me to be an inmate of the house——

The other night I went into Alvarez's liq. shop and the She-Cat there (upon colonade) was most anxious—as only a cat can be—for a male companion—I saw it—& having learnt some of their calls—I made the endearing noise of a cat and up got her pelvis down on her breast—ready—I repeated it several times to show the owner it was no accident.

A Dog, even a strange dog will generally repeat my Uffh! Uffh! being the first noise of *warning* of distant danger by dogs—My dog could say my friends name, Hoof, when asked and if not saying it plainly repeat it—at my order until well pronounced—he generally knowing when that was by jumping about, pleased, *immediately*

The brain of my clever dog I have carefully compared to that of a childs at one year and three months—not even to 1 – ½ year—— (too long to argue)

[page 6]

Graingers [?] barn-door cock who had two horns [drawing included here] hanging at each side—artificially produced in a cruel way I have seen two—a man near Wolverhampton did it. The bone was laid bare on either side & then partshorn & something else was rubbed over the bone.

A child was lately brought to the King's College that had the tail of a dog which it wagged like a dog. I think the whole lower part was that of a dog. Mr Dawson of King's College was my informant———

[page 7]

Astonishment—simple—head & neck extended mouth open as if the eyes & ears were not sufficient to receive the impress.

Astonishment with wonder—hands showing the palms—raised the eyes, wide open eyebrows arched, the mouth open but rounded for the interjection Oh. head thrown back——

Astonishment—agreeable—hands raised as before, eyes dilated eyebrows raised head rather on one side slightly bent and a suppressed smile.

Contempt: the body averted head turned to the object the eyes looking ascant—corners of the mouth drawn downwards with an expression about the nose as if something smelt disagreeably—

A sneer—a disguised contempt accompanied with an awkward forced smile on one side the face—

The same by a female accompanied by a sudden puff of whispering breath and a sudden jerk of her drapery towards herself (on the side of the dir. object) to blow away the 'pestilent' air and afraid that even the hem of her garment should be touched.

—male= pah! pah!

{Pleading, apologetically, 'as stated'—only add a most defenseless attitude.}

Spitting—defiance—at a person

In contempt—on the ground

(Llamas spit their undigested food that raises a blister on the skin)

Showing the tongue by way of insult children & very low persons show it broadly & far out

Grown up females do it pointedly & quickly

Silly people often play the tongue over the lower lip—

{Hunchbacked—club-footed child much squinting

children more mischievous especially the squinting}

Furtive looks

NOTES

Notes to Chapter 1

1. Charles Darwin, *The Autobiography of Charles Darwin* (1887), Nora Barlow, ed. [New York: W.W. Norton & Company, 1958], p. 23. Although the first edition of his *Autobiography* was not published until 1887, Darwin wrote this passage in 1876.

2. Darwin searched for photographs most intensely from 1869 to 1871, but his interest in such material must have begun earlier, because at least one of his pictures is noted on the reverse as having been purchased in 1866 (Henry Hering, *Insane Woman—from Bethlem Hospital, Darwin Papers folio 197.*) In total, the Darwin Archive contains forty-one photographs of unknown provenance, which were probably acquired on one or more of Darwin's photographic shopping trips.

3. Charles Darwin, *The Expression of the Emotions in Man and Animals* (1872), Paul Ekman, ed. [London: HarperCollins, 1998], pp. 17–18.

4. Bruno Latour, "Visualization and Cognition: Thinking with Eyes and Hands," *Knowledge and Society: Studies in the Sociology of Culture Past and Present*, Vol. 6, 1986, pp. 1–40.

5. Charles Darwin, *The Descent of Man, and Selection in Relation to Sex* [London: John Murray, 1871], p. 2.

6. Darwin, *Autobiography*, p. 131.

7. The degree to which one accepts *Expression* as primary in this area is, however, largely a matter of definition. Several private-press scientific portfolios were published before it, e.g., Anna Atkins's sumptuous *Photographs of British Algae* (1843–53), but these were neither produced in quantity nor widely available. In addition, a number of "how-to" books were produced, notably

with regard to photographic processes, which might be considered "scientific." Helmut Gernsheim's *Incunabula of British Photographic Literature, 1839–1875* [London: Scolar Press, 1984], the most comprehensive listing of early photographically illustrated books, lists only six works predating *Expression* with a natural science focus. Two are astronomical in nature, and four are medical, ranging from the treatment of clubfoot to the pathology of cancer in rats. Gernsheim's catalogue does not include works published in other countries, e.g., France and Germany, which were equally advanced in photographic publishing. However, it does confirm the general rarity of photographically illustrated books on scientific topics at the time. Those that did exist were generally of narrow specialist interest. *Expression* was the first science book to achieve wide circulation in the more than thirty years since the invention of photography.

8. Desmond, Adrian, and James Moore, *Darwin: The Life of a Tormented Evolutionist* [New York: Warner Books, 1991], p. 594.

9. Unpublished correspondence, Charles Darwin to Robert Cooke, 17 August 1872, John Murray Archive, item 283.

10. Unpublished correspondence, Charles Darwin to John Murray, 8 January 1872, John Murray Archive, item 271.

11. Chapter 16 of Guillaume-Benjamin Duchenne de Boulogne's *Mécanisme de la physionomie humaine ou analyse electro-physiologique de l'expression des passions applicable à la pratique des arts plastiques* [Paris: Jules Renouard, 1862], titled "A Critical Study of Several Antiquities from the Point of View of *m. corrugator supercillii* and *m. frontalis*," is devoted to this type of analysis.

12. Oscar Rejlander, "An Apology for Art Photography," *British Journal of Photography*, 16 February 1863, p. 78.

13. Dry-plate collodion negatives were available from 1856 but would not have been used, because they required much longer exposure times. There were, of course, other processes that did not require the use of collodion, but these were also slow and even less suited to motion photography.

14. A more comprehensive survey of instantaneous photography in the 1860s and 1870s is contained in Phillip Prodger, *Time Stands Still: Muybridge and the Instantaneous Photography Movement* [New York: Oxford University Press with the Iris & B. Gerald Cantor Center for Visual Arts at Stanford University, 2001].

15. Rejlander published details of his own, specially designed, light-enhancing studio in 1863. See "Mr. Rejlander's Glass Room," *British Journal of Photography*, 2 March 1863, pp. 96–97.

16. Oscar Gustave Rejlander, "Rejlander in Type," *British Journal of Photography*, 8 September 1865, p. 462.

17. Darwin, *Expression*, p. 13.

18. For the whole of Darwin's collection, see Vol. 53.1 of the Darwin Manuscripts Collection at Cambridge University Library.

19. Darwin, *Autobiography*, p. 61.

20. Edmund Burke, *Philosophical Enquiry into the Origin of Our Ideas of the Sublime and Beautiful* (1757) [London: J. Dodsley, 1793], pp. 84–85.

21. Darwin, *Expression*, p. 281.

22. Ibid., p. 238.

23. Ibid., p. 286.

24. Unpublished correspondence, Hensleigh Wedgwood to Charles Darwin, 21 March 1871, Darwin Papers, Vol. 195, item 54.

25. Darwin, *Expression*, p. 14.

26. Ibid., p. 98 n.

27. Ibid., p. 98.

28. Unpublished correspondence, Charles Darwin to James Crichton-Browne, 8 June 1869, Darwin Manuscripts Collection, Cambridge University Libraries, Vol. 143.

29. Unpublished correspondence, London Stereoscopic Company to Thomas Henry Huxley, 11 March 1871, Huxley Papers, Vol. 30, p. 117.

30. Now parts of the Imperial College of Science, Technology, and Medicine, London.

31. Now the Victoria and Albert Museum.

32. The diaries record that Cole formally received Huxley fifteen times at his home and at the South Kensington Museum between 1870 and 1871, and that he also received Wallich at least once during this period. Unpublished diaries of Henry Cole, Manuscripts Collection, National Art Library, South Kensington, London. The dates of Huxley's visits were 10 and 13 January, 15 May, 14 June, 24 and 25 September, and 23 November 1871; 6 January, 23 May, 8 July, and 3 November 1872. Wallich, who was interested in having Cole publish a collection of his photographs of eminent men, called on 15 February 1870.

33. Richard Broke Freeman and P.J. Gautrey, "Charles Darwin's Queries about Expression," *Bulletin of the British Museum (Natural History) Historical Series*, Vol. 5, No. 3, 1972, pp. 207–10.

34. Darwin, *Expression*, p. 16.

35. Ibid., p. 278.

36. Roslyn Poignant, p. 56.

37. For a discussion, see Paul Ekman, *Darwin and Facial Expression: A Century of Research in Review* [New York: Academic Press, 1973].

38. Unpublished correspondence, J.P. Mansell Weale to the editors of *Nature*, January 1873, Darwin Papers, Vol. 181.

39. Michael Pritchard, *A Directory of London Photographers 1841–1908* [Watford, England: PhotoResearch, 1994], p. 19.

40. Unpublished marginalia, figures 30 and 32, from Darwin's copy of Duchenne's *Mécanisme*, Darwin Library, University of Cambridge.

41. Unpublished photographs, *Expressive Pets*, Darwin Papers, Vol. 53.1, Folio 56–58, items 100–9. See chapter 7 for discussion of this work.

42. Studio of Ernst Schultz, *The Actor Portraying Various Character Types*, 1867, Darwin Library, unnumbered folio.

43. ibid.

44. E.g., Oscar Rejlander's *Two Views of a Pouting Boy (Profile)*, Darwin Papers, Vol. 53.1, Folio 30, item 4.

45. Unpublished correspondence, Leonard Darwin to Charles Darwin, undated (c. 1870), Darwin Papers, Vol. 186.

46. This photograph is currently in the collections at his home at Down House. Interestingly, contemporaneous copies of this image bearing the same inscription occasionally surface. I am aware of one, which was presented for

sale to the Photographers' Gallery in London in 1994, and another, which was submitted for auction at the eBay website in June 2000. Both of these were on Colnaghi Studio mounts. In each case, Darwin's signature appeared to be a reproduction of the original at Down House, and not in Darwin's hand. The origin of these facsimile-signed Camerons is mysterious. It is unclear whether their production was authorized by Darwin, or even whether he was aware of their existence. In any case, Cameron does seem to have co-opted Darwin's compliment as a sales ploy, despite his private reservations about the picture.

47. Frederick Burkhardt, Sidney Smith et al., eds., *The Correspondence of Charles Darwin* [Cambridge: Cambridge University Press, 1989], Vol. 17, p. 508.

48. Letter from Julia Margaret Cameron to Charles Darwin, 10 July 1868, Darwin Papers, 161.1:9.

49. Letter from Julia Margaret Cameron to Charles Darwin, n.d., 1868, Darwin Papers, 161.1:8. The identity of Cameron's "Bromley people" is not known.

Chapter 2

1. Phillip Prodger, "Illustration as Strategy in Charles Darwin's *Expression of the Emotions in Man and Animals*," *Inscribing Science: Materiality and Text in Scientific Communication*, Timothy Lenoir, ed. [Stanford, California: Stanford University Press, 1998].

2. Edward Manier, *The Young Darwin and His Cultural Circle: A Study of Influences Which Helped Shape the Language and Logic of the First Drafts of the Theory of Natural Selection* [Boston, Massachusetts: D. Reidel Publishing Company, 1978].

3. Gillian Beer, *Darwin's Plots: Evolutionary Narrative in Darwin, George Eliot and Nineteenth-century Fiction* [London: Ark Editions, 1985], p. 97.

4. Adrian Desmond and James Moore, *Darwin: The Life of a Tormented Evolutionist* [New York: Warner Books, 1991], pp. 579–580.

5. Darwin Papers, Vol. 53.1, Folio 51, object 90, verso.

6. Lewis Carroll, *Alice's Adventures in Wonderland* [London: MacMillan & Company, 1865], p. 1.

7. Bruno Latour, *Science in Action* [Cambridge, Massachusetts: Harvard University Press, 1987], p. 50.

8. Ernst Gombrich, *Art and Illusion: A Study in the Psychology of Pictorial Representation* (1956) [London: Phaidon, 1996], p. 174.

9. Charles Darwin, "Notebook M," *Charles Darwin's Notebooks 1836–1844*, Paul Barrett et al., eds. [Ithaca, New York: Cornell University Press, 1987], p. 559.

10. Francis Darwin, "Reminiscences," in *The Life of Charles Darwin* (1902) [London: Senate Books, 1995], p. 68.

Chapter 3

1. Charles Darwin to Charles Whitley, 15 November 1831, *The Correspondence of Charles Darwin*, Frederick Burkhardt, Sidney Smith, et al., eds. [Cambridge: Cambridge University Press, 1989], Vol. 1, 1821–36, p. 181.

2. Ibid.

3. Charles Darwin to Erasmus A. Darwin, 21 December 1828, *Correspondence*, Vol. 1, 1821–36, pp. 16–17.

4. Janet Browne, *Charles Darwin: Voyaging* [London: Jonathan Cape, 1995], p. 105.

5. J.W. Goodison, ed., *Fitzwilliam Museum Cambridge: Catalogue of Paintings I: Dutch, Flemish, French, German, Spanish* [Cambridge: Fitzwilliam Museum, 1960], p. ix.

6. Charles Darwin, "Notebook M," *Charles Darwin's Notebooks 1836–1844*, Paul Barrett et al., eds. [Ithaca, New York: Cornell University Press, 1987], p. 539. The editors speculate that the "National Institution" to which Darwin refers is either the National Gallery or the Royal Institution.

7. Charles Whitley to Charles Darwin, 5 February 1835, *Correspondence*, Vol. 1, 1821–36, p. 427.

8. Ibid.

9. Charles Darwin, "Notebook N," *Charles Darwin's Notebooks*, p. 567.

10. Admission card to the Plinian Society, undated, Darwin Papers, Vol. 5, object 42.

11. Unpublished notes, "Zoological Walk to Portobello," c. 1826, Darwin Papers, Vol. 5, object 50.

12. Ibid.

13. Martin Rudwick, "The Emergence of a Visual Language for Geological Science, 1760–1841," *History of Science*, Vol. 14, 1976, p. 184, n. 10.

14. Ibid., p. 150.

15. However, occasionally, fully illustrated versions are attempted, with new pictures.

16. Charles Darwin, "Notebook M," *Charles Darwin's Notebooks*, p. 177.

17. Howard Gruber, "Darwin's 'Tree of Nature' and Other Images of Wide Scope," *On Aesthetics in Nature*, Judith Wechsler, ed. [Cambridge, Massachusetts: MIT Press, 1978], p. 127.

18. Charles Darwin, *The Autobiography of Charles Darwin* (1876), Nora Barlow, ed. [London: Norton & Company, 1993], p. 47.

19. Ibid., pp. 77–78.

20. Charles Darwin to Thomas Henry Huxley, 3 September 1855, *Correspondence*, Vol. 5, 1851–55, p. 423.

21. Unpublished correspondence, Charles Darwin to John Collier, 16 February 1882, Darwin Papers, Vol. 143.

22. The illustrator James de Carle Sowerby produced engravings for the work after dissections by Darwin and Huxley.

23. Jeanne Pingree, *Thomas Henry Huxley: List of His Scientific Papers* [London: Imperial College of Science & Technology, 1968], p. 8.

24. Unpublished sketch, "Vascular Tissue in a small leaf of *Drosera rotundifolia*," Darwin Papers, Vol. 60, object 157.

25. *Insectivorous Plants*, Main Ledger, 30 June 1875, John Murray Archive.

26. Charles Darwin to James de Carle Sowerby, 13 April 1850, *Correspondence*, Vol. 4, 1847–50, p. 332.

27. *Variation* and the second edition of *Descent* each contain one engraving after a photograph; *Vegetable Mould* is almost exclusively illustrated with engravings after photographs.

28. A full list of works in the Darwin Library is printed in Mario di Gregorio and Nick Gill, *Charles Darwin's Marginalia* [New York: Garland, 1990], pp. xli–lxi.

29. For a complete description of this collection, see *A Catalogue of a Collection of Plaques, Medallions, Vases, Figures &c., in Coloured Jasper and Basalte . . . Formed by the Late Lord Tweedmouth* [London: Charles Davis, 1905], pp. 27–28. Several of the reliefs are currently in the collections of the National Museums and Galleries on Merseyside. See Robin Emmerson, "Wax Models by Pacetti and Angelini in the Lady Lever Art Gallery," *Ars Ceramica*, Vol. 12, 1995, pp. 22–33. I have been unable to establish how and when the collection passed from Darwin to Lord Tweedmouth.

30. Francis Hobart Herrick, *Audubon: The Naturalist: A History of His Life and Time* [New York: Appleton, 1938], p. 200.

31. Alice Ford, ed., *The 1826 Journal of John James Audubon* [Tulsa: University of Oklahoma Press, 1967], p. 177.

32. Ruskin used to prepare lecture diagrams for Buckland. See E.T. Cook and Alexander Wedderburn, *The Works of Ruskin* [London: George Allen, 1903–5], Vol. 26, p. xx.

33. Sarah Norton and M.A. DeWolfe Howe, eds., *The Letters of Charles Eliot Norton* [Boston, Massachusetts: Houghton Mifflin Company, 1913], p. 309.

34. Darwin, *Autobiography*, p. 61.

Chapter 4

1. Pseudo-Aristotle, "Physiognomica," *Aristotle, Minor Works*, Vol. 4, W.S. Hett, trans. [London: 1936], p. 808b.

2. Quoted in Michael W. Kwakkelstein, "Leonardo da Vinci's Grotesque Heads and the Breaking of the Physiognomic Mould," *Journal of the Warburg and Courtauld Institutes*, Vol. 54, 1991, p. 127.

3. Artists shared Darwin's preoccupation with expression. I have been unable to ascertain whether Darwin was exposed to the work of contemporaneous artists, e.g., Théodore Géricault and Francisco de Goya, whose interests in expression, and insanity in particular, converge with Darwin's work.

4. René Descartes, *The Passions of the Soul* (1650), Stephen Voss, trans. [Indianapolis, Indiana: Hackett Publishing Company, 1989], p. 24.

5. Ibid.

6. Charles Le Brun, *A Method to Learn to Design the Passions* (1667), David S. Rodes, ed., John Williams, trans. [Los Angeles, California: William Andrews Clark Memorial Library, 1980], p. i.

7. Early versions were called *Essays on the Anatomy of Expression in Painting.*

8. Sir Charles Bell, *The Anatomy and Philosophy of Expression as Connected with the Fine Arts*, 3rd edition (1824) [London: G. Bell & Sons, 1912], pp. 2–3.

9. Ibid., p. 194.

10. Charles Darwin, "Notebook C" 243, *Charles Darwin's Notebooks 1836–1844*, Paul Barrett et al., eds. [Ithaca, New York: Cornell University Press, 1987], p. 315.

11. Bell, *Anatomy and Philosophy of Expression*, p. 77.

12. William Paley, *Natural Theology; or, Evidences of the Existence and Attributes of the Deity, Collected from the Appearances of Nature*, 1802.

13. Darwin Library: Bell, *Anatomy and Philosophy*, 3rd edition (1844), p. 64. This is the copy Erasmus Darwin gave to Charles in 1866.

14. Ibid., pp. 106, 118, 119.

15. Ibid., pp. 150, 152, 154, 159.

16. Ibid., pp. 159, 160, 170, 183, 191, 194.

17. John Graham, *Lavater's Essays on Physiognomy: A Study in the History of Ideas* [Berne: Peter Lang Press, 1979], p. 62.

Chapter 5

1. Louis Agassiz, quoted in Frederick Rowe Davis, "Louis Agassiz in Search of Support for a Grand Critique of Darwinian Theory: The South American Thayer Expedition to Brazil," Unpublished thesis, Harvard University Archives, p. 30, as cited in Gwyneira Isaac, "Louis Agassiz's Photographs in Brazil: Separate Creations," *History of Photography*, Vol. 21, No. 1, Spring 1997, p. 5.

2. This although, interestingly, Agassiz had experimented with photography before. See Brian Wallis, "Black Bodies, White Science: Louis Agassiz's Slave Daguerreotypes," *American Art*, Vol. 9, No. 2, 1995, pp. 38–61.

3. Elizabeth Agassiz, *A Journey in Brazil by Professor and Mrs. Louis Agassiz* [Boston, Massachusetts: Fields & Osgood, 1867], p. 296, as cited in Isaac, "Louis Agassiz's Photographs in Brazil."

4. I believe this was the brother of the writer Henry James, although I have been unable to confirm this.

5. Unpublished diary, William James, "Brazilian Expedition Diary and Sketchbook," 1865, William James Papers, Houghton Library, Harvard University.

6. Agassiz did not explain the details of his financial arrangements with the sitters, but it seems likely a formal portrait was partial payment for his models' services. Such photographs would have had a dual purpose: Elizabeth Agassiz recounts that the models, many of whom had never been photographed before, were more comfortable sitting for her husband's photographs once they had seen a sample portrait.

7. Isaac, "Louis Agassiz's Photographs in Brazil," p. 9.

8. Agassiz Album, Peabody Museum Archives, Vol. 2, pp. 8 and 72, respectively.

9. Ibid., p. 22.

10. There is an excellent discussion of this issue in Isaac, "Louis Agassiz's Photographs in Brazil," p. 9.

11. Unpublished correspondence, Louis Agassiz to Charles Darwin, 22 July 1868, Darwin Papers, Vol. 82.1, item 78.

12. Unpublished correspondence, Charles Darwin to Louis Agassiz, 19 August 1868, Darwin Papers, Vol. 94, item 83.

13. Adrian Desmond and James Moore, *Darwin: The Life of a Tormented Evolutionist* [New York: Warner Books, 1991], p. 552.

14. Woodburytype and Albertype photographs, although made from pho-

tographic negatives, are printed in ink and, unlike traditional photographs, do not contain photosensitive salts. Both processes involve the use of gelatin films to transfer the image. In the early 1870s, such photomechanical prints were preferred to common albumen prints for their relative permanence and the consistency with which they could be mass-produced. Darwin considered using the Woodbury process himself but decided it was too expensive. In *Expression*, he used Edwards's newer, heliotype process instead.

15. Alexander Agassiz, "Application of Photography to Illustrations of Natural History. With Two Figures Printed by the Albert and Woodbury Processes," *Bulletin, Museum of Comparative Zoology, Harvard*, Vol. 3, 30 November 1871, p. 47.

16. Alexander Agassiz, letter to Charles Darwin, 9 December 1872, *Letters and Recollections of Alexander Agassiz* [Boston, Massachusetts: Houghton Mifflin, 1913], p. 121.

17. Roslyn Poignant, "Surveying the Field of View: The Making of the RAI Photographic Collection," *Anthropology and Photography: 1860–1920*, Elizabeth Edwards, ed. [New Haven, Connecticut: Yale University Press, 1992], p. 45.

18. Ibid., p. 68. Interestingly, Poignant describes a series of composite prints of Sandwich Islanders that were also presented before the society. Darwin's cousin Francis Galton used a similar composite printing technique in pursuit of his theories on eugenics.

19. Adrian Desmond, *Huxley: The Devil's Disciple* [London: Michael Joseph, 1994], p. 299.

20. Unpublished correspondence and photographs, William Houghton to Thomas Henry Huxley, 28 February 1868, Thomas Henry Huxley Papers, Imperial College of Science, Technology and Medicine, Box H, Folio 85.

21. I have used Johann Caspar Lavater, *Essays on Physiognomy*, [*Physiognomische Fragmente*]Thomas Holcroft, trans. [London: Thomas Tegg, 1840]. In this edition, the plate numbers are 61 and 62.

22. Rosie Mestel, "Gene Sleuth Plays Hunt the Werewolf," *New Scientist*, 3 June 1995, p. 10.

23. Charles Darwin, *The Descent of Man, and Selection in Relation to Sex* [London: John Murray, 1870], p. 25.

24. Unpublished albumen prints, F.R. Bopp Studio, "Vater," "Mutter," "sohn," and "Tochter der Haarigen Familie von Amras," Darwin Papers, Vol. 53.1, Folios 31–33, items 7–10. Darwin was right to be skeptical; it is highly unlikely both mother and father would share the condition, which is inherited. The paintings were formerly hung at Schloss Ambras in Austria; they are currently in the Kunsthistorisches Museum in Vienna.

25. *The People of India* was edited by Dr. John Forbes and Sir John William Kaye, and published by Wm. H. Allen & Co., London in eight volumes from 1868 to 1875. The text was written by Colonel Meadows Taylor. Robertson, Bourne, and Shepherd produced the vast majority of the photographs in the work, although as many as a dozen other photographers may also have contributed (see n. 26). I have been unable to determine which of the eight volumes would have been published prior to Campbell's talk. The first volume featured the people of Hindustan.

26. Bourne is said to have been the photographer, and Shepherd the printer. A number of amateur photographers also participated in the project, including Willoughby Wallace Hooper (1837–1912), Henry Charles Baskerville Tanner (1835–98), and James Waterhouse (1842–1922). The collection had begun informally as a souvenir of the country, initiated by retiring Governor General Canning, but quickly became an official project of the colonial government.

27. Poignant, "Surveying the Field of View," p. 49.

28. Unpublished correspondence, Thomas Henry Huxley to Lord Granville, 12 August 1869, Huxley Papers, Vol. 30, Folio 75, p. 1.

29. Ibid., Folio 76, p. 2.

30. Ibid., Folio 77–78, pp. 3–4.

31. Huxley Papers, Box G, Folio 2.

32. Ibid.

33. Frank Spencer, "Some Notes on the Attempt to Apply Photography to Anthropometry during the Second Half of the Nineteenth Century," *Anthropology and Photography: 1860–1920*, Elizabeth Edwards, ed. [New Haven, Connecticut: Yale University Press, 1992], p. 100.

34. These even though Huxley had not excluded immigrant populations, including white colonists, from his request.

35. Unpublished correspondence, Robert W. Herbert (Colonial Office) to Thomas Henry Huxley, 9 February 1871, Huxley Papers, Vol. 15, Folios 152–54.

36. Unpublished correspondence, Central Board for Aboriginies, Melbourne, to Thomas Henry Huxley, 17 May 1870, Huxley Papers, Vol. 15, Folios 117–18.

37. Unpublished correspondence, A.B. Fryers to Thomas Henry Huxley, 2 March 1872, Huxley Papers, Vol. 15, Folio 155.

38. Unpublished correspondence, Herbert, Huxley Papers, Vol. 15, Folio 154.

39. Charles Darwin, letter to Asa Gray, 11 April 1861, *The Correspondence of Charles Darwin*, Frederick Burkhardt et al., eds. [Cambridge: Cambridge University Press, 1994], Vol. 9, 1861, pp. 88–89.

40. An excellent discussion of Darwin's perception of his own body, including a discussion of photographs made of him, is contained in Janet Browne, "I Could Have Wretched All Night: Charles Darwin and His Body," *Science Incarnate: Historical Embodiments of Natural Knowledge*, Christopher Lawrence and Steven Shapin, eds. [Chicago: University of Chicago Press, 1998], 240–87.

41. Charles Darwin, letter to J.D. Hooker, 27 May 1855, *Correspondence*, Vol. 5, 1855, p. 339. The "Club" to which Darwin refers is the Literary Scientific Portrait Club.

42. Charles Darwin, letter to J.D. Hooker, 17 March 1862, *Correspondence*, Vol. 10, 1862, p. 119.

43. Charles Darwin, *The Autobiography of Charles Darwin* (1887), Nora Barlow, ed. [New York: W.W. Norton & Company, 1958], p. 57.

44. Ibid., p. 72.

45. Charles Darwin, letter to Alphonse de Candolle, 1 January 1863, *Correspondence*, Vol. 11, 1863, p. 74. The son is not identified but is probably William.

46. Charles Darwin, letter to William Darwin, c. 11 September 1857, *Correspondence*, Vol. 6, 1857, p. 451.

47. Ibid.

48. Desmond and Moore, *Darwin*, p. 460.

49. Josef Maria Eder, *History of Photography*, Edward Epstean, trans. [New York: Columbia University Press, 1945], p. 134.

50. See, e.g., Larry Schaaf, *The Photographic Art of William Henry Fox Talbot* [Princeton, New Jersey: Princeton University Press].

51. Desmond and Moore, *Darwin*, p. 214.

52. Darwin, *Autobiography*, pp. 67–68.

53. Ibid., p. 215.

54. John Herschel, *Photographic News*, 11 May 1860, as cited in Mark Haworth-Booth, ed., *The Golden Age of British Photography* [New York: Aperture, 1986], p. 185.

55. Martin Rudwick, *Scenes from Deep Time* [Chicago: University of Chicago Press, 1992], p. 241. In addition to the title, the term "scene from deep time" appears throughout the book.

Chapter 6

1. Guillaume-Benjamin Duchenne de Boulogne, *Mécanisme de la physionomie humaine ou analyse électro-physiologique de l'expression des passions applicable à la pratique des arts plastiques* [Paris: Jules Renouard, 1862].

2. Duchenne's personal copy in the collection of L'École Nationale Supérieure des Beaux Arts, Paris, contains only fifty-eight of the seventy-four photographs originally included in this folio. See Françoise Heilbrun and Philippe Néagu, *Musée d'Orsay: Chefs d'Oeuvre de la Collection Photographique* [Paris: Réunion de Musées Nationaux, 1986], pp. 104–5. Darwin's copy contains seventy-two. The book also contained nine "synoptic" plates comparing expressions to each other, and 150 pages of textual analysis. The scholar Jean-Francois Debord has suggested that the oversized editions of *Mécanisme* may have been intended as a kind of rough draft for a deluxe grand-in-quarto edition, advertised by the publisher Veuve Renouard in 1862 but never completed. If this is true, Darwin must have owned one of the most precious copies of Duchenne's book.

3. See Maria Morris Hambourg and Françoise Heilbrun, *Nadar* [New Haven: Connecticut: Yale University Press, 1994].

4. Strictly speaking, *Mécanisme* contained an astonishing 228 photographic images, because it included nine "synoptic" plates, each containing sixteen miniature photographs that had been masked to point up various aspects of expression that were unclear in the large-format plates. An octavo version of the album was also published, which only contained the synoptic plates. Darwin had both versions of the book in his possession.

5. Unpublished correspondence, G.B. Duchenne de Boulogne to Charles Darwin, 25 March 1871, Darwin Papers, Vol. 162.

6. Unpublished correspondence, Charles Darwin to G.B. Duchenne de Boulogne, 10 March 1871, Darwin Papers, Vol. 96, Folios 103–4.

7. G.B. Duchenne de Boulogne, *The Mechanism of Human Facial Expression*, R. Andrew Cuthbertson, trans. [Cambridge: Cambridge University Press, 1990], p. 42.

8. Ibid., p. 43.

9. Marginalia from Darwin's copy of Duchenne's *Mécanisme* in the Cambridge University Library.

10. Charles Darwin, *The Expression of the Emotions in Man and Animals* (1872), Paul Ekman, ed. [London: HarperCollins, 1998], pp. 13–14.

11. Duchenne, *Mechanism*, p. 103.

12. Ibid.

13. Ibid., p. 106.

14. Ibid., p. 107.

15. A.L. Ashworth, *Stanley Road Hospital, Wakefield: One Hundred and Fifty Years, a History* [Wakefield, England: Wakefield Area Health Authority, 1975], pp. 69–71.

16. Darwin Marable, "Photography and Behaviour in the Nineteenth Century," *History of Photography*, Vol. 9, No. 2, April-June 1985, p. 141.

17. Adrian Desmond and James Moore, *Darwin: The Life of a Tormented Evolutionist* [New York: Warner Books, 1991], p. 32.

18. Unpublished photograph, Darwin Papers, stored with "Anthropological Photographs" folio.

19. Michael Pritchard, *Directory of London Photographers, 1841–1908* [Watford, England: PhotoResearch, 1994], p. 70. Woodcuts after photographs by Henry Hering appeared in the *Illustrated London News* between 1865 and 1871 (Vol. 46, 13 May 1865, p. 464; Vol. 50, 6 April 1867, p. 329; Vol. 58, 17 June 1871, p. 588). These were printed with the Regent Street address.

20. Reproduced in Adrienne Burrows and Iwan Schumacher, *Portraits of the Insane: The Case of Dr. Diamond* [London: Quartet Books, 1990], plate 50.

21. For "Melancholia," see Darwin Papers, Vol. 53.1, Folio 8, items 34 and 35; for "Homicidal Impulses," see item 25.

22. Cover slip: Darwin Papers, Vol. 53.1, Folio 41, object 28.

23. Envelope: Darwin Papers, Vol. 53.1, Folio 6, object 8.

24. Her tartan shawl should have been a dead giveaway. University of Edinburgh Library collections. With thanks to Mike Barfoot for this information.

25. In Darwin, *Expression*, p. 297.

26. Unpublished correspondence, Charles Darwin to James Crichton-Browne, 26 March 1871, Darwin Papers, Vol. 143.

Chapter 7

1. Charles Darwin, "A Biographical Sketch of an Infant," *Mind*, No. 7, July 1877, at p. 285.

2. Charles Darwin, "Notebook N," *Charles Darwin's Notebooks: 1836–1844*, Paul Barrett et al., eds. [Ithaca, New York: Cornell University Press, 1987], p. 582.

3. Darwin, "Notebook M," *Darwin's Notebooks*, p. 545.

4. Ibid., p. 542.

5. Darwin, "Biographical Sketch," p. 288.

6. Ibid., p. 290. This curious remark was probably a reaction to an article that appeared in the *Annals and Magazine of Natural History* in 1852. Darwin cited it in *The Formation of Vegetable Mould through the Action of Worms* [London: John Murray, 1881]. A pike, the author observed, will continue over a course

of months to bruise its head against the wall of an aquarium in hopes of catching a minnow on the opposite side (p. 94).

7. Charles Darwin, *On the Origin of Species by Means of Natural Selection* [London: John Murray, 1859], p. 439.

8. Ibid., p. 450.

9. Janet Browne, "Darwin and the Face of Madness," *The Anatomy of Madness: Essays in the History of Psychiatry*, W.F. Bynum, Roy Porter, and Michael Shepherd, eds. [London: Tavistock, 1985], pp. 154–55. Browne argues that similar interest in instinctive behaviors led him to study photographs of the mentally ill.

10. Charles Darwin, *The Expression of the Emotions in Man and Animals* (1872), Paul Ekman, ed. [London: HarperCollins, 1998], p. 352.

11. Darwin, *Expression*, p. 146. Darwin considered "fear" and "terror" special cases and treated them in a chapter separate from "Suffering and Weeping."

12. Darwin, "Biographical Sketch," pp. 287–88.

13. Darwin, *Expression*, p. 150.

14. Parts of this section are adapted from a short paper by the author that appeared in the journal *History of Photography* under the title "Rejlander, Darwin, and the Evolution of *Ginx's Baby*," Vol. 23, No. 3, Autumn 1999, pp. 260–68.

15. Edgar Yoxall Jones, *O.G. Rejlander: Father of Art Photography: 1813–1875* [London: David & Charles, 1973], p. 38.

16. Darwin, *Expression*, p. 147.

17. The precise number is difficult to quantify, and the history of issues within the first edition is not completely understood. A good discussion appears in Richard Broke Freeman, *The Works of Charles Darwin: An Annotated Bibliographical Handlist*, 2nd edition [Folkestone, England: Wm. Dawson & Sons, 1977].

18. Unpublished letter from Charles Darwin to Robert Cooke, 20 November 1872, Darwin Manuscripts Collection, Cambridge University Library, DAR 143:288.

19. Unpublished letter from Charles Darwin to Robert Cooke, 7 December 1872, Darwin Manuscripts Collection, Cambridge University Library, DAR 143:289.

20. This portrait appears in Ernest Walford's *Representative Men in Literature, Science and Art* (1868).

21. Unpublished correspondence, Charles Darwin to Robert Cooke, 22 April 1872, John Murray Archive, item 285.

22. Unpublished correspondence, Charles Darwin to Robert Cooke, 10 May 1872, John Murray Archive, item 278.

23. Unpublished correspondence, Charles Darwin to Robert Cooke, 28 May 1872, John Murray Archive, item 280.

24. Unpublished correspondence, Charles Darwin to Robert Cooke, 17 August 1872, John Murray Archive, item 283.

25. Darwin wrote, "There is one other element to ascertain & inform me, viz. the cost of Mr. Cooper's engravings. He has done two and is doing a third" (ibid.).

26. Ibid.

27. Unpublished correspondence, Oscar Rejlander to Charles Darwin, 11 November 1871, Darwin Papers, Vol. 176, item 116.

28. Darwin, *Expression*, p. 147.

29. Jones, *O.G. Rejlander*, p. 38. Jones, who is now deceased, did not disclose how these numbers were determined. I think they were probably taken from a nineteenth-century source and are almost certainly an exaggeration. For Rejlander to have produced this many prints would have been a superhuman feat—he suffered from debilitating illness throughout the last years of his life and is not known to have employed any assistants other than his wife, Mary. Nevertheless, it is clear that a large number of prints were made. As of this writing, it is still fairly easy to find copies of *Ginx's Baby* in carte-de-visite sales.

30. Edward Jenkins, *Ginx's Baby, His Birth and Other Misfortunes* [London: Strahan & Co., 1876]. The name Ginx" is clearly a homophone of "jinx," an appropriate name for the luckless father. This dual meaning probably accounts for the variant spellings of the title ("Jinx's Baby" and "Jinks Baby" are the most common) that appear both in critical literature and on the mounts of some copies of the image. These alternative spellings could also have been conceived as an attempt to avoid copyright infringement.

31. Jenkins, *Ginx's Baby*, pp. 62–63.

32. S.F. Spira, "Polka Tot," *History of Photography*, Vol. 11, No. 4, October-December 1987, p. 328.

33. Naomi Salaman and Ronnie Simpson, eds., *Postcards on Photography: The Handmade Copy in Reproduction* [Cambridge: Cambridge Darkroom Gallery, 1998], p. 79.

34. Ibid.

35. Darwin did not refer to the photographer by name; however, Wallich is the only London-based photographer Darwin is known to have commissioned work from at this time.

36. Unpublished correspondence, Charles Darwin to James Crichton-Browne, 8 June 1870, Darwin Papers, Vol. 143.

37. The complete version is an unpublished photograph: George Charles Wallich, "Beatrice Wallich," Darwin Papers, Vol. 53.1, Folio 45, item 50. The archive also includes a Wallich photograph of a "Nun in Rapture," Vol. 53.1, Folio 43, item 38.

38. Darwin, *Expression*, p. 200.

39. Unpublished correspondence, Charles Darwin to Adolph Kindermann, 27 March 1871, Darwin Papers, Vol. 96, item 98.

40. The Darwin Archive includes sixteen Kindermann portraits.

41. Darwin, *Expression*, p. 147.

42. Unpublished correspondence, Charles Darwin to James Crichton-Browne, 8 June 1869, Darwin Papers, Vol. 143.

43. "American Novelties," *Photographic Times*, Vol. 4, No. 44, p. 127.

44. "Landy's Photographic Pets," *Photographic Times*, Vol. 2, 19 July 1872, p. 104.

45. Darwin's copy of *Expressive Pets* appears to be the earlier, twelve-plate version.

46. "Editor's Table—Medals Awarded at the Cincinnati Exhibition," *Philadelphia Photographer*, Vol. 9, No. 107, November 1872, p. 400.

47. "Buy Landy's *Expressive Pets*," *Anthony's Photographic Bulletin*, Vol. 5, No. 6, June 1874, p. 226.

48. Darwin, *Expression*, p. 27.

49. Unpublished correspondence, Charles Darwin to John Murray, 23 April 1871, John Murray Archive, Darwin 240. Murray was Bell's publisher as well as Darwin's.

50. Mario Di Gregorio and Nick Gill, *Charles Darwin's Marginalia* [New York: Garland Publishing, 1990], p. 48.

51. Darwin, *Expression*, p. 2 (footnote).

52. Sir Charles Bell, *The Anatomy and Philosophy of Expression as Connected with the Fine Arts* (1806) [London: G. Bell & Sons, 1880], p. 136.

53. Ibid.

54. Di Gregorio and Gill, *Darwin's Marginalia*, p. 48.

55. Ibid.

56. See, e.g., Beaumont Newhall, *The History of Photography* [New York: Museum of Modern Art, 1982], p. 77.

57. Royal Photographic Society, PF15-2764.

58. Oscar Rejlander, "An Appeal to Photographers," *British Journal of Photography*, 29 July 1864, p. 266.

59. A.H. Wall, "Rejlander's Photographic Art Studies: Photographing Children," *Photographic Times and American Photographer*, Vol. 16, 1887, pp. 672–73.

60. Oscar Rejlander, unpublished photograph, *Ginx's Baby*, Darwin Papers, Vol. 53.1, Folio 63, item 132.

61. Oscar Rejlander, unpublished photograph, "Oscar Rejlander mimicking *Ginx's Baby*," Darwin Papers, Vol. 53.1, Folio 54, item 96.

62. Oscar Rejlander, "Photography as a Handmaid of Art," *British Journal of Photography Almanac*, 1866, as cited in Jones, *O.G. Rejlander*, pp. 33–34.

63. An album of 1862 in the collections of George Eastman House, Rochester, New York, contains a version of the same image in cabinet card size ["Young girl, standing," 72:249:02].

64. Oscar Rejlander, unpublished albumen print, Darwin Papers, Vol. 53.1, Folio 52, item 91.

65. Newhall, *History of Photography*, p. 62.

66. Oscar Rejlander, unpublished photographic prints, "Child Screaming, Held from the Side," and "Child Screaming, Held from Behind," cyanotypes or related process, Darwin Papers, Vol. 53.1, Folio 62, items 129–30.

67. Oscar Rejlander, *Daisy at One*, carte-de-visite albumen print in the collection of the author.

68. Marcia Pointon, *Hanging the Head: Portraiture and Social Formation in Eighteenth-Century England* [New Haven, Connecticut: Yale University Press, 1993], p. 221.

69. Jones, *O.G. Rejlander*, p. 107.

70. Unpublished photographs, "Oscar Rejlander Mimicking *Ginx's Baby*," Oscar Rejlander, Darwin Papers, Vol. 53.1, Folio 53, items 95–96.

71. Unpublished photograph, "Laughing and Crying," London Stereoscopic Company, Darwin Papers, Vol. 53.1, Folio 47, item 64.

72. Darwin, *Expression*, pp. 91–92.

73. Bridget Henisch and Heinz Henisch, *The Photographic Experience 1839–1914: Images and Attitudes* [University Park: Pennsylvania State University Press, 1994], p. 383.

74. *Introducing . . . Mr. Rejlander,* 1871–72, Darwin Papers, Vol. 53.1, Folio 44, object 46.

75. Robert Sobieszek, *Ghost in the Shell: Photography and the Human Soul, 1850–2000,* [Cambridge, Massachusetts: MIT Press, 1999], p. 201.

76. "The Late O.G. Rejlander," *British Journal of Photography,* 29 January 1875, p. 56.

Chapter 8

1. Amy Woolner, "List of Works," *Thomas Woolner R.A.: His Life and Letters* [London: Chapman & Hall, 1917], pp. 336–37.

2. Amy Woolner, "List of Works," indicates that Woolner was dissatisfied with his portraits of Carlyle and Tennyson. He reworked each of these compositions after his return to England.

3. Letter from Edward Lear to Thomas Woolner, 25 June 1860, reprinted in Amy Woolner, "List of Works," p. 196. Lear refers in the same letter to dining with "Mrs. Cameron." Whether this refers to the photographer Julia Margaret Cameron cannot be established. In a subsequent letter, dated 1 May 1870, Lear reveals some of the effect Darwinism was having among the former Pre-Raphaelites. He described William Holman Hunt (who, unlike his brethren, remained a "Pre-Raphaelite" throughout his career) as having "a holy horror of Darwin." See Amy Woolner, "List of Works," p. 284.

4. Woolner's 1869 medallion and 1870 bust are the only contemporaneous portraits of Darwin by him. A marble plaque of Darwin based on Woolner's design was produced posthumously by Josiah Wedgwood and Sons for installation at Christ's College, Cambridge.

5. Letter from W.E. Darwin to T. Woolner, 7 June 1869, reprinted in Amy Woolner, "List of Works," p. 283.

6. Amy Woolner, "List of Works," p. 340.

7. Letter from Charles Darwin to Thomas Woolner, 7 April 1871, reprinted in Amy Woolner, "List of Works," p. 288.

8. Charles Darwin, *The Expression of the Emotions in Man and Animals* (1872), Paul Ekman, ed. [London: HarperCollins, 1998], p. 311.

9. Both Langstaff and Crichton-Browne provided Darwin with photographs of expressions, though neither photographed blushing.

10. Mark Stocker, "Thomas Woolner," *Dictionary of Art,* Jane Turner, ed. [New York: Grove Dictionaries, 1996], Vol. 33, p. 374.

11. He was acquainted, e.g., with the photographer Julia Margaret Cameron. As Sylvia Wolf has noted, Woolner proposed marriage to Cameron's favorite model, Julia Jackson, in the mid-1860s but was rebuffed. Wolf also suggests that Oscar Rejlander photographed Jackson, although the attribution of photographs of her to Rejlander remains tenuous. Woolner is also known to have used Cameron's sister and occasional photographic model Virginia Dalrymple (1825–1892) as models for the body of the mother figure in his work *Civilisation* in 1859. The hair, however, was rendered from a different model,

one Fanny Waugh of Leatherhead. The sculpture, intended to convey the triumph of maternal love, depicts a woman embracing a child on a pedestal. The base of the pedestal, subordinate to the central figures, is decorated with a bas-relief representing acts of cannibalism and human sacrifice. The child is strikingly similar to photographs of cherubim Cameron made during the late 1860s. This similarity is consistent with observations made by the British photohistorian Mark Haworth-Booth, who has observed that the cherubim depicted on the façade of the Henry Cole Wing of the Victoria and Albert Museum may also have been made after Cameron photographs. Cameron, who aggressively cultivated a reputation as a photographic fine artist, is not often associated with the production of studies for artists. Nevertheless, the location of her darkroom in the courtyard of the former South Kensington Museum would have made her a convenient source for such studies.

12. Francis Turner Palgrave, *Essays on Art* [New York: Hurd & Houghton, 1867], p. 279.

13. Palgrave's frustration was directed primarily at new, mechanical methods for making sculpture, e.g., François Willème's briefly popular system of "photosculpture," which was designed to enable a technician to render the profile of a sitter in three dimensions, using a series of photographs taken in the round. Palgrave pulled no punches about Woolner's contemporaries, denouncing Baron Carlo Marochetti's (1805–67) "colossal clumsiness" and Alexander Munro's (1825–71) "grotesque and babyish" compositions.

14. Thomas Woolner, *Silenus* [London: Macmillan & Company, 1884], p. 40.

15. According to Mark Stocker, the sculpture was admired by William Holman Hunt and helped secure Woolner's entry into the Pre-Raphaelite brotherhood in 1848. Since 1982, the original has been in the possession of Woolner's grandson. See Benedict Read, *Victorian Sculpture* [New Haven, Connecticut: Yale University Press, 1982], p. 37.

16. William Shakespeare, *Midsummer Night's Dream*, act II, scene 1.

17. Amy Woolner, "List of Works," p. 4.

18. Charles Darwin, *The Descent of Man, and Selection in Relation to Sex* [London: John Murray, 1870], p. 22.

19. Lavater, by contrast, did not know what to make of them. He dropped a planned chapter on ears from his *Fragments of Physiognomy* as inconclusive.

20. Darwin, *Descent*, p. 22.

21. Letter from Charles Darwin to Thomas Woolner, 10 March 1869, reprinted in Amy Woolner, "List of Works," p. 283.

22. Letter from Charles Darwin to Thomas Woolner, 7 April 1871, reprinted in Amy Woolner, "List of Works," p. 288.

23. Unpublished correspondence, Charles Darwin to Edwin Ray Lankester, 22 March 1871, Darwin Papers, Vol. 146. Darwin also wrote simultaneously to Woolner, advising him of his intention to find additional illustrations.

24. Unpublished correspondence, Hinrich Nitsche to Charles Darwin, 18 April 1871, Darwin Papers, Vol. 87, Folios 41–42. With thanks to the Darwin Papers Project for this translation of the original letter, which is written in German.

25. Darwin, *The Descent of Man and Selection in Relation to Sex*, 2nd edition [London: John Murray, 1874], p. 17.

26. Darwin Papers, Vol. 87, item 67.

27. Unpublished correspondence, William Preyer to Charles Darwin, 8 May 1871, Darwin Papers, Vol. 87, items 49–50.

28. See Michael Jaffé, *Rubens: Catalogo Completo* [Milan: Rizzoli, 1989], p. 308.

29. Christopher White, *Peter Paul Rubens, Man and Artist* [New Haven, Connecticut: Yale University Press, 1987], p. 62.

30. Unpublished correspondence, Charles Darwin to Thomas Lauder Brunton, 25 September 1874, Darwin Papers, Vol. 143, item 162. The correspondence makes clear the discussion began earlier. However, because previous sections of correspondence have been lost, it remains unclear when the discussion of Rubens began.

31. Unpublished correspondence, Charles Darwin to Thomas Lauder Brunton, 29 October 1874, Darwin Papers, Vol. 143, item 163.

32. Unpublished albumen print, José de Ribera's *Mary in Egypt*, Darwin Papers, Vol. 53.1, Folio 9, item 40.

33. A.H. Palmer, *The Life of Joseph Wolf, Animal Painter* [New York: Longmans, Green, & Company, 1895], p. 63.

34. Unpublished drawing, Darwin Papers, Vol. 53.1, item 163.

35. A variant drawing, in which the ears have not been altered, is reproduced in Palmer, *Life of Joseph Wolf.* Titled Laughing Monkey and described as a "duplicate version of a sketch . . . in the Artist's possession," the original of this drawing has been lost.

36. Darwin Papers, Vol. 53.1, Folio 81, object 158.

37. Palmer, *Life of Joseph Wolf*, pp. 192–93.

38. Letter from Charles Darwin to Joseph Wolf, 3 March 1871, reprinted in Palmer, *Life of Joseph Wolf*, pp. 193–94.

39. Palmer, *Life of Joseph Wolf*, pp. 195–96.

40. Darwin Papers, Vol. 53.1, Folios 84–88, items 163–67.

41. Drawing by Joseph Wolf, Darwin Papers, Vol. 53.1, Folio 85, object 164.

42. The notion of a cat scraping the leg of a chair was one that he had carried with him since his days on the *Beagle*, although in a rather different context. In unpublished notes made on the *Beagle* voyage, he wrote: "I conceive this habit of the Jaguar is exactly similar to one which may any day be seen in the common cat, as with outstretched legs & exserted [*sic*] claws it scrapes the leg of a chair." Darwin Papers, Vol. 29.1, Folio 28.

43. Palmer, *Life of Joseph Wolf*, p. 194.

44. Ibid., p. 196. In another instance, the naturalist Philip Henry Gosse (1810–88) asked Wolf for a drawing of a gorilla "throttling a negro with his hind foot; dragging up a tree and strangling him." Wolf was disgusted:

> Such untruth is awful! . . . In the case of animal life far in the world where I have never been, *to a certain extent* I have been obliged to believe what I have been told. I believed it if what I knew and what I could see at the different Zoological Gardens did not contradict it. But as I have said before, there are very few people who can interpret truly what they see in nature.

45. The number of visits is not specified.

46. Palmer, *Life of Joseph Wolf*, pp. 197–98.

47. Ibid., p. 180.

48. Ibid., p. 197.

49. "Mr. Briton Riviere, R.A.," *Connoisseur*, Vol. 57, No. 225, May 1920, p. 119.

50. In 1870 he exhibited the painting *Charity* at the Royal Academy, the first of his paintings to be engraved. The following year he submitted *Circe and the Friends of Ulysses*, which became his first popular triumph. In the same year he began to produce sentimental pictures of children with dogs, beginning with *His Only Friend* (Manchester City Art Gallery), which the *Art Journal* described as "a sort of Oliver Twist asleep under a hedge, with a terrier cuddling into his side."

51. Sister of Henry Bonham-Carter (1827–1921).

52. Unpublished correspondence, Briton Riviere to Elinor Mary Bonham Carter, 26 June 1871, Darwin Papers, Vol. 176.

53. Unpublished correspondence, Charles Darwin to Briton Riviere, 27 June 1871, Darwin Papers, Vol. 147.

54. Unpublished correspondence, Briton Riviere to Charles Darwin, 3 April 1872, Darwin Papers, Vol. 176.

55. Unpublished correspondence, Briton Riviere to Charles Darwin, 16 May 1872, Darwin Papers, Vol. 176.

56. Unpublished correspondence, Briton Riviere to Charles Darwin, 20 May 1872, Darwin Papers, Vol. 176.

57. Unpublished correspondence, Charles Darwin to Briton Riviere, 1 April 1872, Darwin Papers, Vol. 147, item 318.

58. Unpublished correspondence, Charles Darwin to Briton Riviere, 20 May 1872, Darwin Papers, Vol. 147.

59. Joseph Wolf, *Dog in a Humble and Affectionate Frame of Mind*, Darwin Papers, Vol. 53.1, Folio 85, item 164.

60. The painting was originally called *Prisoners*, the idea being that it showed man and dog behind bars.

61. Elsewhere I have argued that the layout of *Expression* helped to make it more palatable to Darwin's intended popular audience. See Phillip Prodger, "Illustration as Strategy in Charles Darwin's *Expression of the Emotions in Man and Animals*," *Inscribing Science: Materiality and Text in Scientific Communication*, Timothy Lenoir, ed. [Stanford, California: Stanford University Press, 1998].

62. Darwin, *Expression*, p. 26.

63. Main ledger for *Expression*, John Murray Archive, 1872, p. 281.

64. Main ledger for *Descent*, John Murray Archive, 1871, p. 398. The precise figure is difficult to ascertain, as several payments were made for drawings and engravings together.

Chapter 9

1. Algernon Graves, *A Dictionary of Artists Who Have Exhibited Works in the Principal London Exhibitions from 1760 to 1893* (1901) [Bath, England: Kingsmead Reprints, 1973], p. 231.

2. Some of this may be gleaned from existing biographies of Rejlander,

e.g., Edgar Yoxall Jones, *O.G. Rejlander: Father of Art Photography: 1813–1875* [London: David & Charles, 1973], and Stephanie Spencer, *O.G. Rejlander: Photography as Art* [Ann Arbor, Michigan: UMI Research Press, 1985]. However, many of the details in this chapter are revealed here for the first time, in part as a result of the wealth of information contained in the Darwin Archive.

3. Accession numbers FM1965001785 and FM1965001787. See Leif Wigh et al., *Oscar Gustave Rejlander, 1813(?)–1875* [Stockholm: Moderna Museet, 1998], which includes reproductions of both portraits. Dating is based on stylistic assessment of the portraits as well as the known dates of other photographs in the collection, which otherwise consists of genre pictures made prior to 1870. However, photographs by Rejlander are notoriously difficult to date on stylistic grounds alone, because his methods changed little over time and he frequently reprinted negatives made early in his career.

4. Francis Darwin, *The Life and Letters of Charles Darwin* (1887) [New York: D. Appleton & Company, 1919], p. 543. Francis identifies the artist as "O.J. Rejlander." Francis's evaluation is especially credible given his intimate knowledge of his father's physical appearance at the time *Expression* was written and his direct involvement in the project, because he edited the posthumous second edition of 1890.

5. Unpublished correspondence, Charles Darwin to James Crichton-Browne, 1 April 1871, Darwin Papers, Vol. 143.

6. Camfield Wills and Deirdre Wills, *History of Photography: Techniques and Equipment* [New York: Exeter Books, 1980], p. 81.

7. Oscar Rejlander, "Hints Concerning the Photographing of Criminals," *British Journal Photographic Almanac*, 1872, pp. 116–17. The *British Journal Photographic Almanac* is a collection of articles published in the *British Journal of Photography*.

8. The association can be confirmed by photographs Rejlander and Cameron took of each other during this period. See, e.g., Eugenia Parry Janis et al., *For My Beloved Sister Mia: An Album of Early Photographs by Julia Margaret Cameron* [Albuquerque: University of New Mexico Press, 1995]. The authors estimate the period of their collaboration c. 1863–65.

9. The circumstances of how Cameron and Darwin met are also unclear, although it is possible they were brought together at least indirectly through their mutual friend John Herschel (1792–1871). Cameron was also associated with the Pre-Raphaelites; the association suggests the possibility that Darwin's connection with Cameron, and thus Rejlander, was either directly or indirectly mediated by sculptor Thomas Woolner. See Sylvia Wolf et al., *Julia Margaret Cameron's Women* [New Haven, Connecticut: Yale University Press, 1998]. See also chapter 11.

10. In fact, Rejlander's studio is still there. Although it does not enjoy formal architectural protection, his specially built glass-roofed studio has survived. At this writing, it may still be seen on the upper story of Albert Mansions.

11. This calculation includes one or two duplications, and the original photograph for figure 4, Small Dog Watching a Cat on a Table, which has been lost. These are all the photographs attributable to Rejlander in the Darwin Archive; Rejlander probably provided others, but they have not been preserved.

12. *Photographic News*, 2 July 1875, as cited in Jones, *O.G. Rejlander*, p. 38.

13. The title, e.g., of Jones's influential *O.G. Rejlander: Father of Art Photography: 1813–1875*.

14. As cited and translated in "An Explanation by Mr. Rejlander," *British Journal of Photography*, Vol. 12, No. 289, 17 November 1865, p. 589. The spelling "Mittheilungen" is as given in the article.

15. This assertion is often voiced; see, e.g., Rolf H. Kraus, "Photographs as Early Scientific Book Illustrations," *History of Photography*, Vol. 2, No. 4, October 1978, p. 302.

16. *A Young Naturalist*, Albumen print, Royal Photographic Society, RPS 5094.

17. Charles Baudelaire, "The Salon of 1859" (Jonathan Mayne, trans.), *Photography in Print*, Vicki Goldberg, ed. [Albuquerque: University of New Mexico Press, 1988], p. 124.

18. Oscar Rejlander, "Rejlander in Type," *British Journal of Photography*, Vol. 12, No. 279, 8 September 1865, p. 462.

19. "The Photographs of Rejlander," *Art Journal*, January 1868, p. 15.

20. The Royal Collections at Windsor, e.g., include several early landscapes that were purchased by or for Prince Albert.

21. This is especially true of photographs he produced during his residence in Wolverhampton, where he worked until 1862.

22. This, at least, was how the London Stereoscopic Company advertised its prints. *Catalogue of Binocular Pictures of the London Stereoscopic Company, 54 Cheapside and 313 Oxford Street*, undated (c. 1867), Victoria & Albert Museum collection.

23. *Photographic Journal*, 15 September 1859, as cited in Jones, *O.G. Rejlander*, p. 47.

24. I have taken all of these examples from Jones, *O.G. Rejlander*, pp. 43, 47, 51.

25. *Journal of the Photographic Society of London*, 21 January 1857, as cited in Jones, *O.G. Rejlander*, p. 55.

26. Tate Gallery, N00099.

27. See the *Cherub* in, e.g., the collections of the Edinburgh Photographic Society.

28. The claim that Rejlander studied in Rome, which seems to have originated with the artist himself, is frequently repeated in literature about Rejlander. While this is certainly possible, there is no proof this was actually the case.

29. Oliver Mathews, *Early Photographs and Early Photographers: A Survey in Dictionary Form* [New York: Pitman Publishing Corporation, 1973], p. 21.

30. Oscar Gustave Rejlander, "An Apology for Art Photography," *British Journal of Photography*, 16 February 1863, pp. 76–77.

31. Published in Frank Howard, "Photography Applied to Fine Art," *Journal of the Royal Photographic Society*, No. 13, 21 January 1854, pp. 154–58. Perhaps the most eloquent expression of this opinion is contained in Rejlander's allegorical photograph *The Infant Photography Hands the Artist a New Brush* (1856). See, e.g., item 84.XP.458.34, J. Paul Getty Museum, Malibu.

32. For an extensive discussion of this issue, see Mark Haworth-Booth, *Photography: An Independent Art* [Princeton, New Jersey: Princeton University Press, 1997].

33. Rejlander, "Apology," p. 77.

34. Oscar Rejlander, "What Photography Can Do in Art," *Yearbook of Photography and Photographic News Almanac*, 1867, p. 50, as cited in Stephanie Spencer, "O.G. Rejlander: Art Studies," *British Photography in the Nineteenth Century: The Fine Art Tradition*, Mike Weaver, ed. [Cambridge: Cambridge University Press, 1989].

35. The historian Edgar Yoxall Jones, in *O.G. Rejlander*, argued that he may have counted Lawrence Alma-Tadema (1836–1912) and Edward Burne-Jones (1833–98) among his clients. Certainly, when the printseller Henry Evans, one of Rejlander's principal stockists, was raided by police in 1870 for selling supposedly obscene photographs, Burne-Jones and Dante Gabriel Rossetti (1828–1882) were among the artists who wrote to the editors of the *Times* in protest. Jones identifies Burne-Jones's magnificent painting *Flamma Vestalis* (1886, Lord Andrew Lloyd-Webber collection) as having been based on the Rejlander photograph *Truly Thankful*, currently in the collection of the Royal Photographic Society. However, an earlier connection can be made in Burne-Jones's oeuvre to the same photographic study. The painting *Wheel of Fortune* (1871–85, Cecil French Bequest) contains another figure clearly derived from the photograph, and unlike *Flamma Vestalis*, it was executed in Rejlander's lifetime.

36. Not thought to be a relation of William Hogarth. Among Joseph Hogarth's accomplishments was the publication of two prints after Turner; see John Gage, *Collected Correspondence of J.M.W. Turner* [Oxford: Clarendon Press, 1980], p. 260. The engravings were James Tibbitts Willmore's (1800–1863) *Old Téméraire* and John Pye's (1782–1874) *Ehrenbreitstein*.

37. Jones, *O.G. Rejlander*, p. 29. According to Jones, the photographer John Hogarth, Jr., ran a photographic studio next door. The illustrator Gustave Doré (1832–1883) sat for a portrait with Rejlander in 1868 and is reputed to have been a close friend. A copy of this portrait is in the collections of the Royal Photographic Society at the National Media Museum in Bradford, England. Rejlander was also connected with the painter and patron Sir Coutts Lindsay (1824–1913), whose influential Grosvenor Gallery was founded shortly after Rejlander's death. Allison Eckardt Ledes, "The Grosvenor Gallery, an English Forerunner of the Modern Museum," *Antiques*, Vol. 149, March 1996, p. 350.

38. Richmond painted Darwin's portrait in 1838, suggesting another potential link between Darwin and Rejlander.

39. The Lawson album is in the Gernsheim Collection, Harry Ransom Humanities Center, University of Texas at Austin (TR 655.11 R355 HRC-P). The photographs in the Lawson album, although assembled by the artist, may not actually have been used as art studies. Lawson was principally a landscapist, while the photographs he collected were all character studies.

40. Oscar Rejlander, "Stray Thoughts," *Yearbook of Photography and Photographic News Almanac*, 1872, p. 65, as cited in Spencer, "O.G. Rejlander: Art Studies," p. 124.

41. Jones, *O.G. Rejlander*, pp. 56–57.

42. Rejlander, "Apology," p. 78.

43. Jones, *O.G. Rejlander*, p. 53.

44. *Photographic News*, 29 October 1875, as cited in Jones, *O.G. Rejlander*, p. 53.

45. Henry Peach Robinson, *Pictorial Effect in Photography: Being Hints on Composition and Chiaroscuro for Photographers* [London: Piper & Carter, 1869], p. 78.

46. Rejlander reported some of his findings to the South London Photographic Society:

> In Titian's *Venus and Adonis*, Venus has her head turned in a manner that no female could turn it and at the same time show so much of her back. Her right leg is also too long. I have proved the correctness of this opinion by photography with variously shaped female models. In *Peace and War*, by Rubens, the back of the female with the basket is painted from a male, as proved by the same test. (Rejlander, "Apology," p. 78)

47. James Borcoman, "Notes on the Early Use of Combination Printing," *One Hundred Years of Photographic History* [Albuquerque: University of New Mexico Press, 1975], p. 17.

48. De Montfort, "Procéde pour grouper plusieurs portraits obtenus isolément afin d'en former un seul tableau héliographique," *La Lumière*, Vol. 1, No. 7, 23 March 1851, p. 27, as described in Borcoman, "Notes."

49. Peter Henry Emerson, "Our English Letter: O.J. Rejlander [*sic*]," *American Amateur Photographer*, Vol. 2, 2 February 1890, p. 73. Much of Emerson's analysis seems to have been conjectural. Some divergence in quantity may be explained by differences in the several versions of the print in existence, and repeated printing of the same negative within the print, e.g., in the columns and curtains that frame the image.

50. Pam Roberts, *The Royal Photographic Society Collection* [Bath, England: Royal Photographic Society, 1994], p. 22.

51. This may not have been unintentional—there is certainly precedent for the repetition of models in painting.

52. "The Late O.G. Rejlander," *British Journal of Photography*, 29 January 1875, p. 54.

53. Painters frequently used photographs to record the exact position of drapery folds in compositions rendered in a lengthy session, or over several sessions. Even slight movements by a sitter could change the appearance of cloth elements of a picture entirely. Once moved, cloth was difficult to reposition to match its earlier aspect. Photographic studies helped to eliminate this problem. Given his alter ego as a painter, it is likely Rejlander developed his technique of superimposing negatives of drapery from the production of studies for use in painting.

54. "Fine Art Gossip," *Athenaeum*, No. 1539, 25 April 1857, p. 538.

55. Kenneth Clark, *The Nude: A Study in Ideal Form* [New York: Doubleday, 1959], p. 27.

56. It was at least purchased for him, though it has been suggested that Queen Victoria initiated the transaction.

57. Michael Pritchard, *A Directory of London Photographers 1841–1908* [Watford, England: PhotoResearch, 1994], p. 19.

58. Rejlander, quoted posthumously in Wall, p. 506.

59. Jones, *O.G. Rejlander*, p. 35.

60. "The Photographs of Rejlander," p. 15. This portfolio is almost certainly

the same described as a group of albums, containing seventy-two whole-plate prints, which his widow Mary advertised for sale in 1890; see "Notes and News," *Photographic Times*, Vol. 20, No. 445, 28 March 1890, p. 155.

61. The collection was sold to the Fry Manufacturing Company of 5 Chandos Street, Charing Cross, but subsequently disappeared. See Pritchard, *Directory*, p. 98. Like the portfolio, the Camera Club exhibition was drawn from material in the possession of Mary Rejlander after the artist's death.

62. Emerson, "Our English Letter," p. 73.

63. Ibid., p. 74. Emerson did not use the proper titles for the works on display. He was extravagantly critical of Rejlander, and one senses that he did not bother to use the names Rejlander had given them. I have deduced the correct titles from his descriptions. In any case, these were among his most famous works and would not have been omitted from his retrospective or, I would argue, from his portfolio.

64. "The Photographs of Rejlander," p. 15. Curiously, the author describes Rejlander's photographs as "not yet very generally known, nor are they easily obtained." This is the only claim I have read discounting the availability of Rejlander's photographs. All other evidence suggests that the opposite was true. As the *Art Journal* featured articles on arts of all media, it may be supposed that the writer was not a specialist in photography. As a result, he may merely have been unaware of other outlets where Rejlander's photographs could be purchased.

65. Charles Darwin, *The Expression of the Emotions in Man and Animals* (1872), Paul Ekman, ed. [London: HarperCollins, 1998], p. 181. It is curious that the work was apparently not signed, nor was it mounted on one of Rejlander's printed studio mounts. It may be that Darwin had in fact found an illegal copy of a Rejlander photograph, because his photographs were widely pirated.

66. Oscar Rejlander, as cited in Jones, *O.G. Rejlander*, p. 26.

67. Oscar Rejlander, quoted posthumously in *Photographic Times*, Vol. 29, No. 2, February 1897, p. 84.

68. See Malcolm Daniel, "Darkroom vs. Greenroom: Victorian Art Photography and Popular Theatrical Entertainment," *Image* (New York), Vol. 33, Nos. 1–2, Fall 1990, pp. 13–19. Peter Henry Emerson was probably the first to apply the greenroom epithet to Rejlander.

69. Elizabeth Heyert, *The Glass-House Years: Victorian Portrait Photography 1839–1870* [Montclair, New York: Allanheld & Schram; London: George Prior Associated Publishers, 1979], p. 134.

70. A.G. Fielding, "Rejlander in Wolverhampton: His Sponsorship by William Parke," *History of Photography*, Vol. 11, No. 1, January-March 1987, p. 20.

71. Helmut Gernsheim, *A Concise History of Photography* [New York: Dover Publications, 1986], p. 63.

72. Ibid., p. 29.

73. "Mr. Rejlander's Glass Room, &c." *British Journal of Photography*, 2 March 1863, p. 97.

74. As cited in Jones, *O.G. Rejlander*, p. 29.

75. Henry Peach Robinson, *The Studio, and What to Do in It* [London: Piper & Carter, 1891], p. 9.

76. A.H. Wall, "Rejlander's Photographic Art Studies—Their Teachings and Suggestions," *Photographic News,* 3 September 1886, reprinted in Peter Bunnell, ed., *The Photography of O.G. Rejlander* [New York: Arno Press, 1979], unpaginated.

77. This line is taken from his obituary: "The Late O.G. Rejlander," p. 55.

78. "Note" (Oscar Rejlander), *Art Journal,* December 1869, p. 382.

79. This information is unpublished. From the archives of John Murray Publishers, London, detailed ledger "Darwin on Expression," August 1872, p. 370.

80. Ibid.

81. Unpublished correspondence, Oscar Rejlander to Charles Darwin, 30 April 1871, Darwin Papers, Vol. 176, item 115.

82. Unpublished correspondence, Charles Darwin to unknown recipient, New York Botanical Gardens Archives, 12 October 1871, No. 1243. By courtesy of the Darwin Papers Project, Cambridge.

83. Francis Darwin, *Life and Letters,* p. 543 (see n. 4). It was reproduced as an engraving by C.H. Jeens.

84. Darwin Papers, Vol. 225, item 114.

85. Janet Browne, *Charles Darwin: The Power of Place* [Princeton, New Jersey: Princeton University Press, 2002], p. 361.

86. Darwin, *Expression,* p. 45.

87. Rejlander, "Hints," p. 117.

88. Ibid., p. 116.

89. George Eastman House, Rochester, New York, *Young Girl Standing with Hand to Mouth,* albumen print (72:249:02).

90. Darwin Papers, Vol. 53.1, Folio 43, items 34–36.

91. J. Paul Getty Museum, Los Angeles, California, *Babylonian Song,* albumen print, 84.XC.873.5973.

92. J. Paul Getty Museum, Los Angeles, California, *Untitled Portrait of a Boy,* albumen print, 84.XC.845.3. With thanks to Julian Cox for this observation.

93. Wall, p. 121.

94. E.g., Fitzwilliam Museum, Cambridge, *Before* and *After,* oils on canvas, PD. 11-1964 and 12-1964.

95. *Lustful Advances* is Darwin Papers, Vol. 53.1, Folio 43, item 33; *Who 'a You* is Folio 47, item 65; *Girl Chewing Her Pinafore* is Folio 46, item 62.

96. Wall, p. 121.

97. Darwin, *Expression,* p. 254.

98. Unpublished correspondence, Oscar Rejlander to Charles Darwin, 11 November 1871, Darwin Papers, Vol. 189, item 116.

99. See notes 5 and 73.

100. Unpublished correspondence, Charles Darwin to Robert Cooke, 28 May 1872, John Murray Archives, item 280.

101. Unpublished correspondence, Charles Darwin to Robert Cooke, 10 May 1872, John Murray Archives, item 278.

102. Unpublished correspondence, Charles Darwin to Robert Cooke, 22 April 1872, John Murray Archives, item 285.

103. Unpublished correspondence, Oscar Rejlander to Charles Darwin, 30 April 1871, Darwin Papers, Vol. 189, item 115.

104. A.H. Wall, "An Hour with Rejlander," *Photographic News*, Vol. 4, No. 112, 26 October 1860, p. 302.

Chapter 10

1. "Odd Odds and Ends by O.G.R." Unpublished correspondence, Oscar Rejlander to Charles Darwin, 1871, Darwin Papers, Vol. 189, item 139.

2. Thanks to John Gage for this suggestion.

3. Unpublished photograph, *At the End of the Brighton Review . . . OGR Himself Again*, 1871–72, diptych of carte-de-visite albumen prints from two wet-plate collodion negatives, laid to card mount. Darwin Papers, Vol. 53.1, Folio 44, object 45. See also appendix.

4. "Odd Odds and Ends," p. 7.

5. Ibid., p. 3.

6. Ibid., p. 2.

7. Ibid., p. 1.

8. Charles Darwin, *The Expression of the Emotions in Man and Animals* (1872), Paul Ekman, ed. [London: HarperCollins, 1998], p. 35.

9. Ibid., p. 241.

10. Unpublished photograph, *Two Views of a Pouting Boy, Profile*, Darwin Papers, Vol. 53.1, Folio 30, item 4. Darwin was puzzled by the double-lipped pout exhibited in adult anger, such as Rejlander described, although he did note it in the text.

11. *Zygomatic major* and *orbicularis oculi*.

12. Paul Ekman, in Darwin, *Expression*, p. 203.

13. Charles Darwin, *The Expression of the Emotions in Man and Animals* [London: John Murray, 1872], p. 248.

14. Ekman, in Darwin, *Expression* [London: HarperCollins, 1998], p. 182.

15. A.H. Wall, "Rejlander's Photographic Art Studies: Their Teachings and Suggestions," *Photographic Times and American Photographer*, Vol. 17, 18 March 1887, p. 132.

16. *Sadness*, 964:0582:0025.

17. "Odd Odds and Ends," p. 6.

18. Ibid., pp. 6, 6, and 7, respectively.

19. Ibid., p. 4. One wonders, on the basis of this observation, what a discussion between Darwin and Rejlander sounded like.

20. Ibid., p. 1.

21. Not counting the experimental photographs of Rejlander with *Ginx's Baby*. For more on these, see chapter 7.

22. Unpublished note, Darwin Papers, Vol. 53.1, Folio 57b, item 108.

23. Darwin, *Expression*, p. 264.

24. Unpublished correspondence, Oscar Rejlander to Charles Darwin, 30 April 1871, Darwin Papers, Vol. 189, item 115.

25. Oscar Rejlander, "An Apology for Art Photography," *British Journal of Photography*, 16 February 1863, p. 78.

26. Benjamin Robert Haydon, *Lectures on Painting and Design* [London: Longman, Brown, Green, & Longmans, 1844], p. 64. This quote neatly parallels Wall's celebration of Rejlander's massive "os frontis" (A.H. Wall, "An

Hour with Rejlander," *Photographic News*, Vol. 4, No. 112, 26 October 1860, p. 302).

27. He resembles the photographer Walter Bentley Woodbury (1834–85), who was a friend of Rejlander, though the similarity is not strong enough to known portraits of Woodbury to state so authoritatively.

28. Unpublished photograph, Oscar Rejlander, *Astonishment*, 1871–72. Albumen print. Darwin Papers, Vol. 53.1, Folio 58, item 112. Interestingly, Darwin also inscribed on this picture, "Gives the earlier photograph an explanation."

29. Unpublished note, Darwin Papers, Vol. 53.1, Folio 58, item 111.

30. Unpublished note, Darwin Papers, Vol. 53.1, Folio 49, item 72.

31. Unpublished note, Darwin Papers, Vol. 53.1, Folio 53, item 94.

32. Unpublished albumen prints, Darwin Papers, Vol. 53.1, Folio 53, items 53–54.

33. Unpublished albumen prints, Darwin Papers, Vol. 53.1, Folio 30, items 4–6.

34. Robert A. Sobieszek, "Sculpture as the Sum of Its Profiles: François Willème and Photosculpture in France, 1859–1868," *Art Bulletin*, Vol. 62, December 1980, pp. 617–30.

35. Oscar Rejlander, "On Photographing Horses," *British Journal Photographic Almanac*, 1873, p. 115.

36. A full account of Muybridge's project is provided in Robert Bartlett Haas, *Eadweard Muybridge: Man in Motion* [Berkeley: University of California Press, 1976].

37. Unpublished correspondence, Oscar Rejlander to Charles Darwin, 15 December 1874, Darwin Papers, Vol. 176, item 118. I have altered the punctuation in this passage slightly to make it easier to read.

38. "The Photographs of Rejlander," *Art Journal*, January 1868, p. 15.

Chapter 11

1. Thanks to Professor Dr. John Michael Krois, of the Institut fuer Philosophie, Humboldt Institute, Berlin, for raising these connections and for providing the citations in this paragraph.

2. As reported by Krois, ibid.

3. See, e.g., Ernst Cassirer, *The Philosophy of Symbolic Forms*, Vol. 4 [New Haven, Connecticut: Yale University Press, 1996], p. 39.

4. Main ledger, 1878–82, John Murray Archive, p. 432.

5. Unpublished correspondence, Charles Darwin to Lewis Carroll, 10 December 1872, Houghton Library, Harvard University.

6. Carroll Diaries are in the collections of the British Library. This entry comes from Vol. 10, which has not been published. Thanks to Edward Wakeling for supplying this reference.

7. See Roger Taylor and Edward Wakeling, *Lewis Carroll, Photographer: The Princeton University Library Albums* [Princeton, New Jersey: Princeton University Press, 2002]. The image is in Carroll Album A(II), British Library.

8. Hargreaves Album, on deposit at Christ Church Library. This album was formerly in the possession of Alice Hargreaves (née Liddell), one of Car-

roll's favorite subjects. The Huxley photograph was probably made immediately after his celebrated British Institution debate on evolution with Samuel Wilberforce at the new Oxford University Museum in June 1860. Carroll photographed both participants in the debate, as well as a number of celebrities present.

9. Unpublished correspondence, Charles Darwin to Joseph Hooker, 18 January 1874, Darwin Papers, Vol. 95, Folio 311.

10. Unpublished correspondence, Charles Darwin to Thomas Henry Huxley, 29 January 1874, Thomas Henry Huxley Archive, Imperial College London.

11. Huxley Archive, Vol. 28, items 221–23. They are labeled "F. Hudson, 177 Holloway Rd., London."

12. The originals are now in the *Archivi Alinari* in Florence.

13. Unpublished correspondence, Paolo Mantegazza to Charles Darwin, 23 December 1872, Darwin Papers, Vol. 171. Original is in French; translation by the author.

14. In a dizzying non sequitur, he also recalled how Paraguayan howler monkeys (*Mycetes casaya* he called them) have a tendency to hurl their excrement at travelers. Darwin's reply is not preserved.

15. Paolo Mantegazza, *Atlante della espressione del dolore; fotografie prese dal vero e da molte opere d'arte, che illustrano gli studi sperimentali sull'espressione del dolore,* [Firenze: Brogi, 1876].

16. The Howditch portraits are item 158/1A; the Noyes photograph is item 158/1B, Galton Papers, University College London.

17. Unpublished correspondence, Henry Peach Robinson to Francis Galton, 3 May 1882, Galton Papers.

18. Unpublished correspondence, Henry Peach Robinson to Francis Galton, 19 May 1882, Galton Papers.

19. MaryJo Marks, "The Mechanics of Affect: Darwin, Regression, and Photography" [abstract], *Theorizing Gesture: Bodily Signs in Representation,* College Art Association Conference, Boston, 1995, p. 240.

20. Henry Peach Robinson, *Pictorial Effect in Photography: Being Hints on Composition and Chiaroscuro for Photographers,* [London: Piper & Carter, 1869], p. 78.

21. See, e.g., Carol Armstrong, *Scenes in a Library: Reading the Photograph in the Book, 1843–1875* [Cambridge, Massachusetts: MIT Press, 1998], pp. 80–106.

22. Marks, "Mechanics of Affect," p. 241.

23. Jennifer Green-Lewis, *Framing the Victorians: Photography and the Culture of Realism* [Ithaca, New York: Cornell University Press, 1996], p. 159.

24. Armstrong, *Scenes in a Library,* p. 106.

25. See, e.g., Roland Barthes, *Camera Lucida* [New York: Hill & Wang, 1981].

26. Oscar Rejlander, "On Photographic Composition: With a Description of *Two Ways of Life,*" *Journal of the Royal Photographic Society,* 21 April 1858, p. 195.

27. Peter Henry Emerson, *Naturalistic Photography, for Students of the Art* [London: Sampson, Low, Marston, Searle, & Rivington, 1889].

28. The Impressionists' efforts were inspired in part by Hermann von Helmholtz's (1821–94) *Handbook of Physiological Optics* (1856–67), the first volume of which appeared just two years before Rejlander's remarks. Darwin repeatedly cited Helmholtz's *Physiology of Music* (1868) in *Expression*, pp. 88–92.

29. Darwin, *Expression*, pp. 194–95.

SELECTED BIBLIOGRAPHY

Agassiz, Alexander, "Application of Photography to Illustrations of Natural History. With Two Figures Printed by the Albert and Woodbury Processes," *Bulletin of the Museum of Comparative Zoology, Harvard*, Vol. 3, 30 November 1871, pp. 47–48.

Agassiz, Louis, and Elizabeth Agassiz, *A Journey in Brazil by Professor and Mrs. Agassiz* [Boston, Massachusetts: Fields & Osgood, 1867].

Armstrong, Walter, "Briton Riviere, Royal Academician: His Life & Work," *Art Journal*, Annual Supplement, 1891, pp. 1–32.

Ashworth, A.L., *Stanley Royd Hospital, Wakefield: One Hundred and Fifty Years, a History* [Wakefield, England: Wakefield Area Health Authority, 1975].

Barrett, Paul, et al., *Charles Darwin's Notebooks 1836–1844* [Ithaca, New York: Cornell University Press, 1987].

Barrett, Paul, Donald Weinshank, et al., *A Concordance to Darwin's Expression of the Emotions in Man and Animals* [Ithaca, New York: Cornell University Press, 1986].

Batchen, Geoffrey, *Burning with Desire: The Conception of Photography* [Cambridge, Massachusetts: MIT Press, 1997].

Beer, Gillian, *Darwin's Plots: Evolutionary Narrative in Darwin, George Eliot and Nineteenth-Century Fiction* [London: Ark Paperbacks, 1985].

Bell, Sir Charles, *The Anatomy and Philosophy of Expression as Connected with the Fine Arts* (1806) [London: G. Bell & Sons, 1912].

Bewick, Thomas, and Ralph Beilby, *A General History of Quadrupeds* [Newcastle: S. Hodgson, R. Beilby, & T. Bewick, 1790].

Borcoman, James, "Notes on the Early Use of Combination Printing," *One Hundred Years of Photographic History* [Albuquerque: University of New Mexico Press, 1975].

Bourriau, Janine, and Trevor Lamb, *Colour: Art & Science* [Cambridge: Cambridge University Press, 1995].

Browne, Janet, *Charles Darwin: The Power of Place* [Princeton, New Jersey: Princeton University Press, 2002].

Browne, Janet, *Charles Darwin: Voyaging* [London: Jonathan Cape, 1995].

Browne, Janet, "Darwin and the Expression of the Emotions," *The Darwinian Heritage*, David Kohn, ed. [Princeton, New Jersey: Princeton University Press in association with Nora Pacifica, 1985].

Browne, Janet, "Darwin and the Face of Madness," *The Anatomy of Madness*, W.F. Bynum, Roy Porter, and Michael Shepherd, eds. [London: Tavistock, 1985].

Browne, Janet, "I Could Have Wretched All Night: Charles Darwin and His Body," *Science Incarnate: Historical Embodiments of Natural Knowledge*, Christopher Lawrence and Steven Shapin, eds. [Chicago: University of Chicago Press, 1998], pp. 240–87.

Bunnell, Peter, ed., *The Photography of O.G. Rejlander* [New York: Arno Press, 1979].

Burke, Edmund, *Philosophical Enquiry into the Origin of Our Ideas of the Sublime and Beautiful* (1757) [London: J. Dodsley, 1793].

Burkhardt, Frederick, and Sydney Smith, eds., *A Calendar of the Correspondence of Charles Darwin, 1821–1882* [Cambridge: Cambridge University Press, 1994].

Burkhardt, Frederick, Sydney Smith, et al. eds., *The Correspondence of Charles Darwin* [Cambridge: Cambridge University Press, 1985–2000].

Burrows, Adrienne, and Iwan Schumacher, *Portraits of the Insane: The Case of Dr. Diamond* [London: Quartet Books, 1990].

"Buy Landy's Expressive Pets," *Anthony's Photographic Bulletin*, Vol. 5, No. 6, June 1874, p. 226.

Cambridge University Library, *Handlist of Darwin Papers at the University Library, Cambridge* [Cambridge: Cambridge University Press, 1960].

Catalogue of Binocular Pictures of the London Stereoscopic Company, 54 Cheapside and 313 Oxford Street, undated (c. 1867), Victoria & Albert Museum collection.

A Catalogue of a Collection of Plaques, Medallions, Vases, Figures &c., in Coloured Jasper Basalte . . . Formed by the Late Lord Tweedmouth [London: Charles Davis, 1905].

Christ's College, University of Cambridge, *The Portraits, Prints, and Writings of Charles Robert Darwin* [Cambridge: Christ's College, 1909].

Clark, Kenneth, *The Nude: A Study in Ideal Form* [New York, Doubleday, 1959].

Cohen, Morton N., *Lewis Carroll: A Biography* [London: Macmillan, 1995].

Collier, John, *A Manual of Oil Painting* [London: Cassell & Company, 1886].

Cook, E.T., and Alexander Wedderburn, *The Works of John Ruskin* [London: George Allen, 1903–5].

Cronin, Helena, *The Ant and the Peacock: Altruism and Sexual Selection from Darwin to Today* [Cambridge: Cambridge University Press, 1991].

Dance, Peter S., *The Art of Natural History: Animal Illustrators and Their Work* [London: Country Life Books, 1978].

Daniel, Malcolm, "Darkroom vs. Greenroom: Victorian Art Photography and Popular Theatrical Entertainment," *Image* (New York), Vol. 33, Nos. 1–2, Fall 1990, pp. 13–19.

Darwin, Charles, *The Autobiography of Charles Darwin* (1887), Nora Barlow, ed. [New York: W.W. Norton & Company, 1958].

Darwin, Charles, *The Descent of Man, and Selection in Relation to Sex* [London: John Murray, 1871].

Darwin, Charles, *The Different Forms of Flowers on Plants of the Same Species* [London: John Murray, 1877].

Darwin, Charles, *The Effects of Cross and Self Fertilisation in the Vegetable Kingdom* [London: John Murray, 1876].

Darwin, Charles, *The Expression of the Emotions in Man and Animals* [London: John Murray, 1872].

Darwin, Charles, *The Expression of the Emotions in Man and Animals* (1872), Paul Ekman, ed. [London: HarperCollins, 1998].

Darwin, Charles, *The Formation of Vegetable Mould through the Action of Worms* [London: John Murray, 1881].

Darwin, Charles, *The Geology of the Voyage of the Beagle*: Vol. 1, *Coral Reefs* (1842); Vol. 2, *Volcanic Islands* (1844); Vol. 3, *South America* (1846) [London: Smith Elder & Company].

Darwin, Charles, *Insectivorous Plants* [London: John Murray, 1875].

Darwin, Charles, *Journal of Researches into the Geology and Natural History of the Countries Visited by H.M.S. Beagle etc.* [London: Henry Colburn, 1839].

Darwin, Charles, *A Monograph of the Fossil Lepadidae, or Pedunculated Cirripedes of Great Britain* [London: Palaeontographical Society, 1851].

Darwin, Charles, *A Monograph of the Sub-class Cirripedia, with Figures of All the Species [The Lepadidae; or, Pedunculated Cirripedes (1851); Balanidae (or Sessile Cirripedes); the Verrucidae, etc. (1854)]* London: Ray Society].

Darwin, Charles, *Movement and Habits of Climbing Plants* [London: John Murray, 1875].

Darwin, Charles, *On the Origin of Species by Means of Natural Selection* [London: John Murray, 1859].

Darwin, Charles, *On the Origin of Species by Means of Natural Selection* (London: John Murray, 1859), Thinker's Library edition [London: Watts & Company, 1929].

Darwin, Charles, *On the Power of Movement in Plants* [London: John Murray, 1880].

Darwin, Charles, *On the Various Contrivances by which British and Foreign Orchids are Fertilised by Insects* [London: John Murray, 1862].

Darwin, Charles, *The Variation of Animals and Plants under Domestication* [London: John Murray, 1868].

Darwin, Francis, *The Life and Letters of Charles Darwin* (1887) [New York: D. Appleton & Company, 1919].

Darwin, Francis, "Reminiscences," *The Life of Charles Darwin* (1902) [London: Senate Books, 1995].

Daston, Lorraine, and Peter Galison, "The Image of Objectivity," *Representations*, Vol. 40, Fall 1992, pp. 81–129.

Daston, Lorraine, and Peter Galison, *Objectivity* [New York: Zone Books, 2007].

Davis, Frederick Rowe, "Louis Agassiz in Search of Support for a Grand Critique of Darwinian Theory: The South American Thayer Expedition to Brazil," unpublished thesis, Harvard University Archives.

De Montfort, "Procéde pour grouper plusieurs portraits obtenus isolément afin d'en former un seul tableau héliographique," *La Lumière*, Vol. 1, No. 7, 23 March 1851, p. 27.

Descartes, René, *The Passions of the Soul* (1650), Stephen Voss, trans. [Indianapolis, Indiana: Hackett Publishing Company, 1989].

Desmond, Adrian, *Huxley: The Devil's Disciple* [London: Michael Joseph, 1994].

Desmond, Adrian, and James Moore, *Darwin: The Life of a Tormented Evolutionist* [New York: Warner Books, 1991].

Di Gregorio, Mario, and Nick Gill, *Charles Darwin's Marginalia* [New York: Garland Publishing, 1986].

Dorsey, George, *The Evolution of Charles Darwin* [New York: Doubleday, Page & Company, 1927].

Duchenne de Boulogne, Guillaume-Benjamin, *Mécanisme de la physionomie humaine ou analyse électro-physiologique de l'expression des passions applicable à la pratique des arts plastiques* [Paris: Jules Renouard, 1862].

Duchenne de Boulogne, Guillaume-Benjamin, *The Mechanism of Human Facial Expression*, (1862), Andrew Cuthbertson, ed. and trans. [Cambridge: Cambridge University Press, 1990].

Dupree, A. Hunter, *Asa Gray: American Botanist and Friend of Darwin* [Baltimore, Maryland: Johns Hopkins University Press, 1988].

Eder, Josef Maria, *History of Photography*, Edward Epstean, trans. [New York: Columbia University Press, 1945].

Eder, Josef Maria, *La photographie instantanée: son application aux arts et aux sciences* [Paris: Gauthier Villars et Fils, 1888].

Edwards, Elizabeth, *Anthropology and Photography: 1860–1920* [New Haven, Connecticut: Yale University Press, 1972].

Edwards, Gary, *International Guide to Nineteenth-Century Photographers and Their Works* [Boston, Massachusetts: G.K. Hall & Company, 1988].

Ekman, Paul, *Darwin and Facial Expression: A Century of Research in Review* [New York: Academic Press, 1973].

Elkins, James, "Art History and Images That Are Not Art," *Art Bulletin*, Vol. 77, No. 4, December 1995, pp. 553–73.

Emerson, Peter Henry, "Our English Letter: O.J. Rejlander [*sic*]," *American Amateur Photographer*, Vol. 2, 2 February 1890, pp. 73–74.

Emmerson, Robin, "Wax Models by Pacetti and Angelini in the Lady Lever Art Gallery," *Ars Ceramica*, Vol. 12, 1995, pp. 22–33.

"Fine Art Gossip," *Athenaeum*, No. 1539, 25 April 1857, p. 538.

Ford, Alice, *The 1826 Journal of John James Audubon* [Tulsa: University of Oklahoma Press, 1967].

Ford, Brian, *Images of Science: A History of Scientific Illustration* [London: British Library, 1992].

Freeman, Richard Broke, *The Works of Charles Darwin: An Annotated Bibliographical Handlist*, 2nd edition [Folkestone, England: Wm. Dawson & Sons, 1977].

Freeman, Richard Broke, and Gautrey, P.J., "Charles Darwin's Queries about Expression," *Bulletin of the British Museum (Natural History) Historical Series*, Vol. 5, No. 3, 1972, pp. 207–10.

Freeman, Richard Broke, and Gautrey, Peter Jack, "Charles Darwin's Queries about Expression," *Bulletin of the British Museum (Natural History)*, Vol. 4, No. 3, 1972, pp. 205–19.

Gage, John, *Collected Correspondence of J.M.W. Turner* [Oxford: Clarendon Press, 1980].

Gage, John, "Colour and Culture," *Colour, Art and Science* [Cambridge: Cambridge University Press, 1995], pp. 175–93.

Gernsheim, Alison, "Medical Photography in the Nineteenth-Century," *Medical and Biological Illustration*, Vol. 11, 1961, pp. 147–56.

Gernsheim, Helmut, *A Concise History of Photography* [New York: Dover Publications, 1986].

Gernsheim, Helmut, *Incunabula of British Photographic Literature, 1839–1875* [London: Scolar Press, 1984].

Gernsheim, Helmut, *Lewis Carroll: Photographer* [New York: Dover Publications, 1969].

Gilman, Sander L., "Darwin Sees the Insane," *Journal of the History of the Behavioural Sciences*, Vol. 15, 1979, pp. 253–62.

Gilman, Sander L., *The Face of Madness: Hugh W. Diamond and the Origin of Psychiatric Photography* [New York: Brunner-Mazel, 1976].

Gilman, Sander L., *Seeing the Insane* [New York, Chichester: J. Wiley & Sons in association with Brunner-Mazel, 1982].

Goldberg, Vicki, ed., *Photography in Print: Writings from 1816 to the Present* [Albuquerque: University of New Mexico Press, 1981].

Gombrich, Ernst, *Art and Illusion: A Study in the Psychology of Pictorial Representation* (1956) [London: Phaidon, 1996].

Goodison, J.W., ed., *Fitzwilliam Museum Cambridge: Catalogue of Paintings I: Dutch, Flemish, French, German, Spanish* [Cambridge: Fitzwilliam Museum, 1960].

Graham, John, *Lavater's Essays on Physiognomy: A Study in the History of Ideas* [Berne: Peter Lang Press, 1979].

Graves, Algernon, *A Dictionary of Artists Who Have Exhibited Works in the Principal London Exhibitions from 1760 to 1893* (1901) [Bath, England: Kingsmead Reprints, 1973].

Green-Lewis, Jennifer, *Framing the Victorians: Photography and the Culture of Realism* [Ithaca, New York: Cornell University Press, 1996].

Gruber, Howard, "Darwin's 'Tree of Nature' and Other Images of Wide Scope," *On Aesthetics in Nature*, Judith Wechsler, ed. [Cambridge, Massachusetts: MIT Press, 1978], pp. 121–40.

Haes, Frank, "The Early Days of Animal Photography," *Photographic News*, Vol. 36, 1892, pp. 38–39.

Haworth-Booth, Mark, *Photography: An Independent Art* [Princeton, New Jersey: Princeton University Press, 1997].

Haworth-Booth, Mark, ed., *The Golden Age of British Photography* [New York: Aperture, 1986].

Henisch, Bridget, and Heinz Henisch, *The Photographic Experience 1839–1914: Images and Attitudes* [University Park: Pennsylvania State University Press, 1994].

Herrick, Frances Hobart, *Audubon: The Naturalist: A History of His Life and Time* [New York: Appleton, 1938].

Hogarth, William, *The Analysis of Beauty: Written with a View of Fixing the Fluctuating Ideas of Taste* [London: Samuel Bagster, 1753].

Howard, Frank, "Photography Applied to Fine Art," *Journal of the Royal Photographic Society*, No. 13, 21 January 1854, pp. 154–58.

Iles, Chrissie, and Russell Roberts, eds., *In Visible Light: Photography and Classification in Art, Science, and the Everyday* [Oxford: Museum of Modern Art, 1997].

Isaac, Gwyneira, "Louis Agassiz's Photographs in Brazil: Separate Creations," *History of Photography*, Vol. 21, No. 1, Spring 1997, pp. 3–11.

Jacobs, Merle E., *Mr. Darwin Misread Miss Peacock's Mind: A New Look at Mate Selection in Light of Lessons from Nature* [Ashland: Nature Books, 1999].

Janis, Eugenia Parry, et al., *For My Beloved Sister Mia: An Album of Early Photographs by Julia Margaret Cameron* [Albuquerque: University of New Mexico Press, 1995].

Jardine, Nicholas, James Secord, and Emma Spary, *Cultures of Natural History* [Cambridge: Cambridge University Press, 1996].

Jay, Bill, "Charles Darwin, Photography, and Everything Else," *British Journal of Photography*, Vol. 127, No. 45, 7 November 1980, pp. 1116–18.

Jenkins, Edward, *Ginx's Baby, His Birth and other Misfortunes* [London: Strahan & Company, 1871].

Johnson, William Stanley, *Nineteenth-Century Photography: An Annotated Bibliography, 1839–1879* [Boston, Massachusetts: G.K. Hall, 1990].

Jones, Caroline, and Peter Galison, eds., *Picturing Science, Producing Art* [New York: Routledge, 1998].

Jones, Edgar Yoxall, *O.G. Rejlander: Father of Art Photography:1813–1875* [London: David & Charles, 1973].

Jordanova, Ludmilla, *Defining Features: Scientific and Medical Portraits 1660–2000* [London: Reaktion Books, 2000].

Keynes, Richard Darwin, *Charles Darwin's Beagle Diary* [Cambridge: Cambridge University Press, 1988].

Knoepflmacher, U.C., and G.B. Tennyson, eds., *Nature and the Victorian Imagination* [Berkeley: University of California Press, 1977].

Kwakkelstein, Michael W., "Leonardo da Vinci's Grotesque Heads and the Breaking of the Physiognomic Mould," *Journal of the Warburg and Courtauld Institutes*, Vol. 54, 1991, pp. 127–39.

Kraus, Rolf H., "Photographs as Early Scientific Book Illustrations," *History of Photography*, Vol. 2, No. 4, October 1978, pp. 291–314.

"Landy's Photographic Pets," *Photographic Times*, Vol. 2, 19 July 1872, p. 104.

Langstaff, Charles L., *Hospital Hygiene, being the Annual Address to the Southampton Medical Society* [London: 1872].

"The Late O.G. Rejlander," *British Journal of Photography*, 29 January 1875, pp. 54–57.

Latour, Bruno, *Science in Action* [Cambridge, Massachusetts: Harvard University Press, 1987].

Latour, Bruno, "Visualization and Cognition: Thinking with Eyes and Hands," *Knowledge and Society: Studies in the Sociology of Culture Past and Present*, Vol. 6, 1986, pp. 1–40.

Lavater, Johann Caspar, *Essays on Physiognomy*, Thomas Holcroft, trans. [London: Thomas Tegg, 1840].

Le Brun, Charles, *A Method to Learn to Design the Passions*, David S. Rodes, ed., John Williams, trans. [Los Angeles, California: William Andrews Clark Memorial Library, 1980].

Ledes, Allison Eckardt, "The Grosvenor Gallery, an English Forerunner of the Modern Museum," *Antiques*, Vol. 149, March 1996, p. 350.

Longstaff, George Blundell, *The Langstaffs of Teesdale and Weardale: Materials for the History of a Yeoman Family* [London: Wardour Press, 1923].

Lyell, Charles, *Principles of Geology* [London: John Murray, 1830–33].

Manier, Edward, *The Young Darwin and His Cultural Circle: A Study of Influences Which Helped Shape the Language and Logic of the First Drafts of the Theory of Natural Selection* [Boston, Massachusetts: D. Reidel Publishing Company, 1978].

Mantegazza, Paolo, *Atlante della Espressione del Dolore* [Firenze: Giacomo Brogi, 1876].

Mantegazza, Paolo, *Physiognomy and Expression*, Havelock Ellis, ed. [London: Walter Scott, 1890].

Marable, Darwin, "Photography and Human Behaviour in the Nineteenth Century," *History of Photography*, Vol. 9, No. 2, April-June 1985, pp. 141–47.

Mathews, Oliver, *Early Photographs and Early Photographers: A Survey in Dictionary Form* [London: Pitman Publishing Corporation, 1973].

Mathon, Catherine, ed., *Duchenne de Boulogne* [Paris: École nationale supérieure des beaux-arts, 1999].

Meller, Peter, "Physiognomical Theory in Renaissance Heroic Portraits," *International Congress of the History of Art, Studies in Western Art*, Vol. 20 [New York: Princeton University Press, 1963], pp. 53–69.

Mivart, St. George, *Genesis of Species* [London: MacMillan & Company, 1871].

Mozley, Anita, *Eadweard Muybridge: The Stanford Years* [Stanford, California: Stanford University Museum of Art, 1972].

"Mr. Rejlander's Glass Room, &c.," *British Journal of Photography*, 2 March 1863, p. 97.

Newhall, Beaumont, *The History of Photography* [New York: Museum of Modern Art, 1982].

"Note" (Oscar Rejlander), *Art Journal*, December 1869, p. 382.

"Notes and News" (Oscar Rejlander), *Photographic Times*, Vol. 20, No. 445, 28 March 1890, p. 15.

Osborne, Harold, *Abstraction and Artifice in Twentieth-Century Art* [Oxford: Clarendon Press, 1979].

Palgrave, Francis Turner, *Essays on Art* [New York: Hurd & Houghton, 1867].

Palmer, A.H., *The Life of Joseph Wolf, Animal Painter* [New York: Longmans, Green, & Company, 1895].

"The Photographs of Rejlander," *Art Journal*, January 1868, p. 15.

Pingree, Jeanne, *T.H. Huxley: List of His Scientific Papers* [London: Imperial College of Science, Technology & Medicine, 1968].

Pritchard, Michael, *A Directory of London Photographers 1841–1908* [Watford, England: PhotoResearch, 1994].

Prodger, Phillip, *An Annotated Catalogue of Illustrations of Human and Animal Expression from the Collection of Charles Darwin* [Lewiston, New York: Edwin Mellen Press, 1998].

Prodger, Phillip, "Illustration as Strategy in Charles Darwin's *Expression of the Emotions in Man and Animals*," *Inscribing Science: Materiality and Text in Scientific Communication*, Timothy Lenoir, ed. [Stanford, California: Stanford University Press, 1998].

Prodger, Phillip, "Photography and the Expression of the Emotions," in Charles Darwin, *The Expression of the Emotions in Man and Animals*, Paul Ekman, ed. [London: HarperCollins, 1998], pp. 399–410.

Prodger, Phillip, "The Romance and Reality of the Horse in Motion," *Marey/Muybridge, Pionniers du Cinéma* [Beaune, France: Musée Marey, 1996], pp. 44–59.

Prodger, Phillip, *Time Stands Still: Muybridge and the Instantaneous Photography Movement* [New York: Oxford University Press with the Iris & B. Gerald Cantor Center for Visual Arts at Stanford University, 2001].

Prodger, Phillip, and Paul Ekman, "Note on the Orientation of the Plates," in Charles Darwin, *The Expression of the Emotions in Man and Animals*, Paul Ekman, ed. [London: HarperCollins, 1998], pp. 411–15.

Pseudo-Aristotle, "Physiognomica," *Aristotle: Minor Works*, Vol. 4, W.S. Hett, trans. [London: 1936].

Ray, Gordon, *The Illustrator and the Book in England from 1790 to 1914* [New York: Pierpont Morgan Library, 1979].

Rejlander, Oscar Gustave, "An Apology for Art Photography," *British Journal of Photography*, 16 February 1863, pp. 77–79.

Rejlander, Oscar Gustave, "An Appeal to Photographers," *British Journal of Photography*, 29 July 1864, p. 266.

Rejlander, Oscar Gustave, "Hints Concerning the Photographing of Criminals," *British Journal Photographic Almanac*, 1872, pp. 116–17.

Rejlander, Oscar Gustave, "On Photographic Composition; with a Description of *Two Ways of Life*," *Journal of the Royal Photographic Society*, 21 April 1858, pp. 191–97.

Rejlander, Oscar Gustave, "On Photographing Horses," *British Journal of Photography Almanac*, 1873, p. 115.

Rejlander, Oscar Gustave, "Photography as a Handmaid of Art," *British Journal of Photography Almanac*, 1866.

Rejlander, Oscar Gustave, "Rejlander in Type," *British Journal of Photography*, Vol. 12, No. 279, 8 September 1865, p. 462.

Rejlander, Oscar Gustave, "What Photography Can Do in Art," *Yearbook of Photography and Photographic News Almanac*, 1867, p. 50.

Rice, A.L., and A.G.E. Jones, "G.C. Wallich M.D.—Megalomaniac or Misused Oceanographic Genius?" *Journal of the Society for the Bibliography of Natural History*, Vol. 7, No. 4, 1976, pp. 423–50.

Roberts, Pam, *The Royal Photographic Society Collection* [Bath, England: Royal Photographic Society, 1994].

Robinson, Henry Peach, *Pictorial Effect in Photography: Being Hints on Composition and Chiaroscuro for Photographers* [London: Piper & Carter, 1869].

Rogers, Malcolm, *Camera Portraits* [New York: Oxford University Press, 1990].

Rogerson, Brewster, "The Art of Painting the Passions," *Journal of the History of Ideas*, Vol. 14, No. 1, January 1953, pp. 68–94.

Rudwick, Martin, "The Emergence of a Visual Language for Geological Science, 1760–1841," *History of Science*, Vol. 14, 1976, pp. 149–95.

Rudwick, Martin, *Scenes from Deep Time* [Chicago: University of Chicago Press, 1992].

Ruskin, John, *The Elements of Drawing* (1857) [New York: Dover, 1971].

Salaman, Naomi, and Ronnie Simpson, eds., *Postcards on Photography: The Handmade Copy in Reproduction* [Cambridge: Cambridge Darkroom Gallery, 1998].

Schaaf, Larry, *Sun Gardens: The Victorian Photographs of Anna Atkins* [New York: Aperture, 1985].

Severn, Arthur, *The Professor: Arthur Severn's Memoir of John Ruskin*, James Dearden, ed. [London: George Allen & Unwin, 1967].

Smart, Alastair, "Dramatic Gesture and Expression in the Age of Hogarth and Reynolds," *Apollo*, Vol. 82, 1963, pp. 90–97.

Smith, Jonathan, *Charles Darwin and Victorian Visual Culture* [Cambridge: Cambridge University Press, 2006].

Smith, Lindsay, *Victorian Photography, Painting and Poetry: The Enigma of Visibility in Ruskin, Morris and the Pre-Raphaelites* [Cambridge: Cambridge University Press, 1995].

Sobieszek, Robert, *Ghost in the Shell: Photography and the Human Soul, 1850–2000* [Cambridge, Massachusetts: MIT Press, 1999].

Sobieszek, Robert, "Sculpture as a Sum of Its Profiles: François Willème and Photosculpture in France, 1859–1868," *Art Bulletin*, Vol. 62, December 1980, pp. 617–30.

Spencer, Stephanie, *O.G. Rejlander: Photography as Art* [Ann Arbor, Michigan: UMI Research Press, 1985].

Spira, S.F., "Polka Tot," *History of Photography*, Vol. 11, No. 4, October–December 1987, p. 328.

Thomas, Ann, ed., *Beauty of Another Order: Photography in Science* [New Haven, Connecticut: Yale University Press, 1997].

Trevelyan, Raleigh, "Thomas Woolner, Pre-Raphaelite Sculptor: The Beginnings of Success," *Apollo*, No. 105, March 1978, p. 202.

Tucker, Jennifer, *Nature Exposed: Photography as Eyewitness in Victorian Science* [Baltimore, Maryland: Johns Hopkins University Press, 2006].

Turner, Jane, ed., *Dictionary of Art* [New York: Grove's Dictionaries, 1996].

Vasillov, Magda, *Photographs and Other Materials Collected by Charles Darwin for His Study of the Expression of the Emotions* (exhibition pamphlet) [Cambridge: Cambridge University Library, June–August 1986].

Wakeman, Geoffrey, *Victorian Book Illustration: The Technical Revolution* [Newton Abbot: David & Charles, 1973].

Wall, A.H., "An Hour with Rejlander," *Photographic News*, Vol. 4, No. 112, 26 October 1860, p. 302.

Wall, A.H., "Rejlander's Photographic Art Studies: Their Teachings and Suggestions," *Photographic Times and American Photographer*, serialized Vol. 16–17, Nos. 257–287, 20 August 1886 to 18 March 1887; Vol. 16: pp. 435–37, 505–8, 532–33, 555–56, 567–68, 580–81, 672–73; Vol. 17: 72–73, 121, 131–32.

Walsh, William S., *Heroes and Heroines of Fiction* [Philadelphia: J.B. Lippincott Company, 1914].

Weissmann, Gerald, *Darwin's Audubon* [New York: Plenum Books, 1998].

Wigh, Leif, et al., *Oscar Gustave Rejlander, 1813(?)–1875* [Stockholm: Moderna Museet, 1998].

Wills, Camfield, and Deirdre Wills, *History of Photography: Techniques and Equipment* [New York: Exeter Books, 1980].

Wolf, Sylvia, et al., *Julia Margaret Cameron's Women* [New Haven, Connecticut: Yale University Press, 1998].

Woolner, Amy, ed., *Thomas Woolner R.A., His Life and Letters* [London: Chapman & Hall, 1917].

ACKNOWLEDGMENTS

At times, I have known something of how it felt to be Charles Darwin himself, sitting alone in my cluttered office and writing inquisitive letters to experts about things I am interested in but do not know much about. Like him, I am deeply indebted to many kind and generous colleagues for their help.

I owe special thanks to John Gage, formerly of the Department of History of Art at the University of Cambridge, for his assistance at every stage of this research. He provided a peerless blend of expertise, generosity, encouragement, and good humor. I also thank Adam Perkins, Godfrey Waller, and the staff of the Manuscripts Collection at the Cambridge University Library for their tireless assistance. And I gratefully acknowledge the support of the members of the Darwin Correspondence Project at Cambridge and, particularly in the early stages of the project, of Jon Topham. One of the great pleasures of working with this material at the time that I did was the opportunity to consult with Paul Ekman, who probably understands Darwin's *Expression* better than any other living person. I am extremely grateful for all his help.

I am also indebted to the staff at Oxford University Press who helped make this handsome book a reality: Joyce Berry spent many productive hours poring over the manuscript and improved it enormously; Susan Ferber and Lisa Stallings edited the project through the final stages, providing expert advice and support; Brady McNamara designed the interior and the jacket; and Patricia Watson was the copyeditor. Claudia Sorsby provided invaluable comments on early and late drafts of the manuscript. Her enthusiasm for the subject has been indispensable in moving me forward. Dan Hillman, leading by example and with seemingly limitless patience for technical questions, has been a most welcome partner in her endeavor. I am also grateful to Sarah Hammond, who read the manuscript before it was sent to press. April Swieconek deserves special thanks for her advice and encouragement, and for her peerless publicity efforts.

Numerous individuals deserve mention for their advice on various matters, large and small. Members of the Darwin family, including Erasmus Barlow, Philip Darwin, Milo Keynes, Richard Keynes, and Sir Martin Wedgwood, kindly shared their unique knowledge of Charles Darwin's picture collection. Mike Barfoot, Bernard Barryte, Gillian Beer, Janet Browne, Peter Bunnell, Morton Cohen, Charlotte Cotton, Julian Cox, Frances Dimond, Elizabeth Edwards, Robin Emmerson, Kelly George, Helen Gray, Anthony Hamber, Craig Hartley, Mark Haworth-Booth, Steven Joseph, Alex Kidson, John Michael Krois, Timothy Lenoir, Edward Manier, Jean-Michel Massing, Solene Morris, Toby Mundy, Virginia Murray, Mike Proudfoot, Pam Roberts, Martin Rudwick, Larry Schaaf, Roger Taylor, and Edward Wakeling are but a few of the distinguished scholars who lent their assistance. My Ph.D. examiners, Paolo Morello and Sam Smiles, also provided invaluable comments and advice.

I am also indebted to the following institutions that provided assistance and allowed me to consult relevant material in their collections: Archivi Alinari, Florence; Bodleian Library, Oxford; Brantwood House, Consiston, England; Cambridge University Libraries; Christ Church College Library, Oxford; Christie's; Down House, Kent; George Eastman House, Rochester, New York; English Heritage; Ernst Mayr Library, Harvard University; Fitzwilliam Museum, Cambridge; Frick Art Reference Library; J. Paul Getty Museum, Malibu, California; Houghton Library, Harvard University; Thomas Henry Huxley Archive, Imperial College of Science and Technology; Lady Lever Gallery, Port Sunlight, England; John Murray Archive; Linnean Society; National Art Library, London; National Registry of Archives, London; National Art Library, London; Natural History Museum, London; Peabody Museum of Archeology and Ethnology, Harvard University; Department of Manuscripts and Special Collections, Princeton University; Marquand Library and Princeton University Libraries; Providence Atheneum; Harry Ransom Humanities Research Center, University of Texas at Austin; Royal Academy; Royal Archives, Windsor; Royal

Photographic Society, Bath, England; Royal Anthropological Society, London; Library, University College London; Victoria & Albert Museum; the Wakefield Museum; the Wellcome Institute Library, London; and Witt Library, Courtauld Institute.

Dawn Brooks was my sounding board, adviser, and resident evolutionary biologist. She had a hand in every aspect of this project, from the smallest detail to the broadest sweep.

I also thank Brian and Yvonne Prodger, who spoil me with their support, encouragement, and patience.

INDEX

PICTURE CREDITS

Reproductions by kind permission of the following:

British Library: 7-4

Collection of the author: xiv-xix, 1-1, 2-1, 3-3, 3-4, 4-3, 4-4, 4-5, 5-5, 6-5, 6-7, 6-16, 7-1, 7-9, 7-12, 8-2, 8-7, 8-12, 8-14, 9-9, 11-6

Darwin Papers, Syndics of Cambridge University Library: P-1, xx, 1-5, 1-6, 2-2, 2-3, 2-4, 2-5, 3-2, 5-6, 6-1, 6-2, 6-3, 6-4, 6-6, 6-8, 6-9, 6-10, 6-11, 6-12, 6-13, 6-14, 6-15, 7-2, 7-3, 7-6, 7-7, 7-8, 7-10, 7-11, 7-13, 7-14, 7-15, 8-3, 8-4, 8-6, 8-8, 8-9, 8-10, 8-11, 8-13, 9-1, 9-5, 9-6, 9-7, 9-8, 9-11, 9-12, 9-13, 9-14, 10-1, 10-2, 10-3, 10-4, 10-5, 10-6, 10-7, 10-8, 11-2, 11-4, 11-5 and cover

Fitzwilliam Museum, Cambridge: 1-4, 3-1

Harry Ransom Center, University of Texas at Austin: 7-16

J. Paul Getty Museum: 9-3, 9-10

Lady Lever Art Gallery, Port Sunlight: 8-15

Museum of Comparative Zoology, Harvard University: 5-4

© National Gallery, London: 1-2, 1-3

National Museums, Liverpool: 3-5

Peabody Museum of Archeology and Ethnology, Harvard University: 5-1, 5-2, 5-3

Princeton University Library: 4-1, 4-2

Private collection, by permission Conway Library, Courtauld Institute of Art: 8-1

Royal Photographic Society Collection, National Media Museum, Science and

Society Picture Library: 9-2, 9-4

Syndics of Cambridge University Library: 5-7, 7-5, 11-1

Staatliche Kunstsammlungen Dresden, Gemäldegalerie Alte Meister: 8-5

Thomas Henry Huxley Papers, Imperial College London, Records and Archives: 5-8, 5-9, 11-3

Printed and bound by CPI Group (UK) Ltd, Croydon, CR0 4YY